AF356215

Stress Biology of Turfgrass

Stress Biology of Turfgrass

Editors

Zhou Li
David Jespersen

Basel • Beijing • Wuhan • Barcelona • Belgrade • Novi Sad • Cluj • Manchester

Editors

Zhou Li
Department of Turf Science
and Engineering, College of
Grassland Science and
Technology, Sichuan
Agricultural University
Chengdu
China

David Jespersen
Department of Crop and Soil
Sciences, University of
Georgia
Griffin
USA

Editorial Office
MDPI
St. Alban-Anlage 66
4052 Basel, Switzerland

This is a reprint of articles from the Special Issue published online in the open access journal *Plants* (ISSN 2223-7747) (available at: https://www.mdpi.com/journal/plants/special_issues/turfgrass_stress_biology).

For citation purposes, cite each article independently as indicated on the article page online and as indicated below:

Lastname, A.A.; Lastname, B.B. Article Title. *Journal Name* **Year**, *Volume Number*, Page Range.

ISBN 978-3-7258-0769-7 (Hbk)
ISBN 978-3-7258-0770-3 (PDF)
doi.org/10.3390/books978-3-7258-0770-3

Cover image courtesy of Zhou Li

Contents

About the Editors

Zhou Li

Zhou Li is a leading researcher in the field of plant physiology and is mainly involved in studying the physiological and molecular mechanisms of grass species in response to various abiotic stresses such as drought, high temperature, salinity, etc. He received his Ph.D. at Sichuan Agricultural University and had the opportunity to study at Rutgers, The State University of New Jersey as a visiting scholar for two years during this time. Upon completion of his Ph.D., Dr. Zhou Li accepted a position at Sichuan Agricultural University as a professor. He has been engaged in scientific research since 2010 and has published more than 100 peer-reviewed articles worldwide in *The Plant Journal, Journal of Integrative Plant Biology, Crop Science, Plant Cell & Physiology, Plant and Soil, Planta, Physiologia Plantarum*, etc. As a senior reviewer, he reviews more than 50 manuscripts every year. He teaches and has mentored many graduate and undergraduate students since 2016. In 2023, he was ranked as one of the world's top 2% scientists in the field of plant biology and botany.

David Jespersen

David Jespersen is an associate professor in the University of Georgia's Department of Crop and Soil Sciences. Dr. Jespersen received his Ph.D. from Rutgers University in 2016 prior to joining the faculty at the University of Georgia. His lab group focuses on abiotic stress in turfgrasses, combining whole plant physiology with underlying molecular biology in a systems biology approach. The ultimate goal of his research is to understand the mechanisms of abiotic stress tolerance in order to develop more resilient turfgrasses. Dr. Jespersen is active in graduate education and has been a graduate student advisor or committee member for over 20 graduate students and teaches graduate and undergraduate courses including plant physiology and turfgrass physiology and ecology.

Preface

Turfgrass is widely used for landscaping, sports turf, and ecological restoration. However, abiotic stresses decrease turf quality and also increase the costs involved in turf maintenance. This reprint focuses on research related to the physiological, metabolic, and molecular mechanisms of turfgrasses in response to drought, heat, cold, or ionic stress (salt, chromium, and cadmium). This reprint comprises a review article on salinity stress and tolerance in turfgrasses, as well as research articles covering antioxidant defense systems, water homeostasis, and ionic equilibrium in turfgrasses in response to abiotic stress, the effects of plant growth regulators in turfgrasses exposed to abiotic stress, physiological and metabolic differences affected by stress in turfgrasses, and key genes and proteins associated with stress tolerance in turfgrasses.

Zhou Li and David Jespersen
Editors

Review

Turfgrass Salinity Stress and Tolerance—A Review

Haibo Liu [1,*], Jason L. Todd [1] and Hong Luo [2]

1   Department of Plant and Environmental Sciences, Clemson University, Clemson, SC 29634, USA
2   Department of Genetics and Biochemistry, Clemson University, Clemson, SC 29634, USA
*   Correspondence: haibol@clemson.edu; Tel.: +1-864-506-6260

Abstract: Turfgrasses are ground cover plants with intensive fibrous roots to encounter different edaphic stresses. The major edaphic stressors of turfgrasses often include soil salinity, drought, flooding, acidity, soil compaction by heavy traffic, unbalanced soil nutrients, heavy metals, and soil pollutants, as well as many other unfavorable soil conditions. The stressors are the results of either naturally occurring soil limitations or anthropogenic activities. Under any of these stressful conditions, turfgrass quality will be reduced along with the loss of economic values and ability to perform its recreational and functional purposes. Amongst edaphic stresses, soil salinity is one of the major stressors as it is highly connected with drought and heat stresses of turfgrasses. Four major salinity sources are naturally occurring in soils: recycled water as the irrigation, regular fertilization, and air-borne saline particle depositions. Although there are only a few dozen grass species from the *Poaceae* family used as turfgrasses, these turfgrasses vary from salinity-intolerant to halophytes interspecifically and intraspecifically. Enhancement of turfgrass salinity tolerance has been a very active research and practical area as well in the past several decades. This review attempts to target new developments of turfgrasses in those soil salinity stresses mentioned above and provides insight for more promising turfgrasses in the future with improved salinity tolerances to meet future turfgrass requirements.

Keywords: turfgrass; salinity stress; stress; turf soil; salinity tolerance; turfgrass drought stress; turfgrass stress tolerance; turfgrass roots; turf

Citation: Liu, H.; Todd, J.L.; Luo, H. Turfgrass Salinity Stress and Tolerance—A Review. *Plants* **2023**, *12*, 925. https://doi.org/10.3390/plants12040925

Academic Editors: Zhou Li, David Jespersen and Juan Barceló

Received: 16 January 2023
Revised: 4 February 2023
Accepted: 16 February 2023
Published: 17 February 2023

1. Introduction

Turfgrasses are ground cover plants with intensive fibrous roots sensitive to different edaphic stresses. Urban green spaces are turfgrass-dominated and backgrounded landscapes of varying functions and uses that are ubiquitous in areas associated with human population growth and urbanization in the United States. Furthermore, lawns cover 1.9% (approximately 186,800 km^2) of the country's terrestrial areas [1,2], with about 80 million lawns sitting within tropical to subarctic climatic zones. Among the mentioned edaphic stresses, soil salinity is one of the major stressors, and it is highly connected with drought and heat stresses of turfgrasses as well as other stresses [3,4] (Figure 1). This review is focusing on the recent development in turfgrass salinity stress and tolerance literature updates and facing the challenges of shortage of turfgrass irrigation water and climate changes.

Turfgrass is perhaps one of the most important vegetative ground covers in the world as it provides functional, recreational, and ornamental purposes to our landscapes, particularly in urban communities. Approximately 50% of global land area is categorized into one of the four classes of aridity [5], and as the global effects of climate change continue to intensify, the demand for freshwater has drastically increased in the recent years. Along with numerous government regulations restricting freshwater use for turfgrass and crops alike, many managers have turned to non-potable, secondary (recycled), and saline water sources. The presence of dissolved salts in non-potable water in addition to poor soil conditions can lead to saline or sodic soil conditions and reduced plant growth. Plants exhibit many morphological, physiological, and metabolic responses to salinity stress that can significantly decrease crop yield and quality [4,5]. Thus, salinity is one of the most significant soil stress conditions limiting plant growth and productivity all over the world [6].

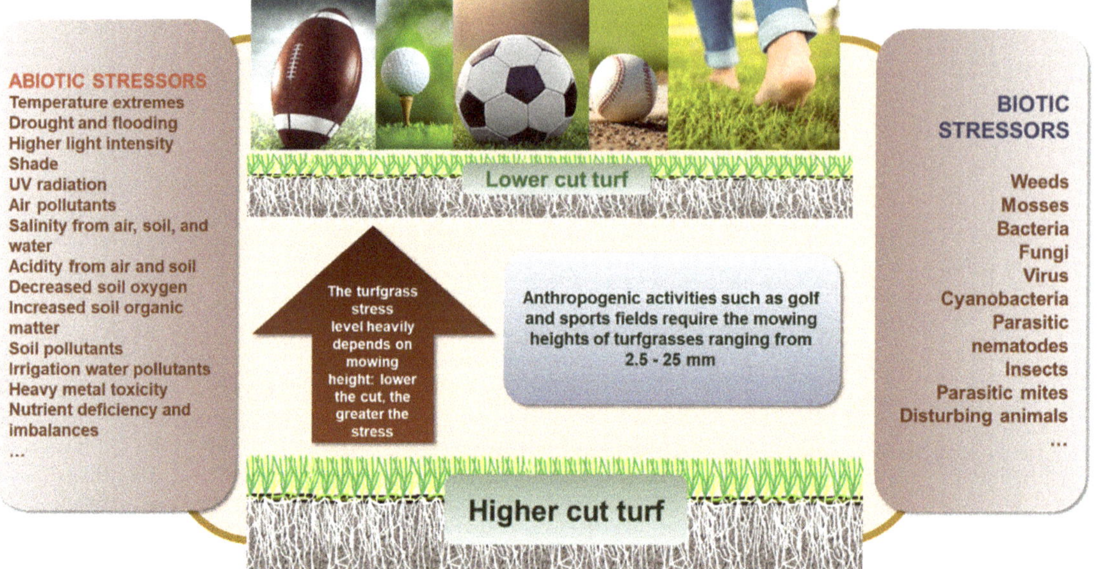

Figure 1. Turfgrasses encounter both abiotic and biotic stresses with mowing height as one of the most critical factors affecting the severity of stresses. A maintained turf area may receive more than one stress at a time or season, and it is very common that a turfgrass must be resistant to multiple stressors in addition to the lower mowing height stress for functional and performance purposes.

Saline soils and salinity stress are often confused with other terminology such as alkaline and sodic soils, and it is important to understand the terminology before reading any literature on salt stress on plants because two terms could imply different meanings. Soils that have high concentrations of neutral soluble salts that are detrimental to plant growth are saline and capable of imposing salinity stress on plants. In contrast, alkalinity and alkaline soils strictly refers to the concentration of hydroxyl (OH^-) ions and soils with a pH > 8.5, respectively (Table 1). Sodic soils are most commonly confused with saline soils, but the difference between the two is that sodic soils have high concentrations of sodium (Na^+) and chloride (Cl^-) salts (ionic soluble salts) that are negative and even detrimental to plant growth [7].

Table 1. Classification of salt-affected soils.

Class	EC	ESP	SAR	Soil pH
	dSm^{-1}	%		
Saline	>4.0	<15	<12	<8.5
Sodic	<4.0	>15	>12	>8.5
Saline-Sodic	>4.0	>15	>12	<8.5

EC: Electrical Conductivity. ESP: Exchangeable Sodium Percentage. SAR: Sodium Absorption Ratio.

2. Causes and Factors Affecting Soil Salinity for Turf

Salts often occur naturally in soil at non-toxic levels, however there are many factors contributing to the cause of salinity stress in plants [8–10]. Soil salinity is caused by a variety of natural processes including rock weathering and fluctuating depth of the water table. Weathering dissolves soluble salts from the composition of rocks causing the solutes to move with water through the soil solution [7,11]. The fluctuating depth of the water table determines where the soluble salts will be in relation to the root zone, and therefore can directly impact plant growth and soil structure. In addition, many environmental, soil physical, and chemical properties all affect the capability of soil to accumulate salts and the susceptibility of turfgrass species to salinity stress. Some environmental factors affecting salinity tolerance include temperature, water status, light, soil depth, soil particle fineness, timing, and depth of irrigation [12]. Hot, dry conditions make turfgrasses more sensitive to saline conditions because of increased evapotranspiration rates favoring salt uptake [4]. Decreases in soil water content show an increase in root zone salinity due to the water potential gradient created by plant roots that pull water towards the root zone. Shallower depths of irrigation increase root zone salinity due to a lack of sufficient water to flush/leach salts out of the root zone. In addition, extended periods of time between irrigation events also increase soil salinity because the groundwater is pulled up the soil profile due to the reduced matric potential in the upper horizons. The rate at which salt accumulates in soil depends on the concentration of total dissolved salts in irrigation water, annual amount of rainfall and irrigation water, the soil's physical and chemical properties, and efficiency of landscape drainage. Salt accumulation occurs more rapidly in soils with thick thatch layers and in clay-dominant soils that have low permeability and are prone to compaction [12].

The quality of irrigation water used is perhaps the most important factor when trying to control soil salinity and reduce plant salinity stress. As available water supplies continue to diminish, turf practitioners are looking for alternative irrigation water sources [13]. Many have turned to recycled and non-potable water sources to irrigate turf landscapes; however, these can have adverse effects on soil's chemical and physical properties including soil salinity [12]. Recycled water, also referred to as recycled wastewater or effluent water, is water that has gone through one cycle of human use and is sent from a home or business through a sewer system to a sewage treatment plant. Recycled wastewater will have a higher concentration of suspended solids and various nutrients, thus resulting in varied soil parameters such as sodium adsorption ratio, exchangeable sodium percentage, and pH. All of these influence soil salinity and nutrient availability. There are no consistent concentrations of substances in recycled wastewater due to seasonal and annual variability of trace contaminants in the initial wastewater [14]. However, recycled wastewater generally contains greater concentrations of sodium, bicarbonates, carbonates, boron, and chloride, which can all significantly impact soil's chemical properties and fertility [15]. Due to the differences in nutrient concentrations between recycled wastewater and potable water, soil infiltration rate and hydraulic conductivity are often negatively impacted. A 50% decrease in infiltration rate after three years of recycled wastewater irrigation was observed on golf course fairways due to interactions of dissolved salts and sodium in the recycled wastewater with soil particles [16]. Suspended solid deposits that are physically clogging soil pore spaces is a common cause of reduced soil porosity while under saline conditions of recycled wastewater irrigation. The use of effluent and non-potable water does have greater effects on horticultural crops than turfgrasses due to the lower amount of above-ground biomass that is more sensitive to salinity [17,18]. In addition, many turfgrass species and varieties (particularly C_4 turfgrasses) are relatively tolerant to the salinity of recycled wastewater if applied directly to soil. However, excess foliar absorption of salts accumulates much quicker in leaf margins and shoot tips when compared to root absorption.

Salts are added in standard application of fertilizers and soil amendments, even including some pesticide applications [12,19]. Naturally, coastal areas that have evaporated and left salt deposits at the surface or in layers in the soil profile suffer more turfgrass salinity stresses than inland areas [19]. Although there is a lack of studies specifically focusing on turfgrasses, airborne saline particles are a heterogeneous group of chloride (Cl)-containing airborne materials both from natural as well as anthropogenic origins [20]. These are air pollution problems primarily relating to source and dispersion, but these airborne saline particles can cause the accumulation of Cl on a vegetation canopy, negatively affecting the growth and performance of turfgrasses (Figure 2).

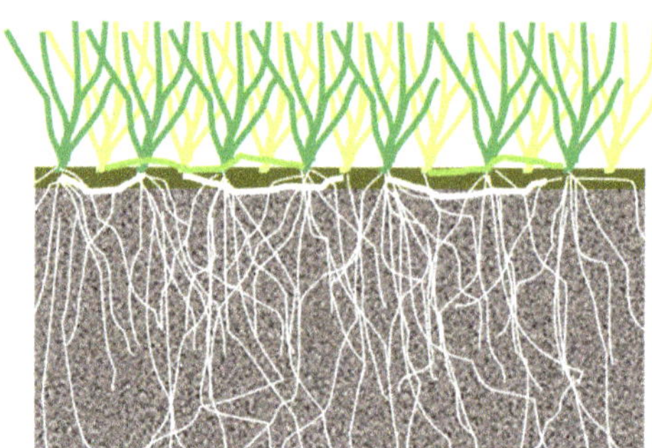

Figure 2. Multiple sources contribute to the salinity stresses of a turfgrass community, and these sources vary depending on location conditions.

3. Turfgrass Responses to Saline Stresses

Individual plant species will respond differently to salinity. Ornamental and vegetable plants are known to be more sensitive to saline conditions, while most grain crops and turfgrasses are considered more tolerant to salinity. Saline conditions trigger a wide variety of physical and chemical responses inside the plant. Most of the detrimental effects of salinity are due to symplastic osmotic pressure fluctuations causing cellular water loss and reduced water uptake. Additionally, the disruption of ion balances, various ion toxicities, and typical chlorotic leaf tissue creates adverse effects on plants metabolic processes. Therefore, plant growth and development are often limited or inhibited while under salinity stress [8–10].

Shoot and root growth responses are perhaps the most documented and recognizable plant responses to any environmental stress, especially soil salinity, in addition to yields as crop plants. A negative linear relationship between shoot growth and salinity is often reflected in salt-sensitive and moderately salt-tolerant species. Salt-tolerant species such as *Cynodon* spp., *Paspalum vaginatum*, *Puccinellia distans*, and *Zoysia matrella* can show stimulated shoot growth during exposure to moderately saline conditions [21–23]. Turfgrass quality and leaf width and length significantly decreased in four species of cool-season turfgrasses (*Lolium perenne* 'Belida', *Festuca rubra commutata* 'Casanova', *Poa pratensis* 'Evora', and *Festuca rubra trichophylla* 'Smyrna') when irrigated with gradually increased salinity levels from 0.54 mM to 200 mM NaCl [24]. The same observations were found in two centipedegrass (*Eremochloa ophiuroides*) accessions, E092 and E092-1, as salinity increased from 0 mM to 75 mM NaCl [25]. Seed germination is also affected by soil salinity, as four cool-season turfgrass species (perennial ryegrass, slender creeping red fescue, tall fescue, and Kentucky bluegrass) had an 11% to 25% decrease in field germination at the highest salinity level of 200 mM NaCl [26]. Certain management practices may also influence a species' salinity tolerance, such as mowing height. Lower mowing heights of 'L-93' creeping bentgrass (*Agrostis stolonifera*) severely decreased salinity tolerance as determined by visual turf quality. When mowed at 6.4, 12.7, and 25.4 mm, the turfgrass reached an unacceptable visual quality at soil EC values of 4.1, 12.5, and 13.9 dS m^{-1}, respectively [27]. These results suggest that a higher shoot biomass increases turfgrass salinity tolerance due to the presence of more plant tissue able to sequester saline ions in vacuoles. The same study found that as soil salinity levels increased, the treatments with the highest mowing height of 25.4 mm retained a higher percentage of original root biomass under saline conditions when compared to the lowest mowing height of 6.4 mm [27]. Most turfgrass species that are considered moderately tolerant to tolerant of salinity stress often demonstrate an increase in root/shoot ratio while under salinity stress [21]. This is due to the plant increasing or maintaining rooting, while shoot growth decreased in response to salinity and may act as a tolerance mechanism. All species (alkaligrass, saltgrass, tall fescue, and Kentucky bluegrass) exhibited an increase in root/shoot ratio at all salinity levels. Of the four species, only saltgrass (*Distichlis spicata*) demonstrated a relatively similar shoot mass as soil EC increased from 2.5 to 25 dS m^{-1}. An initial increase in relative root mass was observed for both alkaligrass and saltgrass up to 15 dS m^{-1}, while a maintenance of root mass was observed for tall fescue at the same EC [21]. The same stimulated root growth was observed in a halophytic grass species, *Sporobolus virginicus*, as NaCl concentrations increased up to 1.0 M [28].

Figure 3 summarizes the metabolic processes associated with salinity tolerances and sensitivities among turfgrass species and cultivars. However, the primary mechanism behind shoot and root growth responses to salt stress is due to the inhibition of photosynthesis. Salinity stress causes a reduction in CO_2 assimilation to the chloroplasts due to stomatal closure. Guard cells are responsible for controlling stomatal aperture to regulate water and gas exchange. Due to the structure of guard cells, they respond to changes in turgor pressure, therefore stomates close because of decreased osmotic pressure and reduced K^+ content in leaves [29]. Increased shoot concentrations of Na^+ and Cl^- cause increased osmotic pressure, resulting in cellular water outflow or a lower turgor pressure. As a result, the osmotic potential inside the cell decreases [30]. Salt-sensitive plants often exhibit decreased chlorophyll content through inhibition of chlorophyll synthesis and enzyme-activated chlorophyll degradation resulting from oxidative stress [31]. Total chlorophyll concentrations of three cool-season turfgrass species (*Lolium perenne* 'Belida', *Poa pratensis* 'Evora', and *Festuca rubra trichophylla* 'Smyrna') exhibited a progressive decrease in total chlorophyll content as salinity conditions became more intense [24,32]. In addition to decreased chlorophyll content, salt stress has been reported to cause chloroplast structural damage. As documented in tall fescue and many other higher plants subjected to salinity stress, the arrangement of chloroplasts in mesophyll cells may have become disordered, the connection between grana loosened, and the lipid-bilayer of chloroplasts were damaged [33,34].

Cellular membrane structural damage and lipid peroxidation are common responses to oxidative stress. Many, if not all, edaphic stresses on turfgrasses cause increased production and accumulation of reactive oxygen species. The presence of a higher concentration of reactive oxygen species can cause further plant damage and even death by inactivating certain proteins and enzymes and destroying cell membrane structure and permeability via lipid peroxidation. Plants defend against reactive oxygen species through antioxidant defense systems that respond to an environmental stress. Stimulated activity of certain antioxidant enzymes, such as superoxide dismutase, catalase, hydrogen peroxidase, and ascorbate peroxidase, is used to scavenge reactive oxygen species to reduce oxidative damage. For example, superoxide dismutase converts oxygen radicals into hydrogen peroxide, which is then scavenged into molecular oxygen and water [35,36]. Antioxidant enzyme activity under saline conditions has been described for some turfgrass species. Kurup et al. [36] measured leaf firing percentage (visual turf quality), and activity of superoxide dismutase, catalase, hydrogen peroxidase, and ascorbate peroxidase of three seashore paspalum genotypes, four bermudagrass cultivars, and tall fescue at five different salinity levels (0, 5000, 10,000, 15,000, and 20,000 mg L^{-1} NaCl). Results indicate that increased antioxidant enzyme activity in a genotype correlated to a higher salinity tolerance through a lower leaf firing %. Bermudagrass cvs. 'Princess 77' and 'Tifgreen' as well as tall fescue were shown to be the least salinity tolerant throughout all parameters. These cultivars exhibited the highest leaf firing at all salinity levels, and bermudagrass cvs. exhibited amongst the lowest activity for all enzymes at the highest salinity level. All seashore paspalum cultivars ('Salam', 'Seadwarf', and 'Seaisle 2000') exhibited minimal leaf firing at all salinity levels when compared to bermudagrass cvs. 'Salam' showed the lowest leaf firing % of paspalum genotypes and the highest hydrogen peroxide activity [36]. These results suggest that any one of the tested paspalum genotypes is viable for high salinity conditions.

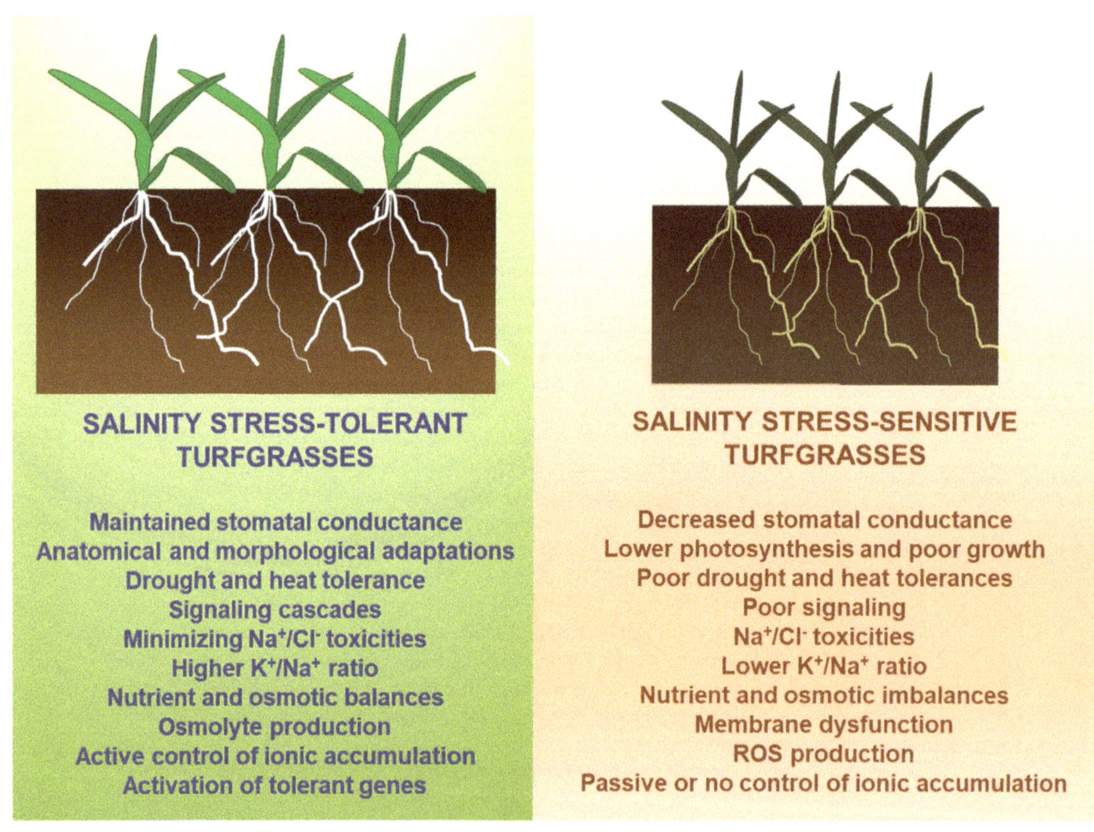

Figure 3. Metabolic processes associated with salinity tolerances and sensitivities among turfgrass species and cultivars.

Due to enhanced nutrient competition as a result of excessive NaCl uptake in saline conditions, ion imbalances and toxicities are common effects of salt stress on plants. Sodium (Na$^+$), chloride (Cl$^-$), and boron (B^{3+}) toxicities pose the greatest threat to plant growth, whereas deficiencies of calcium (Ca^{2+}), potassium (K$^+$), and iron (Fe^{2+}) also create adverse effects on plant growth while under salt stress.

3.1. Sodium (Na$^+$) Toxicity

In saline soils, sodium is the most common toxic element due to the large amounts of soluble salts in the soil solution. High concentrations of sodium in plant tissue negatively impact cell division and enzyme activity through a variety of mechanisms such as disturbance in osmoregulation; uptake of other ions such as K$^+$, Ca^{2+}, and NO$_3{}^-$; and increased production of reactive oxygen species [37]. Such mechanisms lead to a decrease in chlorophyll content and destruction of the cellular membrane and organelle structure, thus impeding plant growth and development at all stages [38]. Guo et al. [39] found that no correlation was found between leaf firing % and Na$^+$ shoot content while under salt stress. However, a significant negative correlation existed between the Na$^+$ shoot content and the relative shoot weight and relative root weight while under salt stress [39]. The results of this study indicate that sodium accumulation in leaf tissue does not have a direct impact on leaf firing/scorch, which is a common symptom associated with salt stress. As a result of increasing NaCl content in the plant and the soil, K$^+$ and Ca^{2+} often become deficient and may restrict plant growth. A large external influx of Na$^+$ inhibits root uptake and decreases the relative binding affinity of K$^+$ and Ca^{2+} for binding sites on enzymes due to their similar hydrated radii and hydration energy to Na$^+$ [38].

Salts used in road de-icing during the winter season inhibit the growth and development of utility turfgrass species. Germination capacity was evaluated at five salt solutions of NaCl (0, 50, 100, 150 and 200 mM), and physiological parameters were measured during the tillering phase at salinity levels of 0, 150, and 300 mM of NaCl [40]. Seeds of Kentucky bluegrass were found to have no germination under any level of salinity. During the tillering phase, salinity negatively affected the length, area, and dry mass of roots as well as the relative water content of plants. The maximum chlorophyll fluorescence yield, quantum yield of photosystem II, and electron transport rate were negatively impacted at the early period of stress [40].

3.2. Chloride (Cl$^-$) Toxicity

Both boron and chlorine are considered essential plant micronutrients for plant growth and metabolism; however, in some instances they can become toxic to plants either naturally or anthropogenically. Chlorine toxicity, as well as most micronutrients such as boron and zinc, most commonly occur when using recycled wastewater on soils with low percolation rates in arid and low-rainfall areas due to the ions' high mobility in soil [41]. As an anion, chloride plays essential roles in stomatal regulation, photosynthetic activity, and osmoregulation. Stomatal regulation is mediated by the fluxes of potassium and accompanying anions such as chloride to act as a stabilizer. Chloride (Cl$^-$) and manganese (Mn^{2+}) work harmoniously in the water splitting reaction, or Hill reaction, in photosystem II. Excessive accumulation of chloride in the chloroplasts can cause chlorophyll degradation through a chloride-induced disturbance of photosystem II's structural organization because of the highly permeable chloroplast envelope [42,43]. Perhaps most importantly, chloride helps regulate osmotic pressure in vacuoles at higher concentrations and more specialized tissues, such as apical meristems and root tips, when at low concentrations [44]. However, too much Cl$^-$ can accumulate in leaf tissue, resulting in a bleached or burned appearance of mature leaf tips and margins, thereby negatively impacting photosynthetic efficiency and potentially decreasing a species' salt tolerance by disrupting regular osmoregulatory practices [41].

When two popular C$_3$ cool-season turfgrass species of Kentucky bluegrass (*Poa pratensis* L.) and tall fescue (*Festuca arundinacea* Schreb.) were compared, Kentucky bluegrass accumulated more Na$^+$ and Cl$^-$ in its shoots, although Na$^+$ and Cl$^-$ concentrations in shoots and roots increased with increasing salinity in both turfgrasses treated with 0, 50, 100, 150, and 200 mM NaCl growing for 40 days [45]. The total soluble sugar (TSS) concentration of the tall fescue was significantly higher than that of the Kentucky bluegrass under elevated NaCl concentrations, with significantly lower nitrate (NO$_3{}^-$) concentrations in the shoots of the Kentucky bluegrass, which indicate either poor nitrate transportation from roots to shoots or a stronger negative competition between the two anion concentrations of nitrate and chloride in Kentucky bluegrass [45].

3.3. Potassium (K^+) Deficiency

Potassium plays important roles in stomatal regulation, CO_2 fixation, and the detoxification of reactive oxygen species [46]. With adequate levels of K, CO_2 fixation is increased, and translocation of photosynthates from source to sink organs is also increased to inhibit the photosynthetic electron transport to O_2 to form excess reactive oxygen species [47,48]. Potassium is also shown to downregulate the activity of NADPH oxidase that converts NADPH to a superoxide radical and $NADP^+$ [49]. Therefore, potassium-deficient plants will show reduced rates of photosynthesis followed by damaged membranes and chlorophyll structures due to the overproduction of ROS [46,50]. *Puccinellia tenuiflora* is a saline–alkali-tolerant plant in the Songnen Plain, one of the three largest soda saline–alkali lands worldwide. The soils from the Songnen Plain were reasonably rich in salts and alkali, and the soils were severely deficient in nitrogen, phosphorus, and potassium. When investigated for its ability to tolerate these adverse soil conditions, unigenes involved in the uptake of N, P, K, and micronutrients were found to be significantly upregulated in the soil, which indicated the existence of an efficient nutrition-uptake system in *P. tenuiflora*. Compared with *P. tenuiflora*, rice (*Oryza sativa*) was hypersensitive to saline–alkali stress [51].

In a 2021 study, perennial ryegrass (*Lolium perenne* L.) seeds were cultured in soil mixtures amended with clinoptilolite zeolite (OZ) and potassium-enriched clinoptilolite zeolite (K-EZ), then exposed to three salinity levels (0, 50, or 100 mM NaCl) for three months. The results show that the application of both types of zeolite significantly decreased Na content by 44.36% and 21.31%, but increased K content by 272.34% and 81.59%, as well as the K/Na ratio by 590.47% and 129.43%, in shoots and roots, respectively. Si content in shoots was increased by 28.33%. The soil mixtures with zeolite, especially K-EZ, enhanced relative water content, membrane stability index, total chlorophyll content, total soluble proteins, peroxidase, and superoxide dismutase activities, but reduced the contents of total soluble carbohydrates, hydrogen peroxide, and malondialdehyde in saline conditions. Shoot and root dry weight, root volume, and root/shoot ratio were also found to be improved by the application of OZ and K-EZ [52].

3.4. Calcium (Ca^{2+}) Deficiency

Calcium is an important factor for cell wall and membrane stability, as well as serving as a secondary messenger to both biotic and abiotic stresses [53]. Ca^{2+} signals help coordinate plant responses and defense mechanisms that are induced by many edaphic stresses including salt stress [54]. This would suggest that deficient levels of Ca decrease the plants' ability to sense and tolerate salt stress, leading to more significant plant damage. Since calcium-deficient plants lack the ability to inhibit salt translocation to the shoots, shoot tip injury, loss of apical dominance, and the development of axillary shoots with necrotic young leaves are common Ca-deficiency symptoms [50].

Tall fescue, after a three-month establishment of canopy and roots, was treated with a combination of 300 mM NaCl and calcium nitrate for five days. The nitrogen content in shoots was highly correlated with Ca^{2+} and K^+ contents in roots. Additionally, high levels of salt increased ATP6E and CAMK2 transcription levels in shoots at day one and day five treated with calcium nitrate. In roots, CAMK2 level was reduced by salinity at day five and exogenous calcium helped to recover it. These findings indicate exogenous calcium plays positive roles in tall fescue to improve salt tolerance [55].

3.5. Boron (B^{3+}) Toxicity

Boron's primary function in plants is to help maintain cell wall structure and function. Borate (boric acid) forms complexes with sugars that stabilize the pectin network to regulate cell wall pore sizes. Boron also participates in metabolic processes that influence vegetative and reproductive growth, such as the metabolism of nucleic acids, phosphorus, nitrogenous compounds, and hormonal regulation [56]. Similar to chloride (Cl^-), excessive accumulation of boron in leaf tissue causes leaf scorch appearing as necrotic and/or chlorotic tissue on the tips and margins of mature leaves [56]. The accumulation of high concentrations of boron can cause osmotic imbalances, therefore reducing the plants' ability to tolerate oxidative damage leading to lipid peroxidation, increases in membrane permeability, and accumulation of proline [56,57]. These responses to boron toxicity can inhibit photosynthesis by damaging thylakoid formation and structure, thus reducing CO_2 absorption and creating conditions that disrupt the transmission of photosynthetic electrons and molecular oxygen to act as an electron acceptor to overproduce reactive oxygen species [57]. Boron toxicities have also been shown to create physiological effects during plant growth, such as reducing root growth in rice and inhibiting seed germination in various plant species [58,59]. This growth and development inhibition is proposed to be caused by boron's ability to bind to the ribose section of several key metabolites such as ATP, NADPH, and NADH [57].

Boron toxicity occurs in some salt-affected soils. This is often observed in arid or semi-arid soils and when the irrigation water is high in salt contents [60]. Buffalograss (*Buchloe dactyloides* (Nutt.) Engelm.) pots in a greenhouse were irrigated with solutions containing 0.5, 1, 2, 4, 6, 8, or 12 mM of boron (B), chlorine (Cl), copper (Cu), iron (Fe), manganese (Mn), molybdenum (Mo), and zinc (Zn) [61]. Boron and Mo induced visual toxicity symptoms more readily than other micronutrients. Boron toxicity of chlorosis was often accompanied by bleached leaf tips. Biomass yield was reduced when the nutrient solution contained 2 mM B, 6 mM Cu, or 2 mM Mo, although the elevated levels of other micronutrients of Cl, Fe, Mn, and Zn did not alter dry matter yield [61].

4. Turfgrass Salinity Resistance Mechanisms

Since salinity resistance of the *Poaceae* family encompasses a broad spectrum of salinity levels, there are a multitude of salinity resistance mechanisms that turfgrass species use (Figure 3). Plant salinity resistance is a combination of two processes: avoidance and tolerance. Avoidance mechanisms aim to decrease excessive translocation of salts to the more sensitive leaf tissues, while tolerance mechanisms aim to increase the plant's ability to survive in the presence of accumulated salts in leaf tissues [62]. This is done through a variety of physiological adaptations such as stimulated root growth, osmotic adjustment and ion exclusion, ion regulation, ion sequestration, and ion excretion. All resistance mechanisms can occur at the same time within a species and are shared across most salinity-tolerant plant species [4].

4.1. Turfgrass Morphological and Anatomical Resistance Mechanisms

Most morphological plant responses to salinity stress occur in the form of leaf scorch and wilting; however, root growth can also be positively or negatively affected by saline conditions. Salt-tolerant turfgrasses often exhibit stimulated root growth under low to moderate salinity levels. Stimulated rooting has been observed in many warm-season moderately salt-tolerant and halophytic turfgrasses [21,63,64]. This mechanism increases the root absorptive area/shoot transpirational area ratio by both increasing root biomass and decreasing or inhibiting relative shoot growth. More root biomass helps counter external osmotic stress and overaccumulation of salts in leaf tissue by allocating more biomass that can safely accumulate salts with minimal adverse effects. This has been observed in *Cynodon dactylon* × *C. transvaalensis* cv. 'Tifway', *Cynodon dactylon* cv. 'Celebration', *Stenotaphrum secundatum* cv. 'Raleigh', *Paspalum vaginatum* cv. 'SeaStar', and *Zoysia japonica* cv. 'Palisades' increased their root biomass by 59%, 29%, 113%, 24%, and 27% compared to the control (2.5 dS m^{-1}), respectively, when exposed to 15 dS m^{-1} salinity level for two years [63]. Most cultivars that exhibited stimulated root growth maintained or increased aboveground visual turf quality (TQ); however, *Cynodon dactylon* × *C. transvaalensis* cv. 'Tifway' had a significant decrease in TQ over the two-year period suggesting that 'Tifway' relies on stimulated root growth as its primary mechanism for salt tolerance rather than ion exclusion or regulation [63].

Salt-sensitive plants over-accumulate salts in the plant tissues beyond the amount needed for osmotic adjustment. This, in effect, raises the leaf osmolarity well beyond the root-media salinity, causing toxicity symptoms such as reduced growth and leaf scorch. Salt-tolerant plants maintain cell turgor pressure and normal physiological functions during salt stress by sufficiently increasing sap osmolarity in the xylem to compensate for external osmotic stress, a process known as osmotic adjustment [65]. However, loading of sodium ions into the xylem and vacuoles are active processes due to the membrane negative charges. Therefore, osmotic adjustment is a relatively inefficient avoidance mechanism in many non-halophytic species, because of higher leakage rates across the tonoplast, when compared to halophytic species [66]. Shoot saline ion exclusion paired with osmotic adjustment has been shown to be perhaps the most prevalent mechanism in salt avoidance in both C_3 and C_4 turfgrass species [66–68]. Ion exclusion refers to the exclusion of saline ions from the shoots to minimize toxic effects by inhibiting translocation of Na^+ to the shoots [69]. Many studies have shown that shoot concentration of Na^+ and Cl^- is negatively associated with relative salinity tolerance of both C_3 and C_4 turfgrass species [33,62,64]. This correlation has been successfully used to predict salinity tolerance of *Cynodon* spp. and *Zoysia* spp. [68,70]. Multiple mechanisms of sodium exclusion have been studied, including Na^+ efflux from roots, Na^+ partitioning in vacuoles of root cells or mesophyll cells, and control of Na^+ loading and unloading by xylem parenchyma cells [69]. Salt-tolerant plant species have been reported to have higher vacuolar and plasma membrane H-ATPase enzyme activity, which is highly correlated with the rate of root Na^+ efflux, when compared to salt sensitive cultivars [71,72]. Higher H-ATPase enzyme activity was shown to generate an electrochemical gradient as the driving force for Na^+ exclusion [73]. The ability of a plant species to control xylem ion loading is one of the most important mechanisms for reducing Na^+ transport to shoots [69]. Multiple passive and active mechanisms are proposed to be responsible for Na^+ loading at the xylem/parenchyma interface, including the salt over sensitive 1-encoded Na^+/H^+ antiporter pathway, the high affinity K^+ transporter pathway, and the cation-chloride cotransporter [74,75].

4.2. Sodium and Potassium Uptake Mechanisms

More recently, salinity tolerance has been suggested to be associated more with the ion-specific component rather than with the osmotic component of stress [62]. Regulation or selectivity of ion uptake, such as Na^+, K^+, and Cl^-, are crucial for the functionality of osmotic adjustment because of their commonality in saline solutions. Higher salt tolerance is positively correlated with a plant's ability to maintain a high K^+/Na^+ ratio when under salt stress [62,64]. This occurs by reducing the transport of Na^+ from the roots to the shoots, increasing the absorption of K^+ in the roots and reducing the leakage of K^+ from cells [64]. Ion regulation primarily occurs at the casparian strip of the root endodermis and low-affinity K^+ channels in plasma membrane of root cells [76]. Maintenance of a high K^+/Na^+ ratio in the cytoplasm is mediated by Na^+/K^+ antiport activity, which further aids in proper enzyme function, cell metabolism, and photosynthetic pathways [69]. In salt-tressed seashore paspalum, the K^+ concentration decreased by 26.20% in the shoots and 69.68% in the roots; however, the Na^+ concentration increased 15-fold in the shoots and 25-fold in the roots. Possibly, the regulation mechanisms of the K^+ and Na^+ for seashore paspalum allow the maintenance of high K^+ concentrations in the shoots to decrease Na^+ translocation from the roots to the shoots under high salinity stress [33]. A significant negative correlation was found between the K^+ concentration in the roots and the Na^+ concentration in the shoots under salinity stress. This means that more K^+ were taken up by roots, there was higher $K^+/Na+$ in the roots, and that more K^+ were transferred from roots to shoots, so the transfer of Na^+ from roots to shoots was inhibited [33]. An overexpression of MicroRNA393 in transgenic creeping bentgrass (*Agrostis stolonifera*) led to significantly higher accumulation of Na in the shoots and K in both the roots and shoots while exposed to 200 mM of NaCl. Despite the high accumulation of Na in the shoots of the transgenic lines, the increased accumulation of K in the shoots helped mitigate further oxidative damage and improve membrane stability in the salt-stressed transgenic lines. This was reflected in an increased relative water content and decreased electrolyte leakage when compared to the wild-type control lines. That said, upon further research, —the authors found that an overexpression of MicroRNA393 aids in maintaining membrane stability and chlorophyll content while under salt stress, thus resulting in less leaf damage and a higher salinity tolerance [77]. It is also proposed that calcium ions play an important role in signaling ion regulation during salt stress. In response to NaCl exposure at the roots, Ca reporter proteins trigger systemic, wave-like fluctuations of cytosolic Ca^{2+} content, which is decoded into downstream responses. This is described as the salt overly sensitive1 signaling pathway. Salt overly sensitive3 is a calcium binding protein that interacts with and activates the CBL-interacting protein kinase. The resulting Ca^{2+} sensor-kinase complex then phosphorylates and activates the Na^+/H^+ antiporter salt overly sensitive1, which functions in Na^+ extrusion and long-distance Na^+ transport in plants. In addition, the Ca^{2+} sensor-kinase complex also regulates ion homeostasis and nutrient uptake by activating specific ion transporters and channels [78].

Ion sequestration or compartmentation is a tolerance mechanism that restricts the number of ions in the cytoplasm by sequestering saline ions in the vacuole. When under high saline conditions, specific types of organic solutes called 'compatible solutes' are accumulated in the cytoplasm to promote saline ion translocation across the tonoplast and into the vacuole. The accumulation of saline ions is metabolically expensive and can inhibit enzyme activity, which would decrease ion movement across the tonoplast. 'Compatible solutes', such as proline and glycine betaine, are accumulated in the cytoplasm under saline conditions to sufficiently maintain cellular structure and osmotic balance. Both proline and glycine betaine have been shown to be positively correlated with salinity tolerance by regulating ion homeostasis, enhancing osmotic adjustment, and scavenging reactive oxygen species [62,79]. Proline's primary function is to increase antioxidant activity to mitigate lipid and protein peroxidation. Proline also improves membrane stability and contributes to intracellular osmotic adjustment to promote ion sequestration and maintain photosynthetic activity while under salt stress [62,80]. In rice, the overexpression of transcription factor OsMADS25 resulted in a greater accumulation of proline in the roots causing an increased carbohydrate content and higher salinity tolerance when compared to the wild type [80]. Similar plant responses to glycine betaine have been reported to improve salt tolerance of certain plant species. In wheat, through the introduction of betaine aldehyde dehydrogenase, transgenic lines overexpressing glycine betaine may elicit enhanced salt tolerance. Glycine betaine was shown to have a protective effect on the components and function of photosystem II (PSII) within the thylakoid membrane, which was reflected in a higher photosynthetic rate when compared to the wild types [79]. Exogenous glycine betaine enhanced the salt tolerance in perennial ryegrass (*Lolium perenne*) through elevated antioxidant enzyme activity that decreased cell membranes and improved ion balances by maintaining a higher K/Na shoot ratio while under salt stress [81].

4.3. Ionic Excretions

Ion excretion is a salinity tolerance mechanism in many higher plant species that promotes active excretion of excess saline ions from the shoots via external storage structures on the leaf epidermis. Bicellular leaf epidermal salt glands are responsible for ion excretion in turfgrass species in the *Chloridoideae* subfamily including *Bouteloua* spp., *Buchloe dactyloides*, *Cynodon* spp., *Distichlis spicata* ssp. stricta, *Sporobolus virginicus*, *Zoysia matrella*, and *Z. japonica* [5]. Salt glands have been found on both abaxial and adaxial leaf surfaces and are arranged longitudinally in parallel rows adjacent to rows of stomata [82,83]. A cutinized basal cell on the leaf epidermis and a cap cell comprises the salt glands. Basal cell appearance varies amongst species, from being fully embedded to fully extruded from the leaf epidermis [5,83]. Multiple structural modifications of the leaf epidermis have been proposed to be responsible for salt partitioning and excretion. A series of parallel invaginated channels going in the direction of ion flow on the plasma membrane (infoldings of the plasmalemma) of the basal cell are suggested to partition and guide saline ions into the salt glands on the leaf epidermis [83]. In addition, the localization of ATPase enzyme activity on the salt gland basal cells suggests that there is also active ion loading occurring at the salt glands to promote excretion [84]. Studies showed that 'Celebration' bermudagrass (*Cynodon dactylon*) and 'DALZ1313' zoysiagrass (*Zoysia matrella* × *Z. japonica*) each developed salt glands on the adaxial leaf surface at the 30 dS m^{-1} salinity level, while DALZ1313 also produced salt glands under control conditions (2.5 dS m^{-1}). In addition, salt crystals were observed on leaf surfaces, indicating that ion excretion was present in both species. Salt gland density increased with increasing salinity (from 2.5 to 30 dS m^{-1}) for DALZ1313 in response to increased salinity, suggesting that increased salt gland density could be one of several responses contributing to enhanced salinity tolerance [5]. However, other studies have indicated *Zoysia matrella* showing no difference in salt gland density and size with increasing salinity levels [85]. This suggests that responses of salt gland density are variable amongst the cultivar and species level. Several environmental factors are suggested to affect the rate of salt secretion in all plant species, but these are not fully understood as results often conflict with each other between species [83]. This would suggest that there is a great number of differences across species in regulation of salt secretion. Many studies indicate that salt excretion is influenced by solar radiation and the rate of salt translocation in plants, suggesting that salt excretion rates maximize during the day [86]. It is proposed that saline ion concentrations in soil significantly affect excretion rates. Sugiura determined that the salt glands of *Zoysia matrella* demonstrated that Na$^+$ excretion increased with increasing salt treatment concentrations [85]. Due to the relation of osmolarity and soil moisture, Sugiura's finding suggests that for some species, salt gland excretion rate may be more dependent on soil moisture status. Furthermore, this causes excretion rates to be variable throughout the day and amongst species, depending on their respective photosynthetic pathway. Other mechanisms of ion excretion within the *Poaceae* family have been studied such as bladder-like structures exclusive to *Paspalum vaginatum*, the most tolerant warm-season perennial turfgrass species. Salt bladders or papillae are unicellular structures while salt glands are bicellular, otherwise the two structures are oriented and function similarly [87]. Seashore paspalum exhibited bladder-like structures under both control and 30 dS m^{-1} salinity, which developed along the vascular bundles on the adaxial leaf surface [5]. Another study indicated dense ridges of papillae on the adaxial leaf surface of seashore paspalum [87]. Differences were observed between tested cultivars. Cultivars 'HI10' and '509018-3' were the more salt-tolerant cultivars, exhibiting larger papillae and a higher concentration of Na$^+$ in the papillae than in the underlying leaf tissue, resulting in a higher K concentration in the leaf tissue when compared to other cultivars [87].

5. Management Practices to Improve Turfgrass Salinity Resistance

Various management and cultural practices may be valuable efforts in enhancing turfgrass salinity resistance without using a potable water source. Cultural practices include selecting a more salt-tolerant turfgrass species or cultivar and increasing the mowing height, which can enhance turfgrass salt tolerance. Management practices such as applications of some soil amendments, silicon, and plant growth regulators have also shown positive effects on saline soil recovery and plant salt tolerance.

5.1. Turfgrass Species and Cultivar Salinity Tolerance Variations

Most turfgrass species fall within the range of slightly tolerant to tolerant categories of relative salinity tolerances. With each species, multiple genetic lines, varieties, and cultivars exist and provide a greater opportunity to select a proper turfgrass for a particular site (Table 2). Proper species and cultivar selection for use in saline soil or when using saline irrigation water is of utmost importance when considering maintenance costs and turf quality. Turfgrasses have great species and varietal differences when it comes to salinity resistance. The majority of cool-season turfgrass species and cultivars are defined as glycophytes, which cannot grow in high salt levels. However, in lower saline conditions, glycophytes can be considered sensitive or moderately sensitive to salt stress by triggering a variety of resistance mechanisms [88]. Some plant and turfgrass species are halophytic in nature, as discussed earlier, meaning that they can withstand higher soil salinity levels up to and beyond that of seawater without experiencing severe reductions in growth or metabolic function [28]. Some examples of halophytic turfgrass species include alkaligrasses (*Puccinellia* spp.), inland saltgrass (*Distichlis spicata* sp. stricta), and seashore paspalum (*Paspalum vaginatum*). Although halophyte and glycophyte classifications provide a general level of salt resistance to certain species, the degree of salinity impact on turfgrasses varies among species and the level of genetic tolerance at the cultivar level [19]. As with any stress tolerance, it is difficult to determine the absolute salinity resistance of a genotype and compare relative resistance between genotypes because of many other factors.

Table 2. Relative salinity tolerance among turfgrass species and their hybrids with estimated cultivar variations within each species (within the same category of the same background color, the order is following the scientific name in alphabetical order).

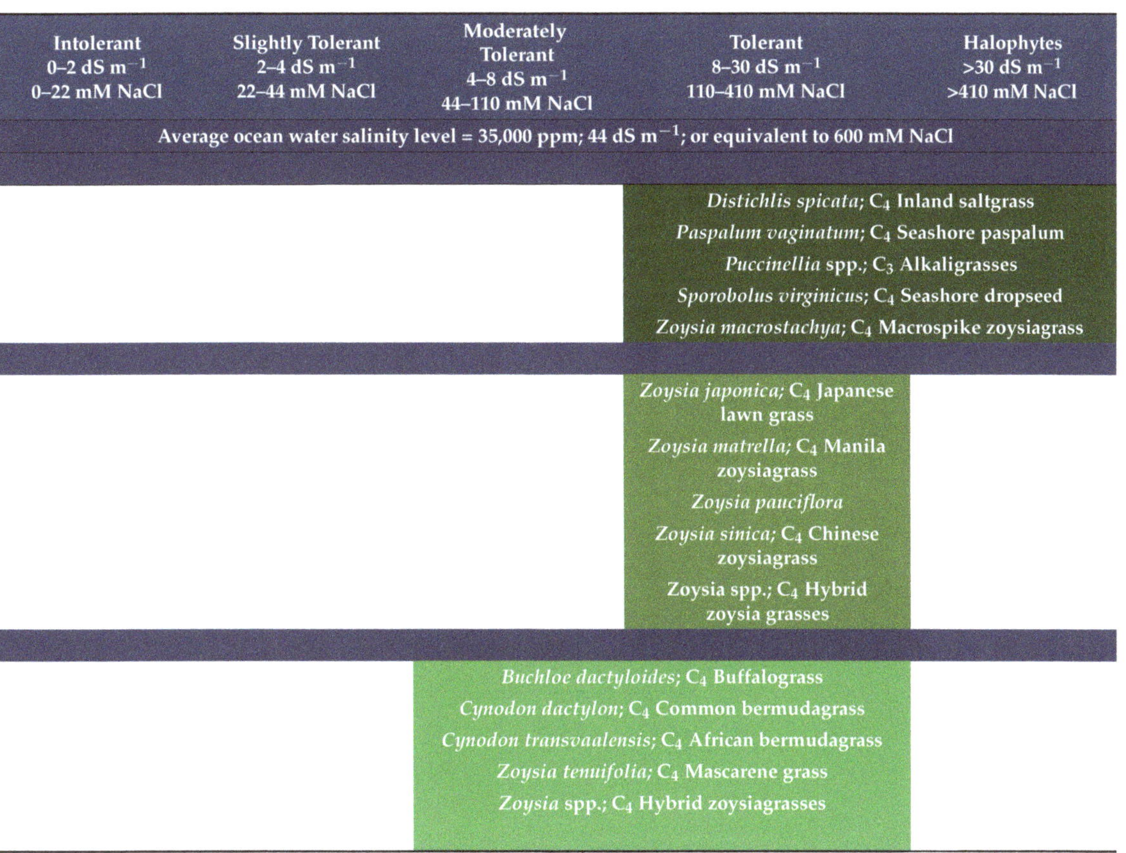

Intolerant 0–2 dS m^{-1} 0–22 mM NaCl	Slightly Tolerant 2–4 dS m^{-1} 22–44 mM NaCl	Moderately Tolerant 4–8 dS m^{-1} 44–110 mM NaCl	Tolerant 8–30 dS m^{-1} 110–410 mM NaCl	Halophytes >30 dS m^{-1} >410 mM NaCl
Average ocean water salinity level = 35,000 ppm; 44 dS m^{-1}; or equivalent to 600 mM NaCl				
				Distichlis spicata; C$_4$ Inland saltgrass
				Paspalum vaginatum; C$_4$ Seashore paspalum
				Puccinellia spp.; C$_3$ Alkaligrasses
				Sporobolus virginicus; C$_4$ Seashore dropseed
				Zoysia macrostachya; C$_4$ Macrospike zoysiagrass
			Zoysia japonica; C$_4$ Japanese lawn grass	
			Zoysia matrella; C$_4$ Manila zoysiagrass	
			Zoysia pauciflora	
			Zoysia sinica; C$_4$ Chinese zoysiagrass	
			Zoysia spp.; C$_4$ Hybrid zoysia grasses	
		Buchloe dactyloides; C$_4$ Buffalograss		
		Cynodon dactylon; C$_4$ Common bermudagrass		
		Cynodon transvaalensis; C$_4$ African bermudagrass		
		Zoysia tenuifolia; C$_4$ Mascarene grass		
		Zoysia spp.; C$_4$ Hybrid zoysiagrasses		

Table 2. *Cont.*

Intolerant 0–2 dS m^{-1} 0–22 mM NaCl	Slightly Tolerant 2–4 dS m^{-1} 22–44 mM NaCl	Moderately Tolerant 4–8 dS m^{-1} 44–110 mM NaCl	Tolerant 8–30 dS m^{-1} 110–410 mM NaCl	Halophytes >30 dS m^{-1} >410 mM NaCl
		Agrostis alba; C$_3$ Redtop bentgrass *Agrostis canina*; C$_3$ Velvet bentgrass *Agrostis capillaris*; C$_3$ Colonial bentgrass *Agrostis* spp.; C$_3$ Hybrid bentgrasses *Agrostis stolonifera*; C$_3$ Creeping bentgrass *Axonopus* spp.; C$_4$ Carpetgrasses *Bouteloua gracilis*; C$_4$ Blue grama grass *Eremochloa ophiuroides*; C$_4$ Centipedegrass *Fescue ovina* var. *ovina*; C$_3$ Sheeps fescue *Festuca arundinacea*; C$_3$ Tall fescue *Festuca longifolia*; C$_3$ Hard fescue *Festuca rubra* subsp. *commutata*; C$_3$ Chewings fescue *Festuca rubra* var. *rubra*; C$_3$ Creeping red fescue *Koeleria macrantha*; C$_3$ prairie Junegrass *Lolium perenne*; C$_3$ Perennial ryegrass *Lolium* spp.; C$_3$ Hybrid ryegrasses *Pennisetum clandestinum*; C$_4$ Kikuyugrass *Poa pratensis*; C$_3$ Kentucky bluegrass *Poa* spp.; C$_{3-4}$ Hybrid bluegrasses *Poa supina*; C$_3$ Supina bluegrass *Poa trivialis*; C$_3$ Rough-stock bluegrass *Stenotaphrum secundatum*; C$_4$ St. Augustinegrass		
	Paspalum notatum; C$_4$ Bahiagrass *Lolium multiflorum*; C$_3$ Annual ryegrass *Poa* spp.; C$_3$ Hybrid bluegrasses			
Poa annua; C$_3$ Annual bluegrass				

5.2. Environmental Factors

Environmental factors such as temperature and light greatly influence plant salinity resistance. Hot and dry climates are often associated with a greater degree of salinity stress because of the increased evapotranspirational demand that favors salt uptake when compared to cool and humid climates [89]. Thus, warm-season C$_4$ turfgrass species are oftentimes more capable to adapt and survive salinity stress when compared to cool-season C$_3$ species because of their natural habitats. Warm-season turfgrass species include *Cynodon dactylon* × *C. transvaalensis* cv. 'Tifway', *Cynodon dactylon* cv. 'Celebration', and *Paspalum vaginatum* cvs. 'Sea Isle 1' and 'SeaStar' maintained the highest normalized difference vegetation index, visual turf quality, lowest growth reduction, and lowest percentage of leaf firing under elevated salinity when compared to other species tested. Increased shoot growth and turf quality were noted in 'Celebration' bermudagrass and both seashore paspalum varieties at the 15 dS m^{-1} treatment in relation to the control treatment (2.5 dS m^{-1}). However, most zoysiagrass ('Palisades' and 'Zeon') and St. Augustinegrass ('Raleigh', 'Floratam', and 'Palmetto') varieties exhibited decreased growth and turf quality in response to any magnitude of elevated salinity in relation to the control treatment. Both bermudagrass and seashore paspalum cultivars recovered back to an acceptable turf quality at the 15 dS m^{-1} treatment, while no cultivars tested in the study recovered to an acceptable level at salinity levels 30 dS m^{-1} and 45 dS m^{-1} [63,90]. A recent study evaluating the salinity tolerance of eight total varieties in four different C$_3$ turfgrass species, indicated that *Lolium perenne* cv. 'Ringles', *Festuca rubra* L. ssp. *trichophylla* cv. 'Abercharm', *Poa pratensis* cv. 'Prafin', *F. arundinacea* cvs. 'Fesnova', and 'Golden Gate' demonstrated salt tolerance when treated with up to 200 mM of NaCl. These varieties exhibited a higher germination percentage, plant growth, photosynthetic efficiency, and K^+/Na^+ ratio under salt-stressed conditions when compared to the other varieties tested [26]. Significant reductions in turfgrass quality and development associated with gradually increasing salinity levels (from 0.54 mM to 200 mM NaCl) were noted for *Poa pratensis* cv. 'Evora', while *Festuca rubra commutata* cv. 'Casanova' was only moderately affected by salinity. *Lolium perenne* cv. 'Belida' and *Festuca rubra trichophylla* cv. 'Smyrna' had the least-affected clipping yield and total root and shoot dry weight values while increasing salinity and are thus considered highly tolerant to salinity stress [32]. When salinity tolerance of two cultivars of Kentucky bluegrass (*Poa pratensis*), 'Limousine' and 'Kenblue', were evaluated, 'Limousine' exhibited a 25% shoot growth reduction at 4.7 dS m^{-1} EC, while 'Kenblue' exhibited a 25% shoot growth reduction occurring at 3.2 dS m^{-1}. 'Limousine' also maintained a higher K^+/Na^+ ratio and relative water content under moderate salinity (8.2 dS m^{-1}), suggesting that *Poa pratensis* cv. 'Limousine' is more salt-resistant than Poa pratensis cv. Kenblue [91].

5.3. Practical Management Approaches

The most common remedy that crop producers and turfgrass managers use to improve soil structure and ameliorate soil salinity is the application of gypsum ($CaSO_42H_2O$). Gypsum is a cheap and relatively easy to use soil amendment that has the potential to reduce soil pH, remove excess sodium ions from the soil's cation exchange sites, and provide calcium to salt-stressed plants. The sulfur component in gypsum provides the ability to form strong acids in the soil, such as sulfuric acid (H_2SO_4), to lower the soil pH. The soil pH of a sandy loam soil when 4 tons ha^{-1} of gypsum was applied experienced a slight decrease over a two-year period [92]. In the same study, the effect that gypsum applications had on saline soil reclamation was greater when paired with a leaching factor. Electrical conductivity, sodium absorption ration, and exchangeable sodium percentage values demonstrated a decreasing trend as application rates of gypsum increased with a leaching factor. The highest crop yield of *Allium cepa* cv. 'Bombe red' was found in the treatment with the highest gypsum rate (4 tons ha^{-1}) plus a leaching factor. Use of only a leaching factor provided non-significant results in saline soil reclamation when compared to treatments including both a leaching factor and gypsum [92]. The calcium in gypsum provides a stronger competition for Na^+ on the soil cation exchange capacity sites, causing the Na^+ to be replaced by the Ca^{2+} and the insoluble sodium sulfate (Na_2SO_4) to be leached out. This helps prevent soil deflocculation and unstable soil structures, which are common issues in saline soils [93]. Applications of gypsum on a mixture of sand-based soil media increased the soil moisture content and calcium content and decreased the sodium adsorption ratio of all soil media mixtures when compared to gypsum-free treatments. However, gypsum had no significant relationship with soil pH and clipping yield of Kentucky bluegrass (*Poa pratensis*) on any soil media [94]. Gypsum treatment had a significant effect on the salinity tolerance of 'Zeon' Zoysiagrass, 'Platinum' Seashore paspalum, and 'TifEagle' bermudagrass after 2 weeks of treatment. Electrolyte leakage of salt-stressed plants decreased with the application of gypsum in all tested turfgrass species [95]. Exogenous calcium supplementation in rice was found to reduce cadmium-induced oxidative stress through a reduction in hydrogen peroxide content, and an increased activity of antioxidant enzyme activity superoxide dismutase, catalase, and glutathione S-transferase when compared to cadmium treatments with no exogenous calcium applications. Furthermore, increased carotenoid content and visual rice quality was associated with calcium-treated rice at all levels of cadmium stress including the control [96]. An increase in calcium concentration in rhizosphere was also noted in Challa's [92] study with applications of gypsum and leaching factors on onion crops, thereby reducing the Na uptake and increasing K^+ uptake, which in turn caused an increased K^+/Na^+ ratio in roots while the turfgrass species were exposed to saline conditions [95].

Recent research has indicated that the use of bioorganic soil amendments has had positive effects on plant salinity tolerance and enhanced physical, chemical, and biological properties of saline soils. Applications of humic acid (0.84 and 2.54 L ha^{-1}) on two golf course fairways, one in Colorado and the other in North Dakota, improved visual turf quality, organic matter content, and microbial biomass in both inherently saline soils and soils irrigated with recycled wastewater. However, humic acid did not significantly reduce soil pH or electrical conductivity over the two-year period but increased other aspects of the soil to yield a better-quality fairway [97]. The use of organic amendments on saline soils also improved many physical properties including porosity, hydraulic conductivity, and soil structure stability, which are all negatively affected by soil salinity [93,97]. Soil microbes are extensively associated with nutrient availability through mineralization and immobilization, thus potentially improving plant tolerance and soil fertility under saline conditions [93]. Plant-growth-promoting rhizobacteria have various direct and indirect mechanisms to mitigate plant growth and quality reduction in responses to salt stress [98–100]. A salt-sensitive rice cultivar (*Oryza sativa* cv. 'GJ17') that was inoculated with the plant-growth-promoting rhizobacterium species *Pseudomonas pseudoalcaligenes* and *Bacillus pumilus* (independently and separately) experienced a reduction in the harmful effects of salinity stress. Treatments including plant-growth-promoting rhizobacteria resulted in increased plant biomass amongst all treatments, and the treatment with both plant-growth-promoting rhizobacterium species exhibited a synergistic effect to create the greatest biomass. Inoculation of 'GJ17' with plant-growth-promoting rhizobacteria reduced the toxicity of reactive oxygen species on plant cells, as reflected through a decrease in lipid peroxidation and caspase-like protease activity. In effect, this enhanced membrane stability, decreased programmed cell death, and increased cell viability of the plants when under salt stress [99].

Salt stress originates from saline solutions remaining in the root zones for a sufficient period for plants to accumulate salts to a toxic level. Therefore, sufficient leaching and adequate drainage are critical when using reclaimed water as an irrigation source [65,101]. Highly maintained turfgrass surfaces such as golf course greens, tees, and fairways are highly prone to salinity stress due to the low mowing height. Since salinity stress largely affects and accumulates salts within leaf tissues, turfgrasses with lower shoot biomasses can exhibit more severe salinity stress symptoms when compared to taller cut landscapes [102–104]. As the mowing heights increased from 15 mm to 45 mm of three bermudagrass cvs. ('Tifway', 'Tifgreen', and 'Tifdwarf'), higher turf quality was noted in the higher-cut treatments (45 mm) when exposed to a 16 dS/m salinity level across all cultivars. In addition to higher turf quality with increased mowing heights, higher photosynthetic rates and total nonstructural carbohydrate contents were also responsible for increased salt resistance of bermudagrass under salinity stress [103]. Lower mowing heights of 'L-93' creeping bentgrass (*Agrostis stolonifera*) putting greens were reported to severely decrease in salinity resistance as determined by visual turf quality. When mowed at 6.4, 12.7, and 25.4 mm the turfgrass reached an unacceptable quality at soil EC 4.1, 12.5, and 13.9 dS/m, respectively [102]. The previous studies suggest that a higher shoot biomass increases turfgrass salinity tolerance by providing more plant tissue to sequester or excrete saline ions and produce more photosynthates.

Silicon has become a renowned beneficial macronutrient that can help mitigate a wide variety of biotic and abiotic stresses [105]. Silicon has been reported to regulate carbohydrate metabolism to improve plant yield and photosystem II photochemical efficiency under salinity stress [106,107]. Chlorophyll content was significantly increased in both salt sensitive species of turfgrass, *Lolium perenne* and *Festuca arundinacea* when silicon was applied while under salt stress, while more salt resistant *Cynodon dactylon* only exhibited a slight increase in chlorophyll content [108]. This would suggest that salt sensitive plant species may benefit from exogenous silicon applications greater than salt resistant species. The mechanisms in which silicon increases photosynthesis and plant growth under abiotic stresses is relatively unknown. However, silicon's effects on ion uptake and oxidative stress have been studied intensively. In the same study, silicon treatments reduced both shoot and root Na^+ concentrations in *Cynodon dactylon, Lolium perenne*, and *Festuca arundinacea*, while no significant increases in root or shoot potassium content were noted across all species [108] indicating that silicon may influence plant ion uptake to improve salinity resistance. Application of silicon overall reduced root and shoot concentration of Na^+ in wheat. The silicon reduced root absorption and translocation to the shoots of Na^+ resulting in a decrease in Na^+/K^+ ratio to mitigate salinity stress [109]. The same mechanism of silicon induced salt stress mitigation was found in maize with applications of 1 mM of Si, in the form of Na_2SiO_3. The applied Si alleviated salt stress through depreciations in Na^+/K^+ ratio, Na^+ ion uptake at the surface of maize roots, and translocation in plant tissues with reduced Na^+ accumulation in leaf tissues [110]. The mechanism in which the silicon reduced Na^+ uptake and transport within the plant is suggested to be a silicon induced stimulation of H^+-ATPase enzymes on the root plasma membranes [111]. H^+-ATPase H^+-pyrophosphatase enzymes play a crucial role in creating a potential gradient to power the tonoplastic Na^+/H^+ antiport that transfers Na^+ from the cytoplasm to the vacuole [112]. Exogenous silicon applications have enhanced antioxidant enzyme activities and decreased oxidative stress in many plant species [113]. In wheat, applications of silicon at 5 mM and 30 mM increased antioxidant enzyme activity of both superoxide dismutase and catalase. Resulting in more efficient reactive oxygen species scavenging and reduced oxidative stress while exposed to 7.6 dS m^{-1} salinity level in the field [106]. Three turfgrass species (*Cynodon dactylon, Lolium perenne*, and *Festuca arundinacea*) demonstrated a similar effect on reducing oxidative stress through maintaining a higher relative water content and mitigating electrolyte leakage [108].

Table 3. Some trials on mitigation of salinity, heat, and drought stresses by exogenous applications of plant metabolites among turfgrass species.

Exogenous Applications	Turfgrasses Tested	The Rates Used	Salinity Stress Level Range Used	Findings and References
24-epibrassinolide (EBL)-a plant hormone	Perennial ryegrass Tall fescue	0.15 mg L^{-1} EBL in 100 mL of the solution per pot (18 cm diameter and 20 cm depth)	0–6 dS m^{-1}	The EBL application under salt stress alleviated loss in clipping yield by 35% and 12%; and reduced leaf firing, through weeks 2–6 post-application, by75–40% and 50–20% for perennial ryegrass and tall fescue, respectively [114].
Chitosan (CTS)-a natural polysaccharide	Creeping bentgrass	0.1, 0.2, 0.5, 1, and 2 g L^{-1}	0–200 mM NaCl	The application of CTS increased antioxidant enzyme activities, thereby reducing oxidative damage to roots and leaves [115].
Glycinebetaine (GB)-an osmoprotectant	Creeping bentgrass Kentucky bluegrass Perennial ryegrass Tall fescue	50, 100, 150, or 200 mM solution of GB of seed priming	0.1 and 14.6 dS m^{-1}	Seeds primed with GB showed a higher germination rate (11.0% to 13.9% increase) and seedling growth (19.3% to 20.7% increase) in mannitol or NaCl solution than in distilled water [116].
Melatonin (ME)-a natural hormone	Creeping bentgrass	0 or 20 μM ME	Drought stress	ME-alleviation of drought-induced leaf senescence in creeping bentgrass was associated with the down-regulation of chlorophyll catabolism and the synergistically interaction with CK-synthesis gene and signaling pathways [117].
Proline-a amino acid	Creeping bentgrass	Proline (10 mM)	Heat stress	Proline-enhanced heat tolerance of creeping bentgrass [118].
Spermidine (Spd)-a polyamine compound	Zoysiagrass (*Zoysia japonica* Steud) cultivars, 'Z081' and 'Z057'	Spd 0, 0.15, 0.30, 0.45, 0.60 mM	0–200 mM NaCl	H_2O_2 and malondialdehyde (MDA) levels significantly decreased in both cultivars treated with Spd [119].

Table 3. *Cont.*

Exogenous Applications	Turfgrasses Tested	The Rates Used	Salinity Stress Level Range Used	Findings and References
γ-Aminobutyric acid (GABA)-a neurotransmitter, a chemical messenger in human brain	Creeping bentgrass cv. Penncross	0.5 mM GABA	0–250 mM NaCl	GABA application is an efficient approach to enhance salt tolerance of creeping bentgrass during a prolonged period of salt stress and also provides valuable information to better understand key candidate genes and regulatory pathways of GABA-induced salt tolerance in plants [120].

5.4. Exogenous Applications of Plant Metabolites to Turfgrass Species

For many years, several groups of plant growth regulators have been used for improving turf quality and enhancing environmental stress tolerances. Propiconozole, a triazole fungicide that now can be used as a plant growth regulator, has shown to ameliorate salt stress in many plant species including annual vinca (*Catharanthus roseus*), perennial ryegrass (*Lolium perenne*), and Kentucky bluegrass (*Poa pratensis* cv. 'Plush') [121–123]. Annual vinca (*Catharanthus roseus*) exhibited higher root biomass and antioxidant enzyme activity under a salinity and propiconozole treatment (80 mM NaCl and 20 mg propiconozole) when compared to control (80 mM NaCl). Although the propiconozole caused reduced stem length and leaf surface area, applications of plant growth regulators on annual vinca enhanced salinity tolerance [121]. In addition to propiconozole, a fortified seaweed extract of seaweed (*Ascophyllum nodosum*), humic acid, thiamin, and L-ascorbic acid were shown to enhance salinity tolerance of perennial ryegrass. In Yan's [123] dissertation, applications of both propiconozole and fortified seaweed extract had positive effects on leaf water potential, total lipid concentration, and nutrient content in perennial ryegrass while under salt stress. Propiconozole provided the greatest leaf water content and total lipid concentration, and fortified seaweed extract provided the greatest decrease in Na^+ and Cl^- leaf tissue concentrations while under salt stress [123]. These results suggest that both propiconozole and fortified seaweed extract can be used to enhance salinity tolerance in turfgrass species through reductions in osmotic pressure and lipid peroxidation. A similar enhancement of salinity tolerance by foliar seaweed extract applications was noted on seashore paspalum (*Paspalum vaginatum* cv. 'Salam'). While under saline conditions, seashore paspalum treated with seaweed extractant exhibited higher turf quality, K and Ca leaf content, photochemical efficiency, and antioxidant enzyme activity when compared to plants not treated with seaweed extractant [124]. In both studies, applications of fortified seaweed extract and propiconozole provided greater benefits only when the plant was salt-stressed [123,124]. Another plant growth regulator researched for turfgrass salinity tolerance is aminoethoxyvinylglycine, which is an ethylene synthesis inhibitor commonly used on fruit to delay ripening [125,126]. Drake's [125] dissertation studied the effects of many plant growth regulator groups, including aminoethoxyvinylglycine, on creeping bentgrass' (*Agrostis stolonifera*) salt tolerance. Applications of aminoethoxyvinylglycine at 10 μM exhibited the highest fresh root weight, tiller count, and total chlorophyll content of creeping bentgrass, as well as the lowest electrolyte leakage amongst all tested plant growth regulators at 50 mM NaCl.

Both fungal and bacterial endophytes typically colonize and live within an individual host plant throughout its life cycle without causing any significant parasitic symptoms as symbiotic relationships. Among turfgrasses, several species have been identified to mitigate salinity stresses [127]. The mechanisms are highly related to the host plant water and nutrient uptake, and enhanced plant hormone productions to minimize salinity, drought, and heat stresses. Turf quality and carotenoid content were positively correlated with the incidence of the phyla *Chloroflexi* and *Fibrobacteres* in rhizosphere soil, and indole acetic acid level was positively correlated with the phyla *Actinobacteria* and *Chloroflexi* in the roots. The salt-tolerant bacterium Enterobacter ludwigii B30 was isolated from *Paspalum vaginatum*. A salt tolerance test showed that B30 could grow normally in a 500 mM NaCl medium [127].

5.5. Symbiotic Relationship with Soil Microorganisms and Genetic Modifications

Responses of *Puccinellia distans* (Table 2), a halophytic grass with low (50 mM) and high (200 mM) NaCl salinity, were studied in a sand culture experiment with and without inoculation by the arbuscular mycorrhizal fungus *Claroideoglomus etunicatum* isolated from its saline habitat [128]. Plant biomass was found to be uninfluenced by salinity levels, but a tendency to develop a higher biomass was observed in arbuscular mycorrhizal fungus plants under the lower and higher saline conditions. Interestingly, leaf photosynthesis increased regardless of salinity or arbuscular mycorrhizal fungus inoculation. Arbuscular mycorrhizal fungus-inoculated plants demonstrated a higher water-use efficiency under the higher saline condition. Arbuscular mycorrhizal fungus inoculation significantly increased leaf osmotic potential. Arbuscular mycorrhizal fungus colonization diminished salt-induced malondialdehyde accumulation, although antioxidative enzymes responded differently. K and Ca contents were not affected by the treatments. Arbuscular mycorrhizal fungus inoculation increased salinity tolerance also by improving water relations and protections against oxidative damage in the leaves [128]. Due to variable environmental factors, the influence of arbuscular mycorrhizal fungus on plant biomass was found to be unrelated to plant phylogeny. The greater biomass accumulation in arbuscular mycorrhizal fungus plants is partially influenced by improved water status, photosynthetic efficiency, and uptake of Ca and K in plants irrespective of salinity stress. In many cases, the uptake of N and P was higher in arbuscular mycorrhizal fungus plants as the salinity increased. Additionally, the activities of malondialdehyde, peroxidase, and superoxide dismutase, as well as the proline content, changed due to arbuscular mycorrhizal fungus inoculation under salinity stress [129].

Genetic diversities (Table 2) exist from a morphologic level to a molecular level among turfgrass germplasm resources [4,130–134], and the extents of variation in their salinity tolerance have been intensively studied. A salt-induced CdWRKY50 was isolated and analyzed in wild bermudagrass [135]. The expression of CdWRKY50 was prominently induced by salt, drought, cold, and abscisic acid (ABA) treatments. Subcellular localization analysis revealed its localization in the nucleus. CdWRKY50-silencing bermudagrass conferred enhanced salt tolerance. In this study, authors also found that transgenic *Arabidopsis* plants overexpressing CdWRKY50 showed decreased salt tolerance. Salt- and antioxidant-related genes, including AtRD29A, AtRD29B, AtDREB2A, AtDREB2B, AtSOS1, AtSOS3, AtSOD1, and AtCAT1 were all clearly repressed in CdWRKY50-overexpressed *Arabidopsis* plants. In conclusion, CdWRKY50 plays a key role in the negative regulation of salt stress within bermudagrass, which provides a new perspective for the underlying molecular mechanism of the CdWRKY50 gene involved in salt stress response [134,135].

6. Conclusions

The greatest advantage for improving turfgrass salinity tolerance is the diversity of turfgrass species and cultivars (Table 2). The most significant challenge facing future improvement is the general lack of in-depth research compared to that of common agricultural crops. Two of the main reasons may be that turfgrasses are not food- or fiber-related crops, and there are more severe restrictions on potable water uses for turfgrass irrigation worldwide. Another major challenge is that turfgrass salinity stress is closely associated with other environmental stresses such as drought, heat, soil alkalinity, nutrient imbalances, and other poor soil conditions. Large-scale collaboration with multiple research units seem to promise great potential in facing such challenges. In recent years, soil salinity has become one of the most prominent plant-growth-limiting factors, especially for highly maintained turfgrass landscapes. As many environmental factors, management practices, and soil properties contribute to the cause of salinity stress in plants, the lack of freshwater for irrigation purposes has driven many producers to utilize saline irrigation water sources, thus negatively impacting soil health and fertility. Research over the past few decades has indicated significant metabolic and visual turfgrass responses to soil salinity that negatively affect plant growth and development. Reports have indicated significant differences of salinity tolerance on the varietal level of both cool-season and warm-season turfgrass species. Among these factors, the most important one is that turfgrass salinity tolerance is strongly controlled by a cultivar's specific genome, as many active and passive resistance mechanisms have been observed in turfgrass species and cultivars when exposed to saline conditions. Thus, understanding the effects of soil salinity on turfgrass' physiological processes will allow for improvements in future breeding and management practices to adapt to the worldwide issue. This review has documented the current state of knowledge on intra- and inter-species differences in salinity resistance by describing recent research on specific plant responses and resistance mechanisms to soil salinity. Certain management practices that alleviate salinity stress are critical factors in maintaining plant growth and development in saline conditions.

Turfgrass salinity stress has been well documented for decades. However, as the severity and commonality of saline soils have increased in recent years, there is high demand for new technology to aid turfgrass managers in adapting or combatting saline conditions. In efforts to adapt to soil salinity, further salinity tolerance screening should be conducted on warm-season grasses, especially common (*Cynodon dactylon*) and hybrid bermudagrass (*C. dactylon* × *C. transvaalensis*) species, as they are commonly used in regions that are prone to salinization as the largest single cultivated grass land area (both turfgrass and forage) in the U.S. In addition, future research should include a recovery component after salinity stress to provide more details for consumers regarding a specific turfgrass cultivar's recuperative capacity to salinity stress. As prior studies have indicated, exogenous applications of various chemical products such as PGRs, silicon, various external applicants (Table 3), and even biological agents have elevated abiotic stress resistance and improved carbohydrate metabolism while under stress for many salt-sensitive turfgrass species [108,123,127]. There is still a need for further research on optimal turfgrass nutrition and nutrition homeostasis before and during salinity stress. Overall improvement in turfgrass salinity resistance requires focused breeding efforts as well as extensive screening protocols to develop new cultivars that can grow in saline conditions while maintaining essential management and playability characteristics as three important aspects for the future.

Author Contributions: Conceptualization, H.L. (Haibo Liu), J.L.T. and H.L. (Hong Luo); J.L.T., original draft preparation. All authors have read and agreed to the published version of the manuscript.

Funding: This work was partially supported by Biotechnology Risk Assessment Grant Program competitive grants nos. 2019-33522-30102 and 2021-33522-35342 from the USDA National Institute of Food and Agriculture.

Institutional Review Board Statement: Not applicable.

Informed Consent Statement: Not applicable.

Data Availability Statement: Not applicable.

Acknowledgments: The authors are grateful to Adam Gore for valuable comments and a thorough editing of the manuscript.

Conflicts of Interest: The authors declare no conflict of interest.

References

1. Thompson, G.L.; Kao-Kniffin, J. Applying biodiversity and ecosystem function theory to turfgrass management. *Crop. Sci.* **2017**, *57*, S238–S248. [CrossRef]
2. Ignatieva, M.; Hedblom, M. An alternative urban green carpet. *Science* **2018**, *362*, 148–149. [CrossRef] [PubMed]
3. Liu, H.; Pressarakli, M.; Luo, H.; Menchyk, N.; Baldwin, C.W.; Taylor, D.H. Chapter 35: Growth and physiological responses of turfgrasses under stressful conditions. In *Handbook of Plant and Crop Physiology Expanded*, 4th ed.; CRC Press: Boca Raton, FL, USA, 2021; pp. 713–776.

4. Fan, J.; Zhang, W.; Amombo, E.; Hu, L.; Kjorven, J.O.; Chen, L. Mechanisms of environmental stress tolerance in turfgrass. *Agronomy* **2020**, *10*, 522. [CrossRef]
5. Chavarria, M.R.; Wherley, B.; Jessup, R.; Chandra, A. Leaf anatomical responses and chemical composition of warm-season turfgrasses to increasing salinity. *Curr. Plant Biol.* **2020**, *22*, 100147. [CrossRef]
6. FAO-UN Map of Aridity (Global~19 km) 2021. Available online: https://data.apps.fao.org/map/catalog/static/search?keyword=aridity (accessed on 15 January 2023).
7. Havlin, J.L.; Tisdale, S.L.; Nelson, W.L.; Beaton, J.D. *Soil Fertility and Fertilizers*, 8th ed.; Pearson Education India: Delhi, India, 2016; pp. 51–72.
8. Parida, A.K.; Das, A.B. Salt tolerance and salinity effects on plants: A review. *Ecotoxicol. Environ. Saf.* **2005**, *60*, 324–349. [CrossRef]
9. Zhu, J.-K. Plant salt tolerance. *Trends Plant Sci.* **2001**, *6*, 66–71. [CrossRef]
10. Munns, R.; Tester, M. Mechanisms of salinity tolerance. *Annu. Rev. Plant Biol.* **2008**, *59*, 651–681. [CrossRef]
11. Singh, A. Soil salinity: A global threat to sustainable development. *Soil Use Manag.* **2022**, *38*, 39–67. [CrossRef]
12. Carrow, R.N.; Duncan, R.R. *Salt-Affected Turfgrass Sites: Assessment and Management*; Ann Arbor Press: Chelsea, MI, USA, 1998; ISBN 978-1-57504-091-2.
13. Marcum, K.B. Use of Saline and non-potable water in the turfgrass industry: Constraints and developments. *Agric. Water Manag.* **2006**, *80*, 132–146. [CrossRef]
14. Bondarenko, S.; Gan, J.; Ernst, F.; Green, R.; Baird, J.; McCullough, M. Leaching of pharmaceuticals and personal care products in turfgrass soils during recycled water irrigation. *J. Environ. Qual.* **2012**, *41*, 1268–1274. [CrossRef]
15. Qian, Y.; Lin, Y. Comparison of soil chemical properties prior to and five to eleven years after recycled water irrigation. *J. Environ. Qual.* **2019**, *48*, 1758–1765. [CrossRef]
16. Qian, Y.L.; Mecham, B. Long-term effects of recycled wastewater irrigation on soil chemical properties on golf course fairways. *Agron. J.* **2005**, *97*, 717–721. [CrossRef]
17. Devitt, D.A.; Morris, R.L.; Kopec, D.; Henry, M. Golf course superintendents' attitudes and perceptions toward using reuse water for irrigation in the southwestern United States. *Horttechnology* **2004**, *14*, 577–583. [CrossRef]
18. Younis, A.; Riaz, A.; Mushtaq, N.; Tahir, Z.; Siddique, M.I. Evaluation of the suitability of sewage and recycled water for irrigation of ornamental plants. *Commun. Soil Sci. Plant Anal.* **2015**, *46*, 62–79. [CrossRef]
19. Carrow, R.N.; Duncan, R.R. *Best Management Practices for Saline and Sodic Turfgrass Soils: Assessment and Reclamation*; CRC Press: Boca Raton, FL, USA, 2012; pp. 6–10.
20. McCune, D.C. Effects of airborne saline particles on vegetation in relation to variables of exposure and other factors. *Environ. Pollut.* **1991**, *74*, 176–203. [CrossRef]
21. Alshammary, S.F.; Qian, Y.L.; Wallner, S.J. Growth response of four turfgrass species to salinity. *Agric. Water Manag.* **2004**, *66*, 97–111. [CrossRef]
22. Chang, B.; Wherley, B.G.; Aitkenhead-Peterson, J.A.; West, J.B. Irrigation salinity effects on Tifway bermudagrass growth and nitrogen uptake. *Crop Sci.* **2019**, *59*, 2820–2828. [CrossRef]
23. Karimi, I.Y.M.; Kurup, S.S.; Salem, M.A.M.A.; Cheruth, A.J.; Purayil, F.T.; Subramaniam, S.; Pessarakli, M. Evaluation of bermuda and paspalum grass types for urban landscapes under saline water irrigation. *J. Plant Nutr.* **2018**, *41*, 888–902. [CrossRef]
24. Keyikoglu, R.; Aksu, E.; Arslan, M. Effects of salinity stress on the growth characteristics of four turfgrass species. *Fresenius Environ. Bull.* **2019**, *28*, 2942–2948.
25. Li, J.; Ma, J.; Guo, H.; Zong, J.; Chen, J.; Wang, Y.; Li, D.; Li, L.; Wang, J.; Liu, J. Growth and physiological responses of two phenotypically distinct accessions of centipedegrass (*Eremochloa ophiuroides* (Munro) Hack.) to salt stress. *Plant Physiol. Biochem.* **2018**, *126*, 1–10. [CrossRef]
26. Tenikecier, H.; Ates, E. Impact of salinity on germination and seedling growth of four cool-season turfgrass species and cultivars. *Pol. J. Environ. Stud.* **2022**, *31*, 1813–1821. [CrossRef] [PubMed]
27. Qian, Y.L.; Fu, J.M. Response of creeping bentgrass to salinity and mowing management: Carbohydrate availability and ion accumulation. *HortSci* **2005**, *40*, 2170–2174. [CrossRef]
28. Tada, Y.; Komatsubara, S.; Kurusu, T. Growth and physiological adaptation of whole plants and cultured cells from a halophyte turf grass under salt stress. *AoB Plants* **2014**, *6*. [CrossRef] [PubMed]
29. Qi, J.; Song, C.-P.; Wang, B.; Zhou, J.; Kangasjärvi, J.; Zhu, J.-K.; Gong, Z. Reactive oxygen species signaling and stomatal movement in plant responses to drought stress and pathogen attack: Ros signaling and stomatal movement. *J. Integr. Plant Biol.* **2018**, *60*, 805–826. [CrossRef] [PubMed]
30. Johnson, R.; Vishwakarma, K.; Hossen, M.S.; Kumar, V.; Shackira, A.M.; Puthur, J.T.; Abdi, G.; Sarraf, M.; Hasanuzzaman, M. Potassium in plants: Growth regulation, signaling, and environmental stress tolerance. *Plant Physiol. Biochem.* **2022**, *172*, 56–69. [CrossRef] [PubMed]
31. Taïbi, K.; Taïbi, F.; Ait Abderrahim, L.; Ennajah, A.; Belkhodja, M.; Mulet, J.M. Effect of salt stress on growth, chlorophyll content, lipid peroxidation and antioxidant defence systems in *Phaseolus vulgaris* L. *South Afr. J. Bot.* **2016**, *105*, 306–312. [CrossRef]
32. Nikolaeva, M.K.; Maevskaya, S.N.; Shugaev, A.G.; Bukhov, N.G. Effect of drought on chlorophyll content and antioxidant enzyme activities in leaves of three wheat cultivars varying in productivity. *Russ. J. Plant Physiol* **2010**, *57*, 87–95. [CrossRef]
33. Gao, Y.; Li, D. Assessing leaf senescence in tall fescue (*Festuca arundinacea* Schreb.) under salinity stress using leaf spectrum. *Eur. J. Hortic. Sci.* **2015**, *80*, 170–176. [CrossRef]

34. Hameed, A.; Ahmed, M.Z.; Hussain, T.; Aziz, I.; Ahmad, N.; Gul, B.; Nielsen, B.L. Effects of salinity stress on chloroplast structure and function. *Cells* **2021**, *10*, 2023. [CrossRef]

35. Ozkur, O.; Ozdemir, F.; Bor, M.; Turkan, I. Physiochemical and antioxidant responses of the perennial xerophyte *Capparis Ovata* Desf. to drought. *Environ. Exp. Bot.* **2009**, *66*, 487–492. [CrossRef]

36. Kurup, S.S.; Abdul Mohsen Ali Salem, M.; Cheruth, A.J.; Sreeramanan, S.; Purayil, F.T.; Al Amouri, A.W.; Pessarakli, M. Changes in antioxidant enzyme activity in turfgrass cultivars under various saline water irrigation levels to suit landscapes under arid regions. *Commun. Soil Sci. Plant Anal.* **2017**, *48*, 1989–2001. [CrossRef]

37. Yang, Y.; Guo, Y. Elucidating the molecular mechanisms mediating plant salt-stress responses. *New Phytol.* **2018**, *217*, 523–539. [CrossRef]

38. Hao, S.; Wang, Y.; Yan, Y.; Liu, Y.; Wang, J.; Chen, S. A review on plant responses to salt stress and their mechanisms of salt resistance. *Horticulturae* **2021**, *7*, 132. [CrossRef]

39. Guo, H.; Wang, Y.; Li, D.; Chen, J.; Zong, J.; Wang, Z.; Chen, X.; Liu, J. Growth response and ion regulation of seashore paspalum accessions to increasing salinity. *Environ. Exp. Bot.* **2016**, *131*, 137–145. [CrossRef]

40. Mastalerczuk, G.; Borawska-Jarmułowicz, B.; Kalaji, H.M. How Kentucky bluegrass tolerate stress caused by sodium chloride used for road de-icing? *Environ. Sci. Pollut. Res.* **2019**, *26*, 913–922. [CrossRef] [PubMed]

41. Adams, R. Chlorine Toxicity. Iowa State University Extension TURF Blog 2014. Available online: https://www.extension.iastate.edu/turfgrass/blog/chlorine-toxicity (accessed on 15 January 2023).

42. Geilfus, C.-M. Review on the significance of chlorine for crop yield and quality. *Plant Sci.* **2018**, *270*, 114–122. [CrossRef] [PubMed]

43. Heber, U.; Heldt, H.W. The chloroplast envelope: Structure, function, and role in leaf metabolism. *Annu. Rev. Plant. Physiol.* **1981**, *32*, 139–168. [CrossRef]

44. Chen, W.; He, Z.L.; Yang, X.E.; Mishra, S.; Stoffella, P.J. Chlorine nutrition of higher plants: Progress and perspectives. *J. Plant Nutr.* **2010**, *33*, 943–952. [CrossRef]

45. Xu, R.; Fujiyama, H. Comparison of ionic concentration, organic solute accumulation and osmotic adaptation in Kentucky bluegrass and tall fescue under nacl stress. *Soil Sci. Plant Nutr.* **2013**, *59*, 168–179. [CrossRef]

46. Wang, M.; Zheng, Q.; Shen, Q.; Guo, S. The Critical Role of Potassium in Plant Stress Response. *Int. J. Mol. Sci.* **2013**, *14*, 7370–7390. [CrossRef]

47. Tighe-Neira, R.; Alberdi, M.; Arce-Johnson, P.; Romero, J.; Reyes-Díaz, M.; Rengel, Z.; Inostroza-Blancheteau, C. Role of potassium in governing photosynthetic processes and plant yield. In *Plant Nutrients and Abiotic Stress Tolerance*; Hasanuzzaman, M., Fujita, M., Oku, H., Nahar, K., Hawrylak-Nowak, B., Eds.; Springer: Singapore, 2018; pp. 191–203. ISBN 978-981-10-9043-1.

48. Raza, M.A.S.; Saleem, M.F.; Shah, G.M.; Khan, I.H.; Raza, A. Exogenous application of glycinebetaine and potassium for improving water relations and grain yield of wheat under drought. *J. Soil Sci. Plant Nutr.* **2014**, *14*, 348–364. [CrossRef]

49. Assaha, D.V.M.; Ueda, A.; Saneoka, H.; Al-Yahyai, R.; Yaish, M.W. The role of Na$^+$ and K$^+$ transporters in salt stress adaptation in glycophytes. *Front. Physiol.* **2017**, *8*, 509. [CrossRef] [PubMed]

50. Chen, C.-T.; Lee, C.-L.; Yeh, D.-M. Effects of nitrogen, phosphorus, potassium, calcium, or magnesium deficiency on growth and photosynthesis of *Eustoma*. *Hortscience* **2018**, *53*, 795–798. [CrossRef]

51. Ye, X.; Wang, H.; Cao, X.; Jin, X.; Cui, F.; Bu, Y.; Liu, H.; Wu, W.; Takano, T.; Liu, S. Transcriptome profiling of *Puccinellia tenuiflora* during seed germination under a long-term saline-alkali stress. *BMC Genom.* **2019**, *20*, 589. [CrossRef] [PubMed]

52. Rahimi, E.; Nazari, F.; Javadi, T.; Samadi, S.; Teixeira da Silva, J.A. Potassium-enriched clinoptilolite zeolite mitigates the adverse impacts of salinity stress in perennial ryegrass (*Lolium perenne* L.) by increasing silicon absorption and improving the K/Na ratio. *J. Environ. Manag.* **2021**, *285*, 112142. [CrossRef] [PubMed]

53. Thor, K. Calcium nutrient and messenger. *Front. Plant Sci.* **2019**, *10*, 440. [CrossRef] [PubMed]

54. Manishankar, P.; Wang, N.; Köster, P.; Alatar, A.A.; Kudla, J. Calcium signaling during salt stress and in the regulation of ion homeostasis. *J. Exp. Bot.* **2018**, *69*, 4215–4226. [CrossRef]

55. Wang, G.; Bi, A.; Amombo, E.; Li, H.; Zhang, L.; Cheng, C.; Hu, T.; Fu, J. Exogenous calcium enhances the photosystem II photochemistry response in salt stressed tall fescue. *Front. Plant Sci.* **2017**, *8*, 2032. [CrossRef]

56. García-Sánchez, F.; Simón-Grao, S.; Martínez-Nicolás, J.J.; Alfosea-Simón, M.; Liu, C.; Chatzissavvidis, C.; Pérez-Pérez, J.G.; Cámara-Zapata, J.M. Multiple stresses occurring with boron toxicity and deficiency in plants. *J. Hazard. Mater.* **2020**, *397*, 122713. [CrossRef]

57. Nejad, S.A.G.; Etesami, H. The Importance of boron in plant nutrition. In *Metalloids in Plants*; Deshmukh, R., Tripathi, D.K., Guerriero, G., Eds.; Wiley: Hoboken, NJ, USA, 2020; pp. 433–449. ISBN 978-1-119-48721-0.

58. Bolaños, L.; Lukaszewski, K.; Bonilla, I.; Blevins, D. Why boron? *Plant Physiol. Biochem.* **2004**, *42*, 907–912. [CrossRef]

59. Farooq, M.A.; Saqib, Z.A.; Akhtar, J.; Bakhat, H.F.; Pasala, R.-K.; Dietz, K.-J. Protective role of silicon (si) against combined stress of salinity and boron (b) toxicity by improving antioxidant enzymes activity in rice. *Silicon* **2019**, *11*, 2193–2197. [CrossRef]

60. Carrow, R.N.; Waddington, D.V.; Rieke, P.E. *Turfgrass Soil Fertility and Chemical Problems: Assessment and Management*; Ann Arbor Press: Chelsea, MI, USA, 2001; ISBN 978-1-57504-153-7.

61. Jackson, M.B.; Lee, C.W.; Schumacher, M.A.; Duysen, M.E.; Self, J.R.; Smith, R.C. Micronutrient toxicity in buffalograss. *J. Plant Nutr.* **1995**, *18*, 1337–1349. [CrossRef]

62. Soliman, W.S.; Sugiyama, S.; Abbas, A.M. Contribution of avoidance and tolerance strategies towards salinity stress resistance in eight C3 turfgrass species. *Hortic. Environ. Biotechnol.* **2018**, *59*, 29–36. [CrossRef]

63. Chavarria, M.; Wherley, B.; Thomas, J.; Chandra, A.; Raymer, P. Salinity tolerance and recovery at-tributes of warm-season turfgrass cultivars. *Hortscience* **2019**, *54*, 1625–1631. [CrossRef]
64. Wang, J.; An, C.; Guo, H.; Yang, X.; Chen, J.; Zong, J.; Li, J.; Liu, J. Physiological and transcriptomic analyses reveal the mechanisms underlying the salt tolerance of *Zoysia japonica* Steud. *BMC Plant Biol.* **2020**, *20*, 114. [CrossRef] [PubMed]
65. Pessarakli, M. (Ed.) *Handbook of Turfgrass Management and Physiology*; CRC Press: Boca Raton, FL, USA, 2007; ISBN 978-0-367-38850-8.
66. Munns, R.; Passioura, J.B.; Colmer, T.D.; Byrt, C.S. Osmotic adjustment and energy limitations to plant growth in saline soil. *New Phytol.* **2020**, *225*, 1091–1096. [CrossRef] [PubMed]
67. Marcum, K.B. Salinity tolerance mechanisms of grasses in the subfamily *Chloridoideae*. *Crop Sci.* **1999**, *39*, 1153–1160. [CrossRef]
68. Qian, Y.L.; Engelke, M.C.; Foster, M.J.V. Salinity effects on zoysiagrass cultivars and experimental lines. *Crop Sci.* **2000**, *40*, 488–492. [CrossRef]
69. Naeem, M.; Iqbal, M.; Shakeel, A.; Ul-Allah, S.; Hussain, M.; Rehman, A.; Zafar, Z.U.; Athar, H.-R.; Ashraf, M. Genetic basis of ion exclusion in salinity stressed wheat: Implications in improving crop yield. *Plant Growth Regul.* **2020**, *92*, 479–496. [CrossRef]
70. Marcum, K.B.; Pessarakli, M. Salinity Tolerance and salt gland excretion efficiency of bermudagrass turf cultivars. *Crop Sci.* **2006**, *46*, 2571–2574. [CrossRef]
71. Ghassemi-Golezani, K.; Farhangi-Abriz, S. Foliar sprays of salicylic acid and jasmonic acid stimulate H^+-ATPase activity of tonoplast, nutrient uptake and salt tolerance of soybean. *Ecotoxicol. Environ. Saf.* **2018**, *166*, 18–25. [CrossRef]
72. Keisham, M.; Mukherjee, S.; Bhatla, S. Mechanisms of sodium transport in plants—Progresses and challenges. *Int. J. Mol. Sci.* **2018**, *19*, 647. [CrossRef]
73. Fan, Y.; Wan, S.; Jiang, Y.; Xia, Y.; Chen, X.; Gao, M.; Cao, Y.; Luo, Y.; Zhou, Y.; Jiang, X. Over-expression of a plasma membrane H^+-ATPase SpAHA1 conferred salt tolerance to transgenic *Arabidopsis*. *Protoplasma* **2018**, *255*, 1827–1837. [CrossRef] [PubMed]
74. Ishikawa, T.; Cuin, T.A.; Bazihizina, N.; Shabala, S. Xylem ion loading and its implications for plant abiotic stress tolerance. In *Advances in Botanical Research*; Elsevier: Amsterdam, The Netherlands, 2018; Volume 87, pp. 267–301. ISBN 978-0-12-809390-0.
75. Zarei, M.; Shabala, S.; Zeng, F.; Chen, X.; Zhang, S.; Azizi, M.; Rahemi, M.; Davarpanah, S.; Yu, M.; Shabala, L. Comparing kinetics of xylem ion loading and its regulation in halophytes and glycophytes. *Plant Cell Physiol.* **2020**, *61*, 403–415. [CrossRef] [PubMed]
76. Chen, T.; Cai, X.; Wu, X.; Karahara, I.; Schreiber, L.; Lin, J. Casparian strip development and its potential function in salt tolerance. *Plant Signal. Behav.* **2011**, *6*, 1499–1502. [CrossRef] [PubMed]
77. Zhao, J.; Yuan, S.; Zhou, M.; Yuan, N.; Li, Z.; Hu, Q.; Bethea, F.G.; Liu, H.; Li, S.; Luo, H. Transgenic creeping bentgrass overexpressing Osa-miR393a exhibits altered plant development and improved multiple stress tolerance. *Plant Biotechnol. J.* **2019**, *17*, 233–251. [CrossRef] [PubMed]
78. Seifikalhor, M.; Aliniaeifard, S.; Shomali, A.; Azad, N.; Hassani, B.; Lastochkina, O.; Li, T. Calcium Signaling and Salt Tolerance Are Diversely Entwined in Plants. *Plant Signal. Behav.* **2019**, *14*, 1665455. [CrossRef]
79. Tian, F.; Wang, W.; Liang, C.; Wang, X.; Wang, G.; Wang, W. Overaccumulation of glycine betaine makes the function of the thylakoid membrane better in wheat under salt stress. *Crop J.* **2017**, *5*, 73–82. [CrossRef]
80. Xu, N.; Chu, Y.; Chen, H.; Li, X.; Wu, Q.; Jin, L.; Wang, G.; Huang, J. Rice transcription factor Os-MADS25 modulates root Grgwth and confers salinity tolerance via the ABA-mediated regulatory pathway and ROS scavenging. *PLoS Genet.* **2018**, *14*, e1007662. [CrossRef]
81. Hu, L.; Hu, T.; Zhang, X.; Pang, H.; Fu, J. Exogenous glycine betaine ameliorates the adverse effect of salt stress on perennial ryegrass. *J. Amer. Soc. Hortic. Sci.* **2012**, *137*, 38–46. [CrossRef]
82. Marcum, K.B.; Anderson, S.J.; Engelke, M.C. Salt gland ion secretion: A salinity tolerance mechanism among five zoysiagrass species. *Crop Sci.* **1998**, *38*, 806–810. [CrossRef]
83. Céccoli, G.; Ramos, J.; Pilatti, V.; Dellaferrera, I.; Tivano, J.C.; Taleisnik, E.; Vegetti, A.C. Salt glands in the *Poaceae* family and their relationship to salinity tolerance. *Bot. Rev.* **2015**, *81*, 162–178. [CrossRef]
84. Naidoo, Y.; Naidoo, G. Cytochemical localization of adenosine triphosphatase activity in salt glands of *Sporobolus virginicus* (L.) Kunth. *South Afr. J. Bot.* **1999**, *65*, 370–373. [CrossRef]
85. Sugiura, S. Salt gland characteristics and activity of *Zoysia matrella* Merr. in environments with different levels of salinity. *J. Jpn. Soc. Turfgrass Sci.* **2020**, *48*, 142–148.
86. Imada, S.; Acharya, K.; Yamanaka, N. Short-term and diurnal patterns of salt secretion by *Tamarix ramosissima* and their relationships with climatic factors. *J. Arid. Environ.* **2012**, *83*, 62–68. [CrossRef]
87. Spiekerman, J.J.; Devos, K.M. The halophyte seashore paspalum uses adaxial leaf papillae for sodium sequestration. *Plant Physiol.* **2020**, *184*, 2107–2119. [CrossRef] [PubMed]
88. Kozłowska, M.; Bandurska, H.; Breś, W. Response of lawn grasses to salinity stress and protective potassium effect. *Agronomy* **2021**, *11*, 843. [CrossRef]
89. An, P.; Inanaga, S.; Li, X.J.; Eneji, A.E.; Zhu, N.W. Interactive effects of salinity and air humidity on two tomato cultivars differing in salt tolerance. *J. Plant Nutr.* **2005**, *28*, 459–473. [CrossRef]
90. Jiang, K.; Yang, Z.; Sun, J.; Liu, H.; Chen, S.; Zhao, Y.; Xiong, W.; Lu, W.; Wang, Z.-Y.; Wu, X. Evaluation of the Tolerance and Forage Quality of Different Ecotypes of Seashore Paspalum. *Front. Plant Sci.* **2022**, *13*, 944894. [CrossRef] [PubMed]
91. Qian, Y.L.; Wilhelm, S.J.; Marcum, K.B. Comparative responses of two Kentucky bluegrass cultivars to salinity stress. *Crop Sci.* **2001**, *41*, 1895–1900. [CrossRef]

92. Challa, A.; Kitila, K.; Workina, M. Evaluation of gypsum and leaching application on salinity reclamation and crop yield at Dugada District, East Shoa Zone of Oromia. *Int. J. Environ. Chem.* **2022**, *6*, 1–6. [CrossRef]
93. Bello, S.K.; Alayafi, A.H.; AL-Solaimani, S.G.; Abo-Elyousr, K.A.M. Mitigating soil salinity stress with gypsum and bio-organic amendments: A review. *Agronomy* **2021**, *11*, 1735. [CrossRef]
94. Rahayu, R.; Mo, Y.G.; Soo, C.J. Amendments on salinity and water retention of sand base rootzone and turfgrass yield. *Sains Tanah J. Soil Sci. Agroclimatol.* **2019**, *16*, 103. [CrossRef]
95. Sekar, S. Efficacy of Salinity Mitigation on Warm Season Turfgrasses. Master's Thesis, Cornell University, Ithaca, NY, USA, 2016.
96. Rahman, A.; Mostofa, M.G.; Nahar, K.; Hasanuzzaman, M.; Fujita, M. Exogenous calcium alleviates cadmium-induced oxidative stress in rice (*Oryza sativa* L.) seedlings by regulating the antioxidant defense and glyoxalase systems: Calcium-induced cadmium stress tolerance in rice. *Braz. J. Bot.* **2016**, *39*, 393–407. [CrossRef]
97. Zhu, H.; Li, D. Using humus on golf course fairways to alleviate soil salinity problems. *Horttechnology* **2018**, *28*, 284–288. [CrossRef]
98. Fatima, T.; Arora, N.K. Plant Growth-promoting rhizospheric microbes for remediation of saline soils. In *Phyto and Rhizo Remediation*; Arora, N.K., Kumar, N., Eds.; Microorganisms for Sustainability; Springer: Singapore, 2019; Volume 9, pp. 121–146, ISBN 978-981-329-663-3.
99. Jha, Y.; Subramanian, R.B. PGPR regulate caspase-like activity, programmed cell death, and antioxidant enzyme activity in paddy under salinity. *Physiol. Mol. Biol. Plants* **2014**, *20*, 201–207. [CrossRef]
100. Kumar, K.; Amaresan, N.; Madhuri, K. Alleviation of the adverse effect of salinity stress by inoculation of plant growth promoting rhizobacteria isolated from hot humid tropical climate. *Ecol. Eng.* **2017**, *102*, 361–366. [CrossRef]
101. Manzoor, M.Z.; Sarwar, G.; Aftab, M.; Tahir, M.A.; Zafar, A. Role of leaching fraction to mitigate adverse effects of saline water on soil properties. *J. Agric. Res.* **2019**, *7*, 275–280.
102. Fu, J.M.; Koski, A.J.; Qian, Y.L. Responses of creeping bentgrass to salinity and mowing management: Growth and turf quality. *Hortscience* **2005**, *40*, 463–467. [CrossRef]
103. Shahba, M.A. Interaction effects of salinity and mowing on performance and physiology of bermudagrass cultivars. *Crop Sci.* **2010**, *50*, 2620–2631. [CrossRef]
104. Shahba, M.A.; Alshammary, S.F.; Abbas, M.S. Effects of salinity on seashore paspalum cultivars at different mowing heights. *Crop Sci.* **2012**, *52*, 1358–1370. [CrossRef]
105. Ranjan, A.; Sinha, R.; Bala, M.; Pareek, A.; Singla-Pareek, S.L.; Singh, A.K. Silicon-mediated abiotic and biotic stress mitigation in plants: Underlying mechanisms and potential for stress resilient agriculture. *Plant Physiol. Biochem.* **2021**, *163*, 15–25. [CrossRef]
106. Taha, R.S.; Seleiman, M.F.; Shami, A.; Alhammad, B.A.; Mahdi, A.H.A. Integrated application of selenium and silicon enhances growth and anatomical structure, antioxidant defense system and yield of wheat grown in salt-stressed soil. *Plants* **2021**, *10*, 1040. [CrossRef]
107. Zhu, Y.-X.; Xu, X.-B.; Hu, Y.-H.; Han, W.-H.; Yin, J.-L.; Li, H.-L.; Gong, H.-J. Silicon improves salt tolerance by increasing root water uptake in *Cucumis sativus* L. *Plant Cell Rep.* **2015**, *34*, 1629–1646. [CrossRef]
108. Esmaeili, S.; Salehi, H.; Eshghi, S. Silicon ameliorates the adverse effects of salinity on turfgrass growth and development. *J. Plant Nutr.* **2015**, *38*, 1885–1901. [CrossRef]
109. Pakdel, E.; Daoud, W.A.; Seyedin, S.; Wang, J.; Razal, J.M.; Sun, L.; Wang, X. Tunable photocatalytic selectivity of TiO_2 /SiO_2 nanocomposites: Effect of silica and isolation approach. *Colloids Surf. A Physicochem. Eng. Asp.* **2018**, *552*, 130–141. [CrossRef]
110. Ali, M.; Afzal, S.; Parveen, A.; Kamran, M.; Javed, M.R.; Abbasi, G.H.; Malik, Z.; Riaz, M.; Ahmad, S.; Chattha, M.S.; et al. Silicon mediated improvement in the growth and ion homeostasis by decreasing Na^+ uptake in maize (*Zea mays* L.) cultivars exposed to salinity stress. *Plant Physiol. Biochem.* **2021**, *158*, 208–218. [CrossRef] [PubMed]
111. Liang, Y.; Zhang, W.; Chen, Q.; Ding, R. Effects of silicon on H^+-ATPase and H^+-PPase activity, fatty acid composition and fluidity of tonoplast vesicles from roots of salt-stressed barley (*Hordeum vulgare* L.). *Environ. Exp. Bot.* **2005**, *53*, 29–37. [CrossRef]
112. Silva, P.; Gerós, H. Regulation by Salt of Vacuolar H^+-ATPase and H^+-Pyrophosphatase Activities and Na^+/H^+ Exchange. *Plant Signal. Behav.* **2009**, *4*, 718–726. [CrossRef]
113. Rizwan, M.; Ali, S.; Ibrahim, M.; Farid, M.; Adrees, M.; Bharwana, S.A.; Zia-ur-Rehman, M.; Qayyum, M.F.; Abbas, F. Mechanisms of silicon-mediated alleviation of drought and salt stress in plants: A review. *Environ. Sci. Pollut. Res.* **2015**, *22*, 15416–15431. [CrossRef]
114. Sever Mutlu, S.; Atesoglu, H.; Selim, C.; Tokgöz, S. Effects of 24-epibrassinolide application on cool-season turfgrass growth and quality under salt stress. *Grassl. Sci.* **2017**, *63*, 61–65. [CrossRef]
115. Geng, W.; Li, Z.; Hassan, M.J.; Peng, Y. Chitosan regulates metabolic balance, polyamine accumulation, and Na^+ transport contributing to salt tolerance in creeping bentgrass. *BMC Plant Biol.* **2020**, *20*, 506. [CrossRef] [PubMed]
116. Zhang, Q.; Rue, K. Glycinebetaine seed priming improved osmotic and salinity tolerance in turfgrasses. *Hortscience* **2012**, *47*, 1171–1174. [CrossRef]
117. Ma, X.; Zhang, J.; Burgess, P.; Rossi, S.; Huang, B. Interactive effects of melatonin and cytokinin on alleviating drought-induced leaf senescence in creeping bentgrass (*Agrostis stolonifera*). *Environ. Exp. Bot.* **2018**, *145*, 1–11. [CrossRef]
118. Rossi, S.; Chapman, C.; Yuan, B.; Huang, B. Improved heat tolerance in creeping bentgrass by γ-aminobutyric acid, proline, and inorganic nitrogen associated with differential regulation of amino acid metabolism. *Plant Growth Regul.* **2021**, *93*, 231–242. [CrossRef]

119. Li, S.; Jin, H.; Zhang, Q. The effect of exogenous spermidine concentration on poly-amine metabolism and salt tolerance in zoysiagrass (*Zoysia japonica* Steud) subjected to short-term salinity stress. *Front. Plant Sci.* **2016**, *7*, 1221. [CrossRef]
120. Li, Z.; Cheng, B.Z.; Peng, Y.; Zhang, Y. γ-Aminobutyric acid induces transcriptional changes contributing to salt tolerance in creeping bentgrass. *Biol. Plant* **2020**, *64*, 744–752. [CrossRef]
121. Jaleel, C.A.; Gopi, R.; Manivannan, P.; Gomathinayagam, M.; Murali, P.V.; Panneerselvam, R. Soil applied propiconazole alleviates the impact of salinity on *Catharanthus roseus* by improving antioxidant status. *Pestic. Biochem. Physiol.* **2008**, *90*, 135–139. [CrossRef]
122. Nabati, D.A.; Schmidt, R.E.; Parrish, D.J. Alleviation of salinity stress in Kentucky bluegrass by plant growth regulators and iron. *Crop Sci.* **1994**, *34*, 198–202. [CrossRef]
123. Yan, J. Influence of Plant Growth Regulators on Turfgrass Polar Lipid Composition, Tolerance to Drought and Salinity Stresses, and Nutrient Efficiency. Ph.D. Thesis, Virginia Polytechnic and State University, Blacksburg, VA, USA, 1993.
124. Elansary, H.O.; Yessoufou, K.; Abdel-Hamid, A.M.E.; El-Esawi, M.A.; Ali, H.M.; Elshikh, M.S. Seaweed extracts enhance salam turfgrass performance during prolonged irrigation intervals and saline shock. *Front. Plant Sci.* **2017**, *8*, 830. [CrossRef]
125. Drake, A. Effect of Plant Growth Regulators on Creeping Bentgrass Growth and Health during Heat, Salt, and Combined Heat and Salt Stress. Ph.D. Thesis, The Ohio State University, Columbus, OH, USA, 2019.
126. Khan, A.S.; Ali, S. Preharvest sprays affecting shelf life and storage potential of fruits. In *Preharvest Modulation of Postharvest Fruit and Vegetable Quality*; Elsevier: Amsterdam, The Netherlands, 2018; pp. 209–255. ISBN 978-0-12-809807-3.
127. Wei, H.; He, W.; Li, Z.; Ge, L.; Zhang, J.; Liu, T. Salt-tolerant endophytic bacterium *Enterobacter ludwigii* B30 enhance bermudagrass growth under salt stress by modulating plant physiology and changing rhizosphere and root bacterial community. *Front. Plant Sci.* **2022**, *13*, 959427. [CrossRef]
128. Dashtebani, F.; Hajiboland, R.; Aliasgharzad, N. Characterization of salt-tolerance mechanisms in mycorrhizal (*Claroideoglomus etunicatum*) halophytic grass, *Puccinellia distans. Acta Physiol. Plant* **2014**, *36*, 1713–1726. [CrossRef]
129. Dastogeer, K.M.G.; Zahan, M.I.; Tahjib-Ul-Arif, M.; Akter, M.A.; Okazaki, S. Plant salinity tolerance conferred by arbuscular mycorrhizal fungi and associated mechanisms: A meta-analysis. *Front. Plant Sci.* **2020**, *11*, 588550. [CrossRef]
130. Huang, B.; DaCosta, M.; Jiang, Y. Research advances in mechanisms of turfgrass tolerance to abiotic stresses: From physiology to molecular biology. *Crit. Rev. Plant Sci.* **2014**, *33*, 141–189. [CrossRef]
131. Hu, X.; Hao, J.; Pan, L.; Xu, T.; Ren, L.; Chen, Y.; Tang, M.; Liao, L.; Wang, Z. Genome-Wide Analysis of tandem duplicated genes and their expression under salt stress in seashore paspalum. *Front. Plant Sci.* **2022**, *13*, 971999. [CrossRef]
132. Wang, R.; Wang, X.; Liu, K.; Zhang, X.-J.; Zhang, L.-Y.; Fan, S.-J. Comparative transcriptome analysis of halophyte *Zoysia macrostachya* in response to salinity stress. *Plants* **2020**, *9*, 458. [CrossRef]
133. Liu, A.; Hu, Z.; Bi, A.; Fan, J.; Gitau, M.M.; Amombo, E.; Chen, L.; Fu, J. Photosynthesis, antioxidant system and gene expression of bermudagrass in response to low temperature and salt stress. *Ecotoxicology* **2016**, *25*, 1445–1457. [CrossRef] [PubMed]
134. Noor, M.; Fan, J.-B.; Zhang, J.-X.; Zhang, C.-J.; Sun, S.-N.; Gan, L.; Yan, X.-B. Bermudagrass responses and tolerance to salt stress by the physiological, molecular mechanisms and proteomic perspectives of salinity adaptation. *Agronomy* **2023**, *13*, 174. [CrossRef]
135. Huang, X.; Amee, M.; Chen, L. Bermudagrass CdWRKY50 gene negatively regulates plants' response to salt stress. *Environ. Exp. Bot.* **2021**, *188*, 104513. [CrossRef]

Article

Genome-Wide Identification of *Hsp90* Gene Family in Perennial Ryegrass and Expression Analysis under Various Abiotic Stresses

Charlotte Appiah [1], Zhong-Fu Yang [1], Jie He [1], Yang Wang [1], Jie Zhou [1], Wen-Zhi Xu [2], Gang Nie [1,*] and Yong-Qun Zhu [2,*]

[1] Department of Forage Science, College of Grassland Science and Technology, Sichuan Agricultural University, Chengdu 611130, China; charlotte27appiah@gmail.com (C.A.); yangzf211@163.com (Z.-F.Y.); he13440090515@163.com (J.H.); wangyll5123@163.com (Y.W.); zhoujie1@stu.sicau.edu.cn (J.Z.)

[2] Institute of Agricultural Resources and Environment, Sichuan Academy of Agricultural Sciences, Chengdu 610066, China; xuwenzhi_herb@126.com

* Correspondence: nieg17@sicau.edu.cn (G.N.); zyq80842@hotmail.com (Y.-Q.Z.); Tel.: +86-139-8091-1407 (G.N.)

Abstract: The heat shock protein 90 (*Hsp90*) is a protein produced in plants in response to stress. This study identified and analyzed *Hsp90* gene family members in the perennial ryegrass genome. From the results, eight *Hsp90* proteins were obtained and their MW, pI and number of amino acid bases varied. The amino acid bases ranged from 526 to 862. The CDS also ranged from 20 (*LpHsp0-4*) to 1 (*LpHsp90-5*). The least number of CDS regions was 1 (*LpHsp90-5*) with 528 kb amino acids, while the highest was 20 (*LpHsp90-4*) with 862 kb amino acids, which showed diversity among the protein sequences. The phylogenetic tree revealed that *Hsp90* genes in *Lolium perenne*, *Arabidopsis thaliana*, *Oryza sativa* and *Brachypodium distachyon* could be divided into two groups with five paralogous gene pairs and three orthologous gene pairs. The expression analysis after perennial ryegrass was subjected to heat, salt, chromium (Cr), cadmium (Cd), polyethylene glycol (PEG) and abscisic acid (ABA) revealed that *LpHsp90* genes were generally highly expressed under heat stress, but only two *LpHsp90* proteins were expressed under Cr stresses. Additionally, the expression of the *LpHsp90* proteins differed at each time point in all treatments. This study provides the basis for an understanding of the functions of *LpHsp90* proteins in abiotic stress studies and in plant breeding.

Keywords: perennial ryegrass; *Hsp90*; abiotic stress; expression profiles; phylogenetic analysis

Citation: Appiah, C.; Yang, Z.-F.; He, J.; Wang, Y.; Zhou, J.; Xu, W.-Z.; Nie, G.; Zhu, Y.-Q. Genome-Wide Identification of *Hsp90* Gene Family in Perennial Ryegrass and Expression Analysis under Various Abiotic Stresses. *Plants* **2021**, *10*, 2509. https://doi.org/10.3390/plants10112509

Academic Editor: Biancaelena Maserti

Received: 5 October 2021
Accepted: 16 November 2021
Published: 19 November 2021

Publisher's Note: MDPI stays neutral with regard to jurisdictional claims in published maps and institutional affiliations.

1. Introduction

Improving stress tolerance is among the major efforts of breeding advancement in cool season grass species [1]. Perennial ryegrass (*Lolium perenne*) is one of the major species of forage and turf grasses extensively planted in warm temperate to subtropical regions around the world, because it is to plant, has better tolerance to abiotic stresses, and requires low maintenance [2]. However, achieving the potential yield after cultivation of perennial ryegrass is limited due to exposure to abiotic stresses in cultivated lands [3]. Abiotic stresses such as cold, drought, salinity, freezing, high light intensity and heat cause cell injury resulting in secondary stresses such as osmotic and oxidative stresses that critically impact the quality and yield of the perennial ryegrass plant [4–6]. Moreover, the impact of stresses in *Lolium perenne* is evident in limiting cultivation, leaf appearance, seed emergence, reducing dry matter by up to 25% and causing plant death [7–10]. The response of plants to heat shock is similar to that of other organisms when exposed to adverse stress conditions, producing highly conserved stress proteins called heat shock proteins (*Hsp*) [11–13].

Heat shock proteins are expressed in response to stresses and are highly conserved at both the cellular and organismic levels [14,15]. Generally, plants have five major classifications of *Hsp*s based on their molecular size: *Hsp100*, *Hsp90*, *Hsp70*, *Hsp60* and small *Hsp* (*smHsp*) [16]. One of the most numerous proteins in the cytoplasm of prokaryotic and eukaryotes belongs to the *Hsp90* family, constituting 1–2% of the total protein level [17]. *Hsp90* is an important constituent of cells at developmental stages under normal conditions, aiding in protein translocation, folding and degradation [18]. Furthermore, the heat shock protein transcript is significantly upregulated under heat stress [19], and also some stresses are known to induce synthesis and expression of heat shock proteins in cells [20]. Under stress conditions, *Hsp90* functions as a homeostatic agent by reestablishing the damaged normal protein structure [5,21]. *Hsp90* has important functions in both animals and plants. *Hsp90* is abundantly expressed in the cytoplasm in soluble form at normal temperatures in yeast, fruit flies, and vertebrate families, while it accumulates rapidly in the nucleus under heat shock conditions [22]. *Hsp90* has been well characterized in *Oryza sativa*, *Arabidopsis thaliana* and *Solanum lycopersicum* [18–20]. *Hsp90* was also recently identified in *Aeluropus littoralis*, *Hordeum vulgare*, *Camellia sinensis* and *Cucumis sativus* [23–26].

Hsp90 is an ATP-regulated dimeric chaperone mainly consisting of three highly conserved domains: C-terminal domain of about 25 kDa that binds to the substrate, the 35 kDa intermediate domain, and the 12 kDa N-terminal domain of the ATP-binding site. *Hsp90* cooperates with other chaperones to form a multiprotein chaperone complex in order to play its role [27]. *Hsp90* can be further divided into five subgroups in accordance to the source and subcellular localization. The subgroups are namely *Hsp90A*, *Hsp90B*, *Hsp90C*, TRAP (TNF receptor-associated protein) and HTPG (high temperature protein G) [28]. *Hsp90A* contains no signal peptide and is located in the cytoplasm. The main subtypes include *Hsp90α* (inducible form) and *Hsp90β* (constitutive form), which are the result of gene duplications about 500 million years ago [25,26]. *Hsp90B*, *Hsp90C* and TRAP are located in the endoplasmic reticulum, chloroplast and mitochondria (Animalia), respectively, because they contain signal peptides. HTPG refers to the *Hsp90* of prokaryotes and is distributed in most bacteria [24,28]. Several biological variants of the *HSP90* gene have also been well characterized in some plants. In *Arabidopsis*, *AtHsp90-5* is important for chloroplast biogenesis and embryogenesis [29,30]. In *Brassica napus*, *Hsp90* plays a vital role in the processes of seed development and germination, while in cotton *Hsp90* has been found to play a crucial role in cotton fiber differentiation and development by maintaining cellular homeostasis [31–36]. *Hsp90*s, as housekeeping proteins in plants, can be induced by various abiotic and biotic stresses [37]. Expression of *Hsp90* in *Arabidopsis thaliana* is developmentally regulated and is responsive to abiotic stresses, phytohormones, and light and dark transitions [19,21,38–40]. The overexpression of *AtHsp90-2*, *AtHsp90-5*, and *AtHsp90-7* reduces tolerance to salt and drought stresses whiles improving tolerance to high concentrations of Ca^{2+} [39]. Moreover, the overexpression of *Hsp90-2* in *Arabidopsis thaliana* may inhibit the transcription of HsfA2, and HsfA2 expressed under the inhibition of *Hsp90-2* contributes to the resistance to oxidative stress [41]. Similarly, an *Hsp90* inhibitor produced by root-peripheral fungi may inhibit plant growth and development, but also increase the resistance of *Arabidopsis* to high temperatures [42]. The *Hsp90* complex in *Arabidopsis* directly regulates the activity of resistance proteins and plays a key role in disease resistance as well [43]. The induction of ABA responsive genes is delayed by overexpression of cytosolic *AtHsp90-2*, but is hardly affected by overexpression of *AtHsp90-5* and *AtHsp90-7* under conditions of salt and drought stress, which implies that different cellular compartment localized *Hsp90*s in *Arabidopsis thaliana* might contribute to responses to abiotic stresses by different functional mechanisms, probably through ABA- or Ca^{2+} dependent pathways [39]. *OsHsp90-2* and *OsHsp90-4* were also found to be up-regulated to drought, cold, heat and salt stresses [38]. The resistance of tobacco leaves to mosaic virus increases because of the interaction of *Hsp90* with RAR1 and TIR-NB-LRR in tobacco leaves [44]. In tobacco, *NtHsp90*s were strongly induced by heat stress, while weakly activated by ABA treatment, with expression pattern analysis indicating that *NtHsp90-4*, *NtHsp90-5*, and *NtHsp90-9* were induced by various abiotic stresses. The expression level of *UpHsp90* in *Ulva pertusa* is notably positively regulated by the change in temperature difference between day and night, but it was almost unaffected under long-term treatment with heavy metal stress [15]. In potatoes, *Hsp90*s may be related to the color of potato tuber chip [46]. *Hsp90*s play vital roles in the growth of tumor cells. For example, geldanamycin can specifically interact with the ATPase active site of *Hsp90*, preventing the binding of *Hsp90* and ATP, and finally achieve the purpose of inhibiting tumor [47]. Using interference technology, the expression level of *Hsp90* was reduced, and it was found that the division rate of U937 cells was significantly reduced [48].

According to previous studies, eight *Hsp90* genes were identified in *Oryza sativa*, seven in *Arabidopsis thaliana*, eight in *Brachypodium distachyon*, ten in *P. trichocarpa*, 21 in *Nicotiana tabacum*, seven in *Solanum lycopersicum*, and twelve in *Zea mays* [40,49–51]. However, the identification of the perennial ryegrass *Hsp90* gene family has not been studied yet. The completion of genome-wide sequencing of perennial ryegrass will provide the necessary information for data mining of *Hsp90* at the whole genome level. In this study, we performed a genome-wide survey of *Hsp90* in the perennial ryegrass genome database, and a complete overview was reported on gene structure and phylogenetic and conserved motif characteristics. Additionally, the expression levels of the *LpHsp90* genes under various abiotic stresses were studied. The results will be helpful for further study of the functional characteristics of *Hsp90* genes in response to abiotic stress in perennial ryegrass.

2. Results

2.1. Identification of LpHsp90 Genes in Perennial Ryegrass

Eight *LpHsp90* genes were identified after the removal of redundant sequences from the genome database of perennial ryegrass. LpHps90 proteins were renamed according to their chromosomal locations; that is, *LpHsp90-1* to *LpHsp90-8*. *LpHsp90* sequences obtained varied in length, which ranged from 528 (*LpHsp90-5*) to 862 (*LpHsp90-4*) amino acids, with an average of 779. The pI values ranged from 4.89 (*LpHsp90-5*) to 5.57 (*LpHsp90-4*) and with a MW ranging from 61214.61 kd (*LpHsp90-5*) to 96712.15 kd (*LpHsp90-4*). The *LpHsp90*s were highly cytoplasmic (*LpHsp90-1, 2, 3, 5, 7 and 8*) with the exception of *LpHsp90-4* and *LpHsp90-6*, which were nuclear and ER subcellular localized, respectively (Table 1). Moreover, the analysis of the cis-acting elements of perennial ryegrass *Hsp90* genes showed that plant hormone responsiveness was identified, implying that *LpHsp90* genes might be involved in various plant stress-responsive pathways and closely related to the function of plant hormones such as abscisic acid, gibberellin and methyl-jasmonate (Table S1).

Table 1. The biophysical characteristics and subcellular localization of *Lphsp90* proteins.

Gene	Molecular Weight	Theoretical pI	Number of Amino Acids	Instability Index	Predicted Sub-Cellular Location
LpHsp90-1	80,409.24	4.96	700	41.43	Cytoplasmic
LpHsp90-2	89,044.8	5.19	787	43.15	Cytoplasmic
LpHsp90-3	80,947.92	4.95	710	40.22	Cytoplasmic
LpHsp90-4	96,712.15	5.57	862	43.84	Nuclear
LpHsp90-5	61,214.61	5.08	528	44.14	Cytoplasmic
LpHsp90-6	92,834.85	4.89	809	37.71	Endoplasmic reticulum
LpHsp90-7	88,305.63	4.9	779	47.45	Cytoplasmic
LpHsp90-8	88,305.63	4.9	779	47.4	Cytoplasmic

2.2. Phylogenetic Analysis and Multiple Sequence Alignment

The *Hsp90* protein sequence alignments of *Lolium perenne*, *Oryza sativa*, *Arabidopsis thaliana* and *Brachypodium distachyon* were used to construct a phylogenetic tree employing the maximum-likelihood method with 1000 bootstraps to explore the evolutionary relationship among the plant species using MEGA6 (Figure 1). The *Hsp90* protein sequences were classified into two main groups (group I and II), and each group was further divided into two subgroups (Ia, Ib, IIa and IIb). The group IIb (15 members) had the largest number of members, followed by group Ib (8 members). It was also seen that groups Ia and IIa had 4 members each. Additionally, the phylogenetic tree showed that there was high similarity among cytosolic *Hsp90*s and less similarity among the organelle-localized members.

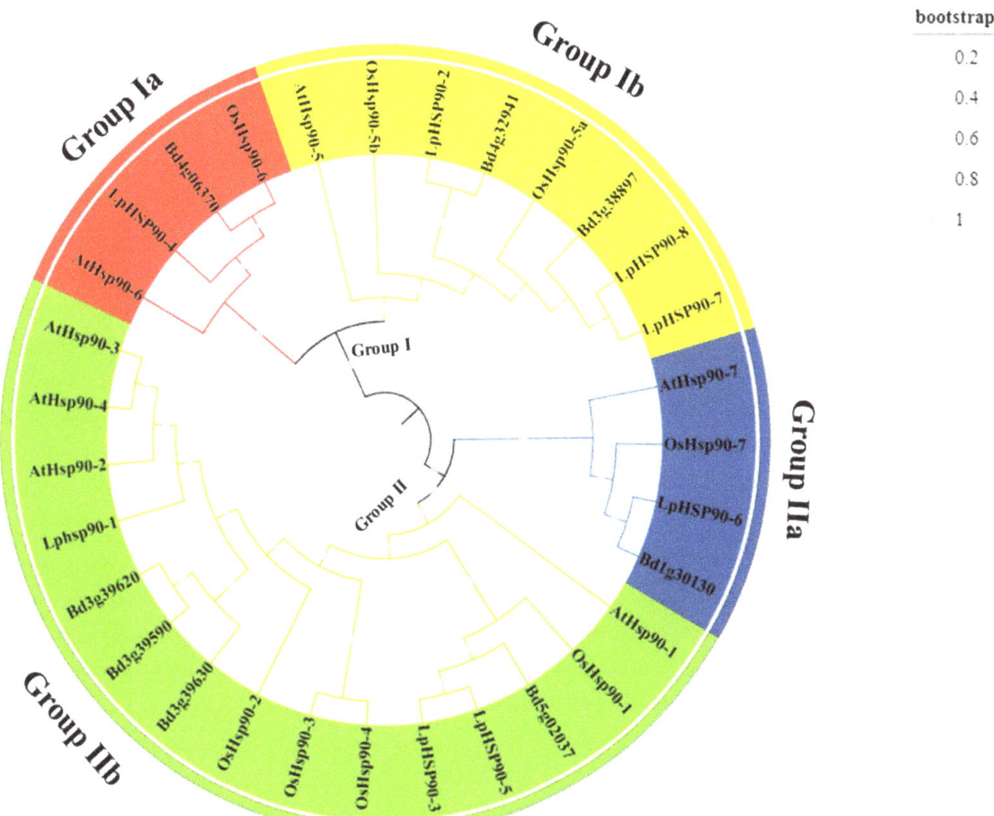

Figure 1. Unrooted phylogenetic tree representing relationships among the *Hsp90* protein sequences of *Lolium perenne*, *Arabidopsis thaliana*, *Brachypodium distachyon* and *Oryza sativa*. The tree was divided into two main groups (I and II) and further divided into subgroups (Ia, Ib, Iia and Iib). Ia had 4 members, and Ib had 8 members; Iia had 4 members and Iib had 15 members.

The phylogenetic tree analysis showed that there were five pairs of paralogs within species, of which two were from *Lolium perenne* (*LpHsp90-7* and *LpHsp90-8*, *LpHsp90-3* and *LpHsp90-5*), one pair in *Oryza sativa* (*OsHsp90-3* and *OsHsp90-4*), one pair in *Brachypodium distachyon* (*Bd3g39620* and *Bd3g39590*) and one pair in *Arabidopsis thaliana* (*AtHsp90-2* and *AtHsp90-3*) (Figure S1) [50]. There were three orthologous gene pairs among the species (*Bd4g06370* and *LpHsp90-4*, *Bd1g30130* and *LpHsp90-6*, *Bd4g32941* and *LpHsp90-2*) [51]. The orthologous and paralog genes may predict the functions and characteristics of the *LpHsp90* genes in the evolutionary relation with *Arabidopsis thaliana*, *Brachypodium distachyon* and *Oryza sativa* [52].

2.3. Conserved Motif and Gene Structure Analysis of LpHsp90 Proteins

A maximum likelihood phylogenetic tree was constructed using eight protein sequences of *LpHsp90* (Figure 2a). The *LpHsp90* sequences were divided into two main subgroups according to the bootstrap values and motif compositions. Ten conserved motifs were identified using the online software MEME (Figure 2b). It was observed that all *LpHsp90* protein sequences in the same group had similar motif compositions and positionings. All motifs were arranged in the order of motif -5, motif-9, motif-8, motif-6, motif-4, motif-10, motif-3, motif-2, motif-7 and motif-1 apart from *LpHsp90-5*. In group Ia, it was observed that *LpHsp90-5* had only seven motifs with the exclusion of motif-5, motif-9 and motif-8, but was similar in motif position with *LpHsp90-1* and *LpHsp90-3*, which may be due to evolutionary change. Additionally, *LpHsp9-2*, *LpHsp90-7* and *LpHsp90-8* in group Iib were similar in motif composition and positioning. Details of motif logo and consensus are listed in supplement file Figure S2. Besides, three heat shock genes *(LpHsp90-1, LpHsp90-3* and *LpHsp90-5)* containing the C-terminal EEVD motif were predicted that they functionally interacted with other family members and were seen to be highly similar (Figure S3).

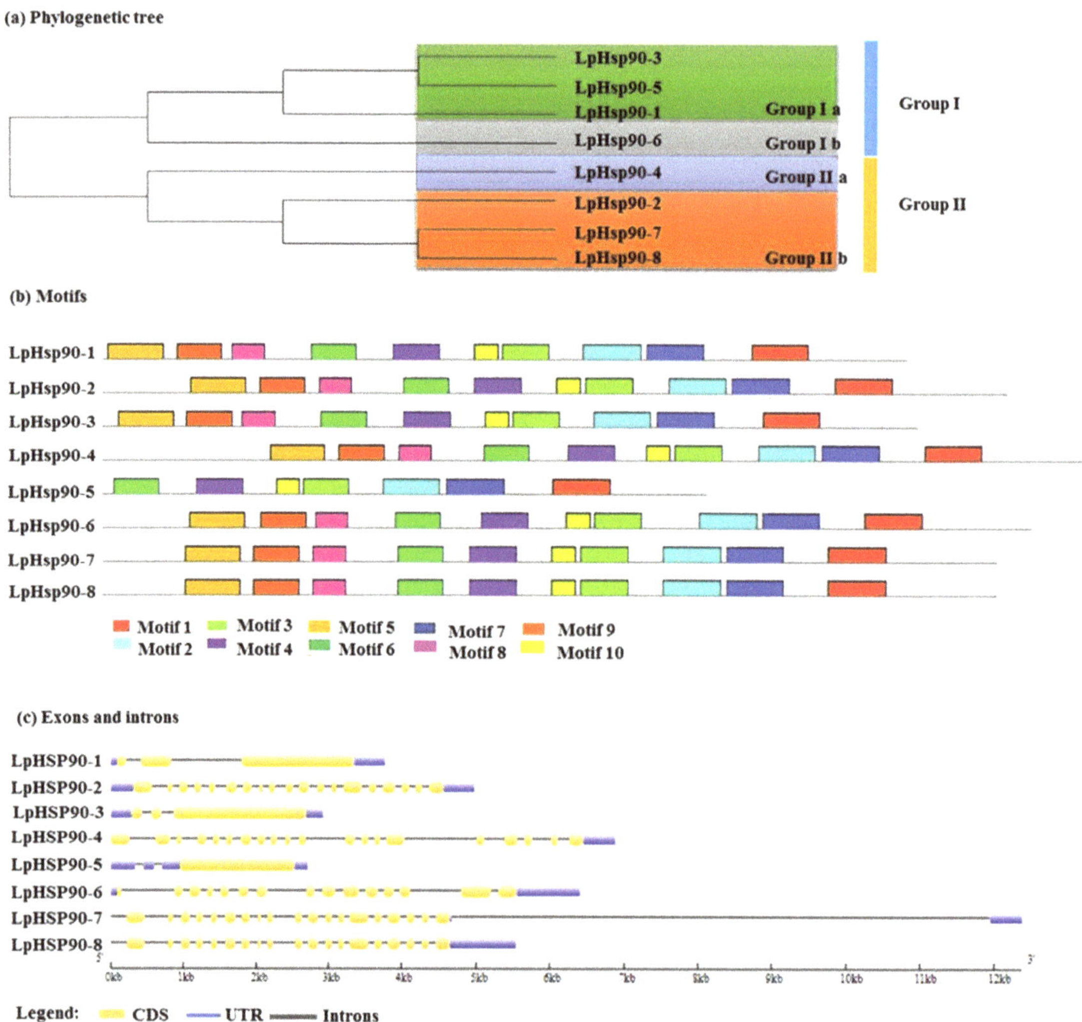

Figure 2. Phylogenetic relationship, gene structure and motif composition of *Hsp90* genes in perennial ryegrass. (**a**) A multiple sequence alignment of the full length of *LpHsp90* protein sequences was executed using Clustal W, and a maximum-likelihood phylogenetic tree with 1000 bootstraps was constructed using MEGA 6.0 software. The four subgroups were marked with different colors. (**b**) Schematic representation of the conserved motifs obtained using MEME online tool in *LpHsp90* proteins. Different motifs are represented by colored boxes of corresponding colors. The grey lines represent the non-conserved. (**c**) The exon/intron structure of the *LpHsp90* protein sequences was obtained employing the online tool GSDS.

The analysis of the exon-intron structure of *Hsp90* protein sequences of perennial ryegrass may provide insights into the evolution of the *LpHsp90* gene family [53]. The online software GSDS tool was used to obtain the exon-intron structure of the *Hsp90* protein sequences. Figure 2c revealed a coding sequence of *LpHsp90* protein interrupted by introns. The least number of CDS regions was one (*LpHsp90-5*), with 528 kb amino acids, while the highest was twenty (*LpHsp90-4*), with 862 kb amino acids, which showed diversity among the protein sequences. *LpHsp90-7* had the longest gene structure due to the length of the intron, although *LpHsp90* had the highest number of CDS regions. Moreover, comparing *LpHsp90-8* and *LpHsp90-7*, they were different in gene structure but had the same MW, pI and number of amino acids.

2.4. Expression Profile of LpHsp90 in Response to Abiotic Stresses

Most plants have mechanisms for defense against stress, and *Hsp90* genes are known to be expressed in response to these abiotic stresses [31]. To analyze the expression pattens of *LpHsp90* under abiotic stress, eight *LpHsp90* proteins were analyzed using the qRT-PCR technique. As shown in Figures 3–5, different expression patterns were observed under ABA, cadmium (Cd), chromium (Cr), salt (NaCl), heat and PEG induced abiotic stresses. It was observed that *LpHsp90* gene regulation was consistent across all stresses at the 0-h time point.

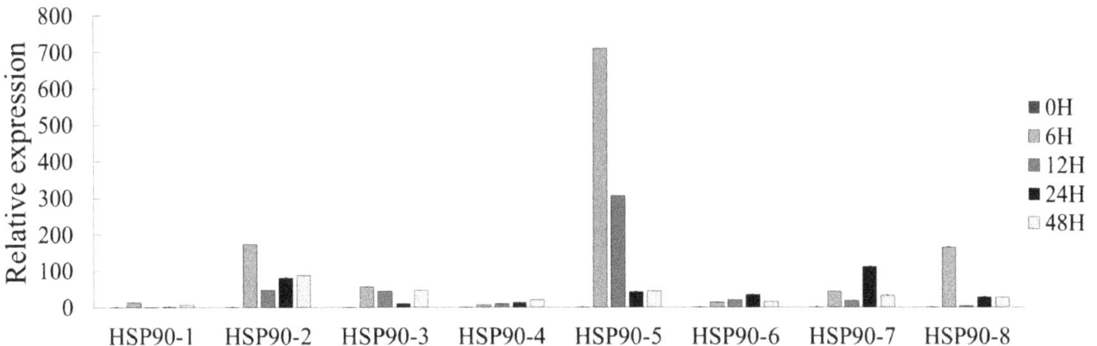

Figure 3. Expression patterns of *LpHsp90s* in response to heat. The values are the mean values of expression ($n = 3$).

LpHsp90 was highly expressed under heat stress as compared to other treatments. *LpHsp90-5* had the highest expression value under heat stress, which was recorded at two time points, 6 h and 12 h (Figure 3). At 6 h, it was also observed that *LpHsp90-2* and *LpHsp90-4* were substantially expressed.

Under salt stress, *LpHsp90-5* was significantly expressed at 12 h (Figure 4a). *Lphsp90-4* was seen to be expressed at 12 h also. Considering drought induced by PEG treatment at 6 h, *LpHsp90-7* was highly induced at 12 h (Figure 4b). *LpHsp90-5* and *LpHsp90-8* were not expressed at any time point. It was observed that the expression patterns of *LpHsp90* proteins were irregular at the various time points, owing to the fact that an increase in expression at 6 h may decrease at 12 h or an increase in expression at 24 h may decrease at 48 h. ABA treatment up-regulated *LpHsp90-5* consistently across all time points, while *LpHsp90-7* was not induced by ABA across all time points (Figure 4c). Some *LpHsp90* genes (*LpHsp90-1*, *LpHsp90-2*, *LpHsp90-3*, *LpHsp90-4*, *LpHsp90-6* and *LpHsp90-8*) were seen to have an undulating expression pattern, thus their expression may be affected by the longevity of exposure to ABA stress.

Under Cr treatment, *LpHsp90-7* was significantly expressed at 6 h but was seen to have reduced with longer time of exposure. *LpHsp90-5* was also observed to be expressed at the 6 h time point. Further, *LpHsp90-2* was highly expressed at 24 h but reduced at 48 h, while *LpHsp90-3* was induced at 48 h under Cd stress (Figure 5a). Cr stress induced the weakest expression, although *LpHsp90-1* and *LpHsp90-3* were induced at the 48 h time point while all of the remaining six *LpHsp90* genes were fairly activated compared to other treatments.

Generally, *LpHsp90-1*, *LpHsp90-3*, *LpHsp90-4* and *LpHsp90-6* were induced under five stresses, while *LpHsp90-7* and *LpHsp90-8* were induced under only two stresses. *LpHsp90* showed the highest expression at 6 h (*LpHSP90-5*, under heat stress), and the lowest expression was recorded at 48 h (*LpHsp90-8*, under Cd stress). A heatmap showing the expression of each *LpHsp90* gene was drawn under all six abiotic treatments (Figure S4).

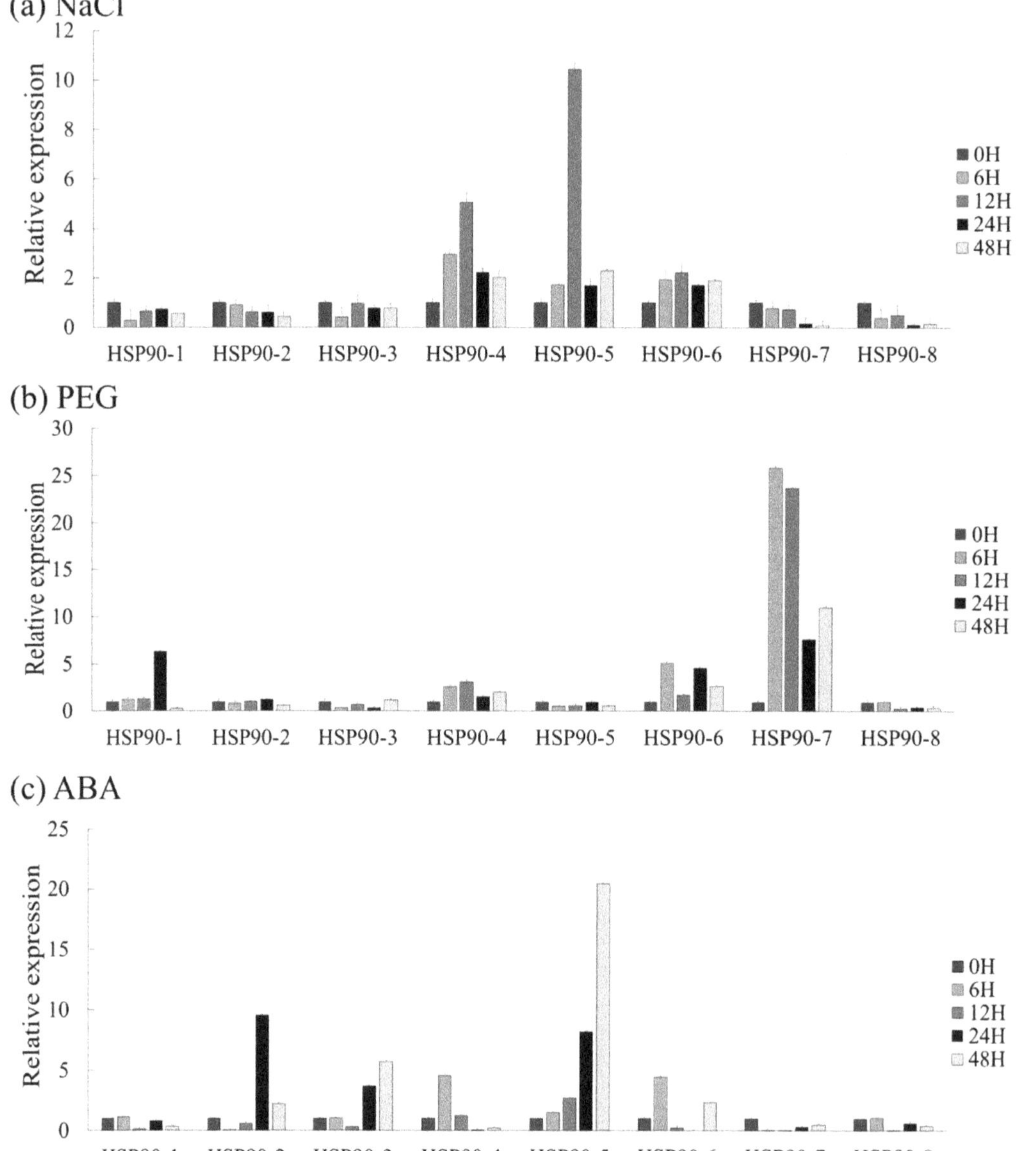

Figure 4. Expression patterns of *LpHsp90s* in response to (**a**) NaCl (**b**) PEG and (**c**) ABA treatment. The values are the mean values of expression (*n* = 3).

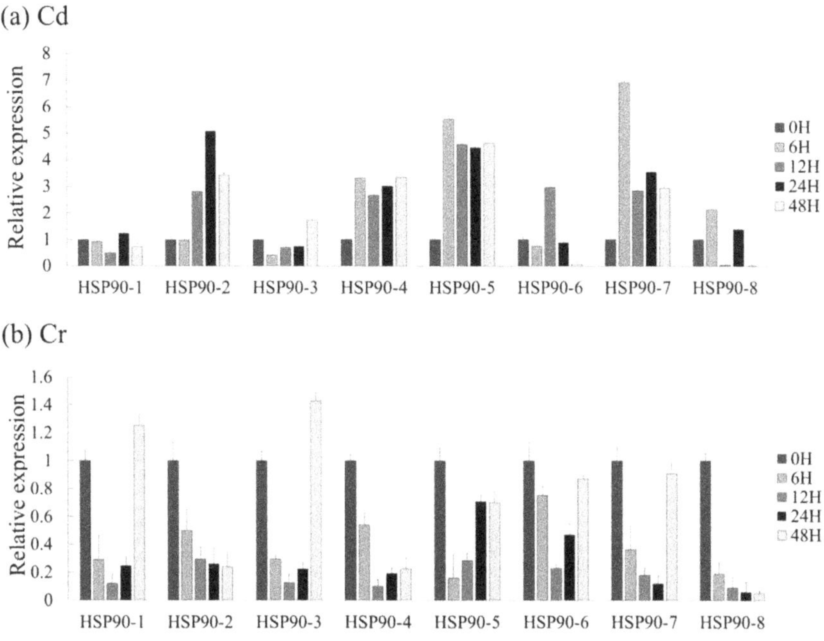

Figure 5. Expression patterns of *LpHsp90s* in response to (**a**) Cd and (**b**) Cr treatment. The values are the mean values of expression ($n = 3$).

3. Discussion

The heat stress proteins have been classified based on their molecular weights into *Hsp100*, *Hsp90*, *Hsp70*, *Hsp60* and *smHsp* [54]; among them, *Hsp90* was known to be important and highly conserved [55,56]. *HSPs* were first identified in the salivary gland chromosomes of Drosophila larva [57]. Later, other studies found organisms that produced a series of proteins of different sizes, known as *HSPs*, in response to increased temperature [58]. The study of *Hsp90* genes proved that they were not only related to stress signal transduction in plants, folding of receptors, transcription factors and kinases and physiological processes [51–53,58], but also to assisting cell survival under stresses [59]. Studies have found out that besides high temperatures, abiotic stresses such as drought, salinity, heavy metals and ABA could induce the production of *Hsp90* in plants [60,61]. *Hsp90* proteins have been identified in plants and differences in their numbers are attributable to their genome sizes. Seven *Hsp90* proteins were identified in *Arabidopsis thaliana* [38], 10 in *P. trichocarpa* [62], 21 in *Nicotiana tabacum* [63], seven in *Solanum lycopersicum* [64], seven in *Oryza sativa* [65], eight in *Brachypodium distachyon* and 12 in *Zea mays* (Maize genome database, http://www.maizegdb.org, accessed on 1 August 2020). As perennial ryegrass is one of the major species of forage and turf grasses extensively planted in warm temperate to subtropical regions around the world, it was expedient to obtain genes that would further aid in its improvement. Therefore, it was necessary to identify *Hsp90* genes related to various abiotic stresses in perennial ryegrass. The identification of Lolium perenne *Hsp90* will provide more insights and fundamental information into genetic improvements in response to stressful conditions in other plants.

In this study, eight *Hsp90* protein sequences were identified in the perennial ryegrass genome database. The *Hsp90*s may play a role in the physiological and environmental stability of perennial ryegrass. The *LpHsp90*s identified had different biophysical and chemical properties, which indicated diversity among the *LpHsp90* protein sequences. The isoelectric point of the *LpHsp90* protein sequences ranged from 4.89 to 5.57, making them acidic, which is consistent with *Hsps* found in *Arabidopsis thaliana*, *Solanum lycopersicum* and other plants [66]. Phylogenetic analysis between *Hsp90s* of *Lolium perenne*, *Arabidopsis thaliana*, *Oryza sativa* and *Brachypodium distachyon* divided the *Hsps* into two groups, consistent with other studies performed [67]. Additionally, phylogenetic analysis aided in the identification of five paralog gene pairs among the plant species. This may imply that most species expanded according to their own species-specific approach during evolution of the *Hsp90* family [50,68]. This finding is consistent with gene families found in cereals such as rice and also in *Nicotiana tabacum* [50,69]. The structure of proteins is known to determine the function they may perform [50,67]. According to the study, *LpHsp90* protein sequences had different gene structure. The number of introns is mostly related to the sensitivity of gene transcription regulation, thus the lesser the number of introns, the more likely that a plant has the ability to adapt to different developmental and environmental stimuli [70]. Theoretical pI and number of amino acids between *LpHsp90-7* and *LpHsp90-8* do not imply they may perform the same function due to the difference in their gene structure. Furthermore, the predicted cis-acting elements from Plantcare stated the likelihood of *LpHsp90-7* to be involved in responsiveness to MeJA and gibberellin.

The study of the expression of *Hsp90* proteins in response to abiotic stresses has been undertaken in various plant species. In this study, it was seen that most *LpHsp90* protein genes were induced under most of the stresses. Heat stress recorded the highest expression in all *LpHsp90* proteins, which is consistent with studies in populous and *CsHSP90* genes, although the expression levels of certain genes were brief and slightly decreased at individual time points [25,62]. The comparatively lowest level of expression was recorded under Cr stress. *LpHsp90-5* was observed to be highly expressed under ABA, heat and salt stress, but these were at different time points. In ABA, there was a rapid rise in the expression of *LpHsp90-5*, with the highest expression level recorded at 48 h. Unlike ABA, NaCl stress induced high expression in *LpHsp90-5* at 12 h but with a rapid reduction at 24 h and 48 h. Under heat stress, there was also a decrease in expression after peaking at 6 h. Additionally, *LpHsp90-7* was highly expressed under PEG and Cd stress, and these were also recorded at different time points. *LpHsp90-1* and *LpHs90-3* were the only *Hsp90* proteins expressed under Cr heavy metal stress. This could imply that *LpHsp90-5* could be an *LpHs90* gene of interest in ABA, heat, and salt stress; *LpHsp90-7* to PEG and Cd; and *LpHsp90-3* and *LpHsp90-1* to Cr. Analysis of *AtHsp90-5* and *AtHsp90-6* expression revealed that the former is mildly induced by heat shock and that the latter is barely induced by heat shock; this was also observed in *LpHsp90-5* in this study. Comparatively, *LpHsp90-2* expression levels increased modestly with heat stress; this was also observed in *AtHsp90-2*, the levels of which were mildly increased after treatment with NaCl or heavy metals [39]. Therefore, further studies may be carried out to explain the functions of *LpHsp90-7*, *LpHsp90-5*, *LpHsp90-3* and *LpHsp90-1* in relation to these stresses. Moreover, the various time points of the expression of the LpHsp90 proteins should also be further investigated, since the effects of stress are influenced by intensity and longevity. It was also observed that the paralogous pairs (*Lphsp90-5* and *LpHsp90-3*, *LpHsp90-7* and *LpHsp90-8*) had different expression profiles under the various abiotic stresses.

These results showed that *LpHsp90s* were induced under ABA, PEG, Cr, Cd NaCl and heat stress, and individual *LpHsp90s* may have different regulatory patterns that reflect their potential roles in the response to different abiotic stresses. Heat stress induced the highest response of all stresses, indicating that the *LpHsp90* protein was very sensitive to heat stress.

4. Materials and Methods

4.1. Identification of LpHsp90 Genes in Perennial Ryegrass

Arabidopsis thaliana Hsp90 protein sequences were downloaded from the TAIR databases (http://www.arabidopsis.org/, accessed on 1 April 2020) [39]. The protein sequences of *Arabidopsis thaliana Hsp90* genes were used as queries to perform BLASTP against the genome resource of perennial ryegrass, which was downloaded from the Perennial Ryegrass Genome Sequencing Project (http://185.45.23.197:5080/ryegrassgenome, accessed on 1 April 2020) [71], followed by the removal of redundant proteins. Subcellular localization of *LpHsp90* proteins was predicted by using CELLO (http://cello.life.nctu.edu.tw/, accessed on 1 August 2020). The physical and chemical parameters of the *LpHsp90* protein sequences were obtained from ProtParam (https://web.expasy.org/protparam, accessed on 1 August 2020).

4.2. Phylogenetic Analysis and Multiple Sequence Alignment

A multiple sequence alignment of *Lolium perenne, Arabidopsis thaliana, Brachypodium distachyon* and *Oryza sativa Hsp90* protein sequences was analyzed using ClustalW [72]. *Oryza sativa Hsp90s* were obtained from the Rice Genome Annotation Project (http://rice.plantbiology.msu.edu/, accessed on 1 April 2020). Brachypodium distachyon were obtained from the Phytozome link in Brachypodium distachyon Assembly and Gene Annotation in ensemble (http://www.phytozome.net/, accessed on 1 April 2020). A maximum-likelihood (ML) phylogenetic tree was constructed with MEGA version 6.0 employing 1000 bootstraps in order to examine the evolutionary relationships among the *Hsp90* family of *Lolium perenne, Arabidopsis thaliana, Brachypodium distachyon* and *Oryza sativa* [73,74].

4.3. Exon-Intron Structure, Conserved Motif, Characteristics Analysis of LpHsp90

The exon-intron structure of *Hsp90* genes was obtained using the online Gene Structure Display Server v2.0 (GSDS: http://gsds.cbi.pku.edu.cn, accessed on 1 October 2020) [75] with the giff3 file. The MEME program (Multiple Expectation Maximization for Motif Elicitation (http://memesuite.org/tools/meme/, accessed on 1 October 2020) [76] was used to identify the conserved motifs in the *LpHsp90* protein sequences.

4.4. Plant Materials, Growth Conditions and Stress Application

Viable seeds of *Lolium perenne* cv 'Mathilde' were grown in a container with dimensions 20 cm × 15 cm × 10 cm. The container was filled with quartz sand and distilled water to one-third of the volume, after which the seeds were evenly spread on the sand to avoid clustering of plants. The container with the seeds was moved to the growth chamber. Plants were grown at a temperature of 20 °C and 15 °C (12 h day/12 h night) at 70% relative humidity and 750 $\mu mol \cdot m^{-2} \cdot s^{-1}$ PAR illumination. After 14 days of germination, Hoagland solution dissolved in distilled water following the manufacturer's protocol was used to supply plants with nutrients for 46 days. The plants were separated into groups of five for the various stress applications.

For salt stress, sodium chloride (NaCl) was used at a concentration of 250 mM by dissolving NaCl in Hoagland solution. For heavy metal stresses, chromium (Cr) (K2Cr2O7) and cadmium (Cd) ($CdCl_3 \cdot 6H_2O$) were used at a concentration of 300 mg/L and 200 mg/L, respectively. Twenty-percent polyethylene glycol 6000 (PEG) was used to stimulate drought stress after dissolving in Hoagland solution. For heat stress, plants were transferred to a growth chamber and subjected to a temperature of 38 °C/30 °C (day/night) with 12 h light/ 12 h dark photoperiod and all other conditions remaining constant, as per the early growth stage. Plants were subjected to 100 mM ABA for the exogenous abscisic acid (ABA) application. Collection of plant leaf samples was performed in time frames of 0 h, 6 h, 12 h, 24 h, 48 h and 72 h after each stress application with three biological replications. Plant leaf samples were collected into a 1.5 mL tube using a pair of surgical scissors, which were intermittently cleaned with ethanol after every stress sample collection to avoid contamination. Approximately 1 g of collected leaves was quickly frozen using liquid nitrogen and stored in a freezer at −80 °C.

4.5. RNA Isolation, cDNA Synthesis and Quantitative Real-Time PCR Expression Analysis

For all RNAs isolated from leaf samples, HiPure Plant RNA Mini Kit (Magen Biotech Co. Ltd., Guangzhou, China) was used following the manufacturer's protocol. A Nanodrop ND-2000 spectrophotometer (Nano-Drop Technologies, Wilmington, DE, USA) was used to determine RNA concentration, purity, and integrity, followed by 1% agarose gel electrophoresis. M5 Super plus qPCR RT kit with gDNA remover (Mei5 Biotechnology, Co., Ltd. Beijing, China) were used for the RNA reverse transcript. The qRT-PCR technique was used to validate the expression of *LpHSP90* genes in the various stress treatments and the specific primers obtained as designed by Premier 3.0 (Table 2). A 10 µL mixture was prepared for each sample, containing 5 µL of abm® EvaGreen 2X qPCR Master Mix (Applied Biological Materials Inc, Richmond, Canada), 1.5 µL of synthesized cDNA product, 0.3 µL of each primer and 2.9 µL of ddH$_2$O. The qRT-PCR reaction protocols were as follows: an enzyme activation step at 95 °C for 10 min with 1 cycle, denaturation at 95 °C for 15 s, and anneal/ extension at 60 °C for 60 s, for a total of 35 cycles. Technical samples and biological samples were used for all qRT-PCRs. Three biological replicates and technical repeats were used for each gene. The relative gene expression level was analyzed according to the $2^{-\Delta\Delta Ct}$ method [77].

Table 2. Primer information for *LpHsp90* genes.

Gene	Forward-Primer (5′-3′)	Reverse-Primer (5′-3′)
Lphsp90-1	ATCGTCTCTGACCGTGTTGT	AAGCATCACCAGGTCCTTGA
Lphsp90-2	GCACACTTCACAACAGAGGG	CTCGCCATCAAAGTCATCCG
Lphsp90-3	CATCATGGACAACTGCGAGG	GGCGTAGTCCTCCTTGTTCT
Lphsp90-4	AGGAGGTGTTTCTTCGGGAG	TGCAATGGTCCCAAGAGAGT
Lphsp90-5	TCGGAGTTCATCAGCTACCC	GCTCACCTCCTTCACCTTCT
Lphsp90-6	GCAAGGACTCGAAGCTCAAG	TTGATCTCCAGGACACGCTT
Lphsp90-7	GCCAATTGATGAGGTTGCCA	TCGCAGACCAACCAAACTTG
Lphsp90-8	GCGGAGGAGAAGTTCGAGTA	CATCCCAATGCCAGTGTCAG

5. Conclusions

This study was conducted to identify the *Hsp90* gene family in perennial ryegrass; hence, a genome-wide identification and expression analysis of the *LpHsp90* gene family was performed. Additionally, the gene structure, conserved motif, evolutionary relationships and expression patterns were studied. Eight *Hsp90* proteins were identified within the perennial ryegrass whole genome and were named according to their locations on the chromosomes. The sub-cellular localization of *LpHsp90* proteins indicated that they are mostly cytoplasmic. Two pairs of *LpHsp90* paralogous genes were identified (*LpHsp90-7* and *LpHsp90-8*, *LpHsp90-3* and *LpHsp90-5*) along with three orthologous gene pairs (*Bd4g06370* and *LpHsp90-4*, *Bd1g30130* and *LpHsp90-6*, *Bd4g32941* and *LpHsp90-2*). Expression pattens indicated that *LpHsp90-7*, *LpHs90-5*, *LpHs90-3* and *LpHsp90-1* were highly expressed under various stresses. *LpHsp90* proteins were generally highly expressed under heat stress and weakly under Cr stress. The functions of *LpHsp90* proteins remain unknown, and further studies are needed to determine their precise functions. This study provides a basis for future comprehensive studies on the functional analysis of *LpHsp90* proteins. Additionally, treatments such as MeJA and gibberellic acid would be of great interest in the experimental design and should be considered in future studies because they are important cellular regulators.

Supplementary Materials: The following are available online at https://www.mdpi.com/article/10.3390/plants10112509/s1, Table S1: Analysis of cis-acting element of *LpHsp90* genes in perennial ryegrass. Figure S1: Unrooted phylogenetic tree of 8(eight) *LpHsp90* proteins with annotated functions. The green color represented *Hsp90* proteins in *Arabidopsis thaliana*, red for *Oryza sativa*, violet for *Brachypodium distachyon* and blue for *Lolium perenne*. Figure S2: Details of motif logo and consensus. Figure S3: Amino acid sequence alignment of 8(eight) *LpHsp90* and the location of the C-terminal EEVD motif predicted. Figure S4: Heatmap showing the expression pattern of tested *LpHsp90* genes of perennial ryegrass under (a) heat (b) NaCl (c) Cd (d) ABA (e) PEG treatment respectively. The color scale indicates expression values normalized by TB tools formula.

Author Contributions: Funding acquisition and experiment design, G.N. and J.Z.; data curation, C.A., Z.-F.Y., W.-Z.X. and Y.-Q.Z.; writing—original draft, G.N., Y.W., C.A. and J.H.; writing—review and editing, G.N. All authors have read and agreed to the published version of the manuscript.

Funding: This research was funded by Agricultural Science and Technology Achievement Transformation Project in Sichuan Province (21NZZH0035), China Agriculture Research System of MOF and MARA, and the Funding of Outstanding Papers Promotion Project for Financial Innovation in Sichuan Province (2018LWJJ-013).

Institutional Review Board Statement: Not applicable.

Informed Consent Statement: Not applicable.

Data Availability Statement: Not applicable.

Conflicts of Interest: The authors declare no conflict of interest.

References

1. Xu, Y.; Wang, J.; Bonos, S.A.; Meyer, W.A.; Huang, B. Candidate Genes and Molecular Markers Correlated to Physiological Traits for Heat Tolerance in Fine Fescue Cultivars. *Int. J. Mol. Sci.* **2018**, *19*, 116. [CrossRef] [PubMed]
2. Huang, B.; Dacosta, M.; Jiang, Y. Research advances in mechanisms of turfgrass tolerance to abiotic stresses: From physiology to molecular biology. *Crit. Rev. Plant Sci.* **2014**, *3*, 141–189. [CrossRef]
3. Huang, L.; Yan, H.; Jiang, X.; Yin, G.; Zhang, X.; Qi, X.; Zhang, Y.; Yan, Y.; Ma, X.; Peng, Y. Identification of candidate reference genes in perennial ryegrass for quantitative RT-PCR under various abiotic stress conditions. *PLoS ONE* **2014**, *9*, e93724.
4. Holmes, W.; Robson, M.J.; Parsons, A.J.; Williams, T.E.; Gill, M.; Beever, D.E.; Osbourn, D.F.; Murdoch, J.C.; Nix, J.S.; Green, B.H. Grass: Its production and utilization. *N. Z. J. Agric. Res.* **2010**, *33*, 351.
5. Wang, W.; Vinocur, B.; Shoseyov, O.; Altman, A. Role of plant heat-shock proteins and molecular chaperones in the abiotic stress response. *Trends Plant Sci.* **2004**, *9*, 244–252. [CrossRef]
6. Nakashima, K.; Yamaguchi-Shinozaki, K. Regulons involved in osmotic stress-responsive and cold stress-responsive gene expression in plants. *Physiol. Plant.* **2006**, *126*, 62–71. [CrossRef]
7. Chaves, M.M.; Maroco, J.P.; Pereira, J.S. Understanding plant responses to drought—From genes to the whole plant. *Funct. Plant Biol.* **2003**, *30*, 239–264. [CrossRef]
8. Bailey-Serres, J.; Voesenek, L. Flooding stress: Acclimations and genetic diversity. *Annu. Rev. Plant Biol.* **2008**, *59*, 313–339. [CrossRef]
9. Norris, I.B. Relationships between growth and measured weather factors among contrasting varieties of *Lolium*, *Dactylis* and *Festuca* species. *Grass Forage Sci.* **2010**, *40*, 151–159. [CrossRef]
10. Ollerenshaw, J.H. Influence of waterlogging on the emergence and growth of *Lolium perenne* L. shoots from seed coated with calcium peroxide. *Plant Soil* **1985**, *85*, 131–141. [CrossRef]
11. Pearson, A.; Cogan, N.O.I.; Baillie, R.C.; Hand, M.L.; Bandaranayake, C.K.; Erb, S.; Wang, J.; Kearney, G.A.; Gendall, A.R.; Smithet, K.F. Identification of QTLs for morphological traits influencing waterlogging tolerance in perennial ryegrass (*Lolium perenne* L.). *Theor. Appl. Genet.* **2011**, *122*, 609–622. [CrossRef]
12. Xiong, Y.; Fei, S.Z.; Arora, R.; Brummer, E.C.; Warnke, S.E. Identification of quantitative trait loci controlling winter hardiness in an annual x perennial ryegrass interspecific hybrid population. *Mol. Breed.* **2007**, *19*, 125–136. [CrossRef]
13. Queitsch, C.; Sangster, T.A.; Lindquist, S. *Hsp90* as a capacitor of phenotypic variation. *Nature* **2002**, *417*, 618–624. [CrossRef]
14. Robert, J. Evolution of heat shock protein and immunity. *Dev. Comp. Immunol.* **2003**, *27*, 449–464. [CrossRef]
15. Hsu, A.L.; Murphy, C.T.; Kenyon, C. Regulation of aging and age-related disease by DAF-16 and heat-shock factor. *Science* **2003**, *300*, 1142–1145. [CrossRef]
16. Krishna, P. Plant responses to heat stress. In *Plant Responses to Abiotic Stress*; Topics in Current Genetics; Springer: Berlin/Heidelberg, Germany, 2003; Volume 4, pp. 73–101.
17. Czar, M.J.; Galigniana, M.D.; Silverstein, A.M.; Pratt, W.B. Geldanamycin, a Heat Shock Protein 90-Binding Benzoquinone Ansamycin, Inhibits Steroid-Dependent Translocation of the Glucocorticoid Receptor from the Cytoplasm to the Nucleus. *Biochemistry* **1997**, *36*, 7776–7785. [CrossRef]
18. Pratt, W.B.; Krishna, P.; Olsen, L.J. *Hsp90*-binding immunophilins in plants: The protein movers. *Trends Plant Sci.* **2001**, *6*, 54–58. [CrossRef]
19. Lindquist, S. The Heat-Shock Response. *Annu. Rev. Biochem.* **1986**, *55*, 1151–1191. [CrossRef]
20. Iqbal, N.; Farooq, S.; Arshad, R.; Hameed, A. Differential accumulation of high and low molecular weight heat shock proteins in basmati rice (*Oryza sativa* L.) cultivars. *Genet. Resour. Crop. Evol.* **2010**, *57*, 65–70. [CrossRef]
21. Lindquist, S.; Craig, E.A. The heat-shock proteins. *Annu. Rev. Genet.* **1988**, *22*, 631–677. [CrossRef]
22. Tapia, H.; Morano, K. *Hsp90* nuclear accumulation in quiescence is linked to chaperone function and spore development in yeast. *Mol. Biol. Cell* **2010**, *21*, 63–72. [CrossRef] [PubMed]

23. Hashemi-petroudi, S.H.; Nematzadeh, G.; Mohammadi, S.; Kuhlmann, M. Analysis of expression pattern of genome and analysis of *HSP90* gene family in *Aeluropus littoralis* under salinity stress. *J. Crop. Breed.* **2019**, *11*, 134–143. [CrossRef]
24. Chaudhary, R.; Baranwal, V.K.; Kumar, R.; Sircar, D.; Chauhan, H. Genome-wide identification and expression analysis of *Hsp70*, *Hsp90*, and *Hsp100* heat shock protein genes in barley under stress conditions and reproductive development. *Funct. Integr. Genom.* **2019**, *19*, 1007–1022. [CrossRef] [PubMed]
25. Chen, J.; Gao, T.; Wan, S.; Zhang, Y.; Yang, J.; Yu, Y.; Wang, W. Genome-Wide identification, classification and expression analysis of the *HSP* gene superfamily in tea plant (*Camellia sinensis*). *Int. J. Mol. Sci.* **2018**, *19*, 2633. [CrossRef]
26. Zhang, K.; He, S.; Sui, Y.; Gao, Q.; Jia, S.; Lu, X.; Jia, L. Genome-Wide Characterization of *HSP90* Gene Family in Cucumber and Their Potential Roles in Response to Abiotic and Biotic Stresses. *Front. Genet.* **2021**, *12*, 584886. [CrossRef]
27. Zhang, Z.; Quick, M.K.; Kanelakis, K.C.; Gijzen, M.; Krishna, P. Characterization of a plant homolog of hop, a cochaperone of *hsp90*. *Plant Physiol.* **2003**, *131*, 525–535. [CrossRef]
28. Yamano, T.; Mizukami, S.; Murata, S.; Chiba, T.; Tanaka, K.; Udono, H. *Hsp90*-mediated assembly of the 26 S proteasome is involved in major histocompatibility complex class I antigen processing. *J. Biol. Chem.* **2008**, *283*, 28060–28065. [CrossRef]
29. Feng, J.; Fan, P.; Jiang, P.; Lv, S.; Chen, X.; Li, Y. Chloroplast targeted *Hsp90* plays essential roles in plastid development and embryogenesis in *Arabidopsis* possibly linking with VIPP1. *Physiol. Plant.* **2014**, *150*, 292–307. [CrossRef]
30. Oh, S.E.; Yeung, C.; Babaei-Rad, R.; Zhao, R. Cosuppression of the chloroplast localized molecular chaperone *HSP90.5* impairs plant development and chloroplast biogenesis in *Arabidopsis*. *BMC Res. Notes* **2014**, *7*, 643. [CrossRef]
31. Reddy, R.K.; Chaudhary, S.; Patil, P.; Krishna, P. The 90 kDa heat shock protein (*hsp90*) is expressed throughout *Brassica napus* seed development and germination. *Plant Sci.* **1998**, *131*, 131–137. [CrossRef]
32. Sable, A.; Rai, K.M.; Choudhary, A.; Yadav, V.K.; Agarwal, S.K.; Sawant, S.V. Inhibition of heat shock proteins *HSP90* and *HSP70* induce oxidative stress, suppressing cotton fiber development. *Sci. Rep.* **2018**, *8*, 3620. [CrossRef]
33. Hoffmann, T.; Hovemann, B. Heat-shock proteins, *Hsp84* and *Hsp86*, of mice and men: Two related genes encode formerly identified tumour-specific transplantation antigens. *Gene* **1988**, *74*, 491–501. [CrossRef]
34. Rebbe, N.F.; Ware, J.; Bertina, R.M.; Modrich, P.; Stafford, D.W. Nucleotide sequence of a cDNA for a member of the human 90-kDa heat-shock protein family. *Gene* **1987**, *53*, 235–245. [CrossRef]
35. Krone, P.H.; Sass, J.B. *Hsp 90α* and *Hsp 90β* genes are present in the zebrafish and are differentially regulated in developing embryos. *Biochem. Biophys. Res. Commun.* **1994**, *204*, 746–752. [CrossRef]
36. Chen, B.; Zhong, D.; Monteiro, A. Comparative genomics and evolution of the *HSP90* family of genes across all kingdoms of organisms. *BMC Genom.* **2006**, *7*, 156. [CrossRef]
37. Song, H.; Fan, P.; Li, Y. Overexpression of organellar and cytosolic *AtHSP90* in *Arabidopsis thaliana* impairs plant tolerance to oxidative stress. *Plant Mol. Biol. Rep.* **2009**, *27*, 342–349. [CrossRef]
38. Hu, W.; Hu, G.; Han, B. Genome-wide survey and expression profiling of heat shock proteins and heat shock factors revealed overlapped and stress specific response under abiotic stresses in rice. *Plant Sci.* **2009**, *176*, 583–590. [CrossRef]
39. Krishna, P.; Gloor, G. The *Hsp90* family of proteins in *Arabidopsis thaliana*. *Cell Stress Chaperones* **2001**, *6*, 238–246. [CrossRef]
40. Yang, X.; Zhu, W.; Zhang, H.; Liu, N.; Tian, S. Heat shock factors in tomatoes: Genome-wide identification, phylogenetic analysis and expression profiling under development and heat stress. *PeerJ* **2016**, *4*, e1961. [CrossRef]
41. Yamada, K.; Fukao, Y.; Hayashi, M.; Fukazawa, M.; Nishimura, M. Cytosolic *HSP90* regulates the heat shock response that is responsible for heat acclimation in *Arabidopsis thaliana*. *J. Biol. Chem.* **2007**, *282*, 37794–37804. [CrossRef]
42. McLellan, C.A.; Turbyville, T.J.; Wijeratne, E.M.K.; Kerschen, A.; Vierling, E.; Queitsch, C.; Whitesell, L.; Gunatilaka, A.A.L. A rhizosphere fungus enhances Arabidopsis thermotolerance through production of an *HSP90* inhibitor. *Plant Physiol.* **2007**, *145*, 174–182. [CrossRef]
43. Takahashi, A.; Casais, C.; Ichimura, K.; Shirasu, K. *HSP90* interacts with RAR1 and SGT1 and is essential for RPS2-mediated disease resistance in Arabidopsis. *Proc. Natl. Acad. Sci. USA* **2003**, *100*, 11777–11782. [CrossRef]
44. Liu, Y.; Burch-Smith, T.; Schiff, M.; Feng, S.; Dinesh-Kumar, S.P. Molecular chaperone *Hsp90* associates with resistance protein N and its signaling proteins SGT1 and Rar1 to modulate an innate immune response in plants. *J. Biol. Chem.* **2004**, *279*, 2101–2108. [CrossRef]
45. Tominaga, H.; Coury, D.A.; Amano, H.; Miki, W.; Kakinuma, M. cDNA cloning and expression analysis of two heat shock protein genes, *Hsp90* and *Hsp60*, from a sterile *Ulva pertusa* (Ulvales, Chlorophyta). *Fish. Sci.* **2012**, *78*, 415–429. [CrossRef]
46. Sołtys-Kalina, D.; Szajko, K.; Sierocka, I.; Śliwka, J.; Strzelczyk-Żyta, D.; Wasilewicz-Flis, I.; Jakuczun, H.; Szweykowska-Kulinska, Z.; Marczewski, W. Novel candidate genes AuxRP and *Hsp90* influence the chip color of potato tubers. *Mol. Breed.* **2015**, *35*, 224. [CrossRef] [PubMed]
47. Manchado, M.; Salas-Leiton, E.; Infante, C.; Ponce, M.; Asensio, E.; Crespo, A.; Zuasti, E.; Cañavate, J.P. Molecular characterization, gene expression and transcriptional regulation of cytosolic *Hsp90*s in the flatfish Senegalese sole (*Solea senegalensis* Kaup). *Gene* **2008**, *416*, 77–84. [CrossRef] [PubMed]
48. Galea-Lauri, J.; Latchman, D.S.; Katz, D.R. The role of the 90-kDa heat shock protein in cell cycle control and differentiation of the monoblastoid cell line U937. *Exp. Cell Res.* **1996**, *226*, 243–254. [CrossRef] [PubMed]
49. Zhang, H.; Li, L.; Ye, T.; Chen, R.; Gao, X.; Xu, Z. Molecular characterization, expression pattern and function analysis of the *OsHSP90* family in rice. *Biotechnol. Biotechnol. Equip.* **2016**, *30*, 669–676. [CrossRef]

50. Bai, J.; Pennill, L.A.; Ning, J.; Lee, S.W.; Ramalingam, J.; Webb, C.A.; Zhao, B.; Sun, Q.; Nelson, J.C.; Leach, J.E.; et al. Diversity in nucleotide binding site-leucine-rich repeat genes in cereals. *Genome Res.* **2002**, *12*, 1871–1884. [CrossRef]
51. Han, Y.; Chen, Y.; Yin, S.; Zhang, M.; Wang, W. Over-expression of TaEXPB23, a wheat expansin gene, improves oxidative stress tolerance in transgenic tobacco plants. *J. Plant Physiol.* **2015**, *173*, 62–71. [CrossRef]
52. Chory, J.; Wu, D. Weaving the Complex Web of Signal Transduction. *Plant Physiol.* **2001**, *125*, 77–80. [CrossRef]
53. Mach, J. Alternative splicing produces a JAZ protein that is not broken down in response to jasmonic acid. *Plant Cell* **2009**, *21*, 14. [CrossRef]
54. Al-Whaibi, M.H. Plant heat-shock proteins: A mini review. *J. King Saud Univ. Sci.* **2011**, *23*, 139–150. [CrossRef]
55. Genest, O.; Wickner, S.; Doyle, S.M. *Hsp90* and *Hsp70* chaperones: Collaborators in protein remodeling. *J. Biol. Chem.* **2019**, *294*, 2109–2120. [CrossRef]
56. Milioni, D.; Hatzopoulos, P. Genomic organization of *Hsp90* gene family in Arabidopsis. *Plant Mol. Biol.* **1997**, *35*, 955–961. [CrossRef]
57. Ritossa, F. A new puffing pattern induced by temperature shock and DNP in drosophila. *Experientia* **1962**, *18*, 571–573. [CrossRef]
58. Cha, J.; Ahn, G.; Kim, J.Y.; Kang, S.B.; Kim, M.R.; Su'udi, M.; Kim, W.Y.; Son, D. Structural and functional differences of cytosolic 90-kDa heat-shock proteins (*Hsp90s*) in *Arabidopsis thaliana*. *Plant Physiol. Biochem.* **2013**, *70*, 368–373. [CrossRef]
59. Jackson, S.E.; Queitsch, C.; Toft, D. *Hsp90*: From structure to phenotype. *Nat. Struct. Mol. Biol.* **2004**, *11*, 1152–1155. [CrossRef]
60. Shinozaki, F.; Minami, M.; Chiba, T.; Suzuki, M.; Yoshimatsu, K.; Ichikawa, Y.; Terasawa, K.; Emori, Y.; Matsumoto, K.; Kurosaki, T.; et al. Depletion of *hsp90beta* induces multiple defects in B cell receptor signaling. *J. Biol. Chem.* **2006**, *281*, 16361–16369. [CrossRef]
61. Rehn, A.; Moroni, E.; Zierer, B.K.; Tippel, F.; Morra, G.; John, C.; Richter, K.; Colombo, G.; Buchner, J. Allosteric Regulation Points Control the Conformational Dynamics of the Molecular Chaperone *Hsp90*. *J. Mol. Biol.* **2016**, *428*, 4559–4571. [CrossRef]
62. Zhang, J.; Li, J.; Liu, B.; Zhang, L.; Chen, J.; Lu, M. Genome-wide analysis of the Populus *Hsp90* gene family reveals differential expression patterns, localization, and heat stress responses. *BMC Genom.* **2013**, *14*, 532. [CrossRef] [PubMed]
63. Song, Z.; Pan, F.; Yang, C.; Jia, H.; Jiang, H.; He, F.; Li, N.; Lu, X.; Zhang, H. Genome-wide identification and expression analysis of *HSP90* gene family in *Nicotiana tabacum*. *BMC Genet.* **2019**, *20*, 35. [CrossRef] [PubMed]
64. Zai, W.S.; Miao, L.X.; Xiong, Z.L.; Zhang, H.L.; Ma, Y.R.; Li, Y.L.; Chen, Y.B.; Ye, S.G. Comprehensive identification and expression analysis of *Hsp90s* gene family in *Solanum lycopersicum*. *Genet. Mol. Res.* **2015**, *14*, 7811–7820. [CrossRef] [PubMed]
65. Kawahara, Y.; de la Bastide, M.; Hamilton, J.P.; Kanamori, H.; McCombie, W.R.; Ouyang, S.; Schwartz, D.C.; Tanaka, T.; Wu, J.; Zhou, S.; et al. Improvement of the *Oryza sativa* Nipponbare reference genome using next generation sequence and optical map data. *Rice* **2013**, *6*, 4. [CrossRef]
66. Liu, Y.; Wan, H.; Yang, Y.; Wei, Y.; Li, Z.; Ye, Q.; Wang, R.; Ruan, M.; Yao, Z.; Zhou, G. Genome-wide identification and analysis of heat shock protein 90 in tomato. *Yi Chuan* **2014**, *36*, 1043–1052.
67. Jain, M.; Tyagi, A.K.; Khurana, J.P. Genome-wide analysis, evolutionary expansion, and expression of early auxin-responsive SAUR gene family in rice (*Oryza sativa*). *Genomics* **2006**, *88*, 360–371. [CrossRef]
68. Skolnick, J.; Fetrow, J.S. From genes to protein structure and function: Novel applications of computational approaches in the genomic era. *Trends Biotechnol.* **2000**, *18*, 34–39. [CrossRef]
69. Singh, R.K.; Lee, J.K.; Selvaraj, C.; Singh, R.; Li, J.; Kim, S.Y.; Kalia, V.C. Protein Engineering Approaches in the Post-Genomic Era. *Curr. Protein Pept. Sci.* **2018**, *19*, 5–15. [CrossRef]
70. Jin, Z.; Chandrasekaran, U.; Liu, A. Genome-wide analysis of the Dof transcription factors in castor bean (*Ricinus communis* L.). *Genes Genom.* **2014**, *36*, 527–537. [CrossRef]
71. Byrne, S.L.; Nagy, I.; Pfeifer, M.; Armstead, I.; Asp, T. A synteny-based draft genome sequence of the forage grass *Lolium perenne*. *Plant J.* **2015**, *84*, 816–826. [CrossRef]
72. Larkin, M.A. Clustal W and Clustal X version 2.0. *Bioinformatics* **2007**, *23*, 2947–2948. [CrossRef]
73. Guidon, S.; Gascuel, O. A simple, fast and accurate algorithm to estimate large phylogenies by maximum likelihood. *Syst. Biol.* **2003**, *52*, 696–704. [CrossRef]
74. Tamura, K.; Stecher, G.; Peterson, D.; Filipski, A.; Kumar, S. MEGA6: Molecular Evolutionary Genetics Analysis version 6.0. *Mol. Biol. Evol.* **2013**, *30*, 2725–2729. [CrossRef]
75. Hu, B.; Jin, J.; Guo, A.Y.; Zhang, H.; Luo, J.; Gao, G. GSDS 2.0: An upgraded gene feature visualization server. *Bioinformatics* **2014**, *31*, 1296–1297. [CrossRef]
76. Bailey, T.L.; Boden, M.; Buske, F.A.; Frith, M.; Grant, C.E.; Clementi, L.; Ren, J.; Li, W.W.; Noble, W.S. MEME SUITE: Tools for motif discovery and searching. *Nucleic Acids Res.* **2009**, *37*, W202–W208. [CrossRef]
77. Schmittgen, T.D.; Livak, K.J. Analyzing real-time PCR data by the comparative C(T) method. *Nat. Protoc.* **2008**, *3*, 1101–1108. [CrossRef]

Article

Lanthanum Promotes Bahiagrass (*Paspalum notatum*) Roots Growth by Improving Root Activity, Photosynthesis and Respiration

Ying Liu [1] and Juming Zhang [2,*]

[1] Qinghai Provincial Key Laboratory of Adaptive Management on Alpine Grassland/Key Laboratory of Superior Forage Germplasm in the Qinghai-Tibetan Plateau, Qinghai Academy of Animal and Veterinary Sciences, Qinghai University, Xining 810016, China; yingliu629@yahoo.com

[2] Guangdong Engineering Research Center for Grassland Science, College of Forestry and Landscape Architecture, South China Agricultural University, Guangzhou 510642, China

* Correspondence: jimmzh@scau.edu.cn

Abstract: Lanthanum (La), one of the most active rare earth elements, promotes the growth of turfgrass roots. In this study, the mechanisms by which La influences bahiagrass (*Paspalum notatum*) growth were evaluated by the analyses of root growth, root activity, cell wall polysaccharide content, respiration intensity, ascorbic acid oxidase (AAO) and polyphenol oxidase (PPO) activity, the subcellular distribution of mitochondria, transcription in roots, photosynthetic properties, chlorophyll fluorescence parameters, and chlorophyll content. The application of 0.3 mM La^{3+} increased root activity, respiration intensity, AAO activity, and the number of mitochondria in the mature cells of bahiagrass roots. La could significantly improve the net photosynthetic rate, transpiration rate, and chlorophyll fluorescence of bahiagrass. Differentially expressed genes identified by high-throughput transcriptome sequencing were enriched for GO (Gene Ontology) terms related to energy metabolism and were involved in various KEGG (Kyoto Encyclopedia of Genes and Genomes) pathways, including oxidative phosphorylation, TCA (Tricarboxylic Acid) cycle, and sucrose metabolism. These findings indicate that La promotes bahiagrass root growth by improving root activity, photosynthesis, and respiration, which clarifies the mechanisms underlying the beneficial effects of La and provides a theoretical basis for its use in artificial grassland construction and ecological management projects.

Keywords: lanthanum; bahiagrass; root growth; photosynthesis; respiration

Citation: Liu, Y.; Zhang, J. Lanthanum Promotes Bahiagrass (*Paspalum notatum*) Roots Growth by Improving Root Activity, Photosynthesis and Respiration. *Plants* **2022**, *11*, 382. https://doi.org/10.3390/plants11030382

Academic Editor: Barbara Hawrylak-Nowak

Received: 31 December 2021
Accepted: 27 January 2022
Published: 30 January 2022

Publisher's Note: MDPI stays neutral with regard to jurisdictional claims in published maps and institutional affiliations.

1. Introduction

Lanthanum (La) belongs to the group of elements known as rare earth elements. Among the fifteen rare earth elements, La is the most active, the most well-studied, and the most widely used. Its application in agricultural production was first proposed in the 1980s. More than three decades of systematic research have generated a wealth of literature indicating that La is a beneficial element for plant growth. In terms of vegetative growth, it can promote seed germination and root growth [1], boost metabolism by the enhancement of photosynthesis [2], and promote the absorption of soil mineral elements by the plant [3]. With respect to reproductive growth, it can improve flower formation and fruit set rates [4]. It can affect the cell skeleton and the function of the plasma membrane [5,6]. The use of La can improve plant resistance to abiotic stresses, such as heavy metals [7], and enhance disease resistance [8]. In short, La is a beneficial nutrient for agricultural production. Recent studies have focused on characterizing the effects of La on plant growth. However, little is known about the mechanisms underlying these effects, including the mechanisms by which La promotes the growth of plant roots, the most important organs for absorbing water and nutrients [9].

Bahiagrass (*Paspalum notatum*) is widely used for slope protection and soil and water conservation in southern China owing to its extensive and deep root system. Our preliminary work has shown that La could significantly improve bahiagrass root growth [10]. It has been reported that La absorbed by plant roots is mainly attached to the cell wall of root cells and that La alleviates the toxic effects of aluminum by changing the porosity of the cell wall [11]. Therefore, the objective of this study was to investigate how La promotes the growth of the bahiagrass root system, focusing on its root-promoting mechanism. To achieve this aim, this study was divided into three parts: (1) the evaluation of La's effect on cell wall components and root activity; (2) the evaluation of the regulatory effects of La on photosynthesis and respiration, as root growth requires a sufficient energy supply; and (3) the evaluation of the genetic basis of the physiological and phenotypic alterations in the root by high-throughput transcriptome sequencing.

2. Results

2.1. Measurement of the La Content

Before studying the effect of La on bahiagrass root growth, it is necessary to first determine that the La added to the root nutrient solution can be absorbed by bahiagrass. As shown in Figure 1A, La treatment significantly increased the La content in bahiagrass roots in a time-dependent manner ($p < 0.05$). In addition, the greatest increase was observed after 2 days, and the rate of increase decreased gradually over time. As shown in Figure 1B, La treatment also significantly increased the La content in bahiagrass leaves ($p < 0.05$). Similar to the results obtained for the root tissues, the greatest increase in La was observed on the second day; however, the La content in the leaves did not increase as the treatment time increased (i.e., the increments on day 4 and day 6 were not significant).

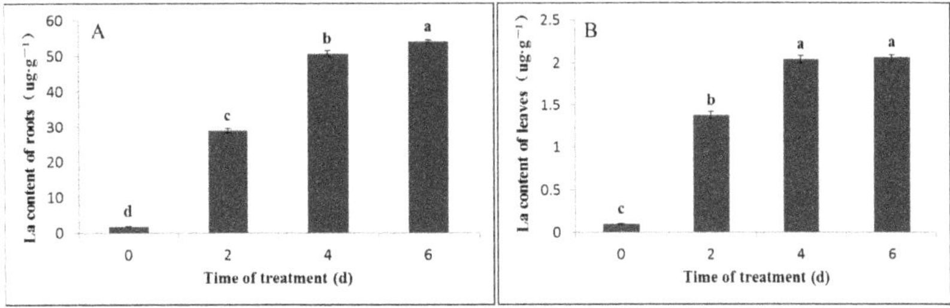

Figure 1. La content in bahiagrass roots and leaves for various La treatment durations. Note: Bars with different letters are significantly different at $p = 0.05$. (**A**) La content of roots, (**B**) La content of leaves.

2.2. Effect of La on Bahiagrass Root Activity

Root activity is the most direct indicator of plant root health. The results of this study showed that La treatment significantly improved the root activity of bahiagrass seedlings ($p < 0.05$) (Figure 2). The application of lanthanum increased root activity nearly fivefold.

2.3. Effect of La on the Cell Wall Polysaccharide Content of Bahiagrass Roots

The cell wall polysaccharide content is the most direct indicator of cell wall functionality. La treatment had no significant effect on the cell wall polysaccharide content of bahiagrass roots ($p > 0.05$) (Figure 3). Cell wall polysaccharide content included pectin, hemicellulose I (HCI), and hemicellulose II (HCII). These results suggest that La does not contribute to the regulation of cell wall components, even though La absorbed by the roots mainly accumulates in the cell walls.

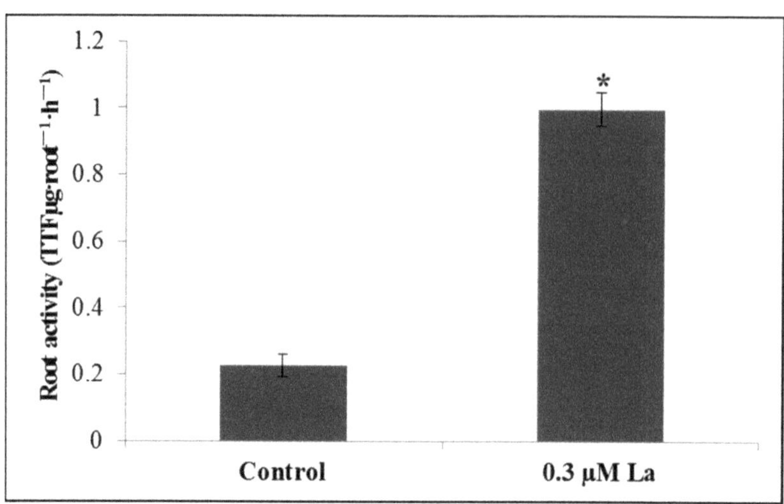

Figure 2. Effect of La on root activity of bahiagrass roots. Note: Asterisks (*) indicate significant differences at $p = 0.05$.

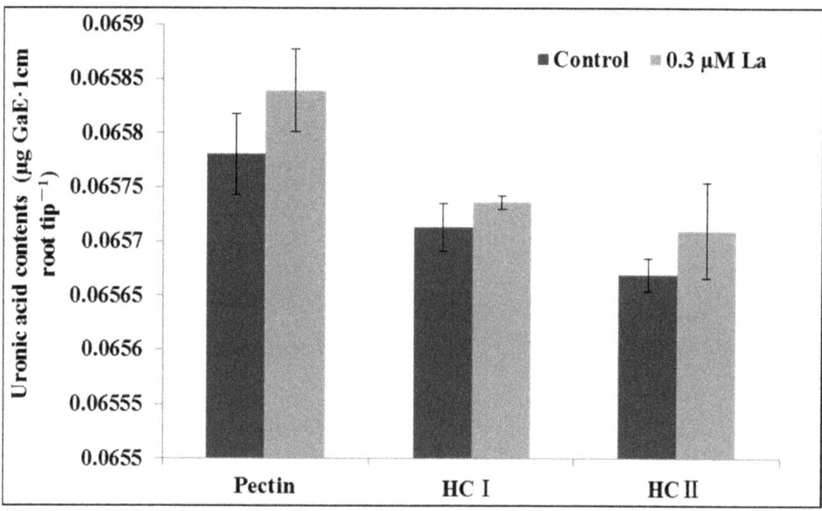

Figure 3. Effect of La on the uronic acid content of bahiagrass.

2.4. Effect of La on Bahiagrass Photosynthesis

2.4.1. Effect of La on the Photosynthetic Properties of Bahiagrass

The net photosynthetic rate is defined as the difference between the rates of photosynthesis and respiration and is often represented as the accumulation of organic matter in the leaf per unit time per unit area. The transpiration rate refers to the amount of transpiration in the leaf per unit time and is one of the main factors affecting the net photosynthetic rate of plants. Stomatal conductance refers to the size of the stomatal opening, which affects the photosynthesis, respiration, and transpiration of plants, and is an important indicator of the impact of stomatal or non-stomatal factors on the photosynthetic rate, along with the intercellular CO_2 concentration [12–14]. Our results showed that La treatment significantly improved the net photosynthetic and transpiration rates of bahiagrass ($p < 0.05$), with no significant effects on stomatal conductance and the intercellular CO_2 concentration ($p > 0.05$) (Figure 4).

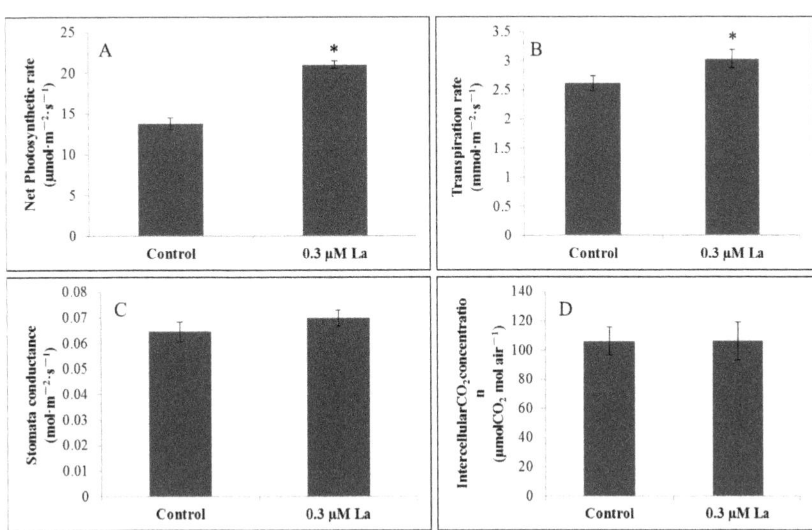

Figure 4. Effect of La on photosynthetic properties of bahiagrass leaf. Note: Asterisks (*) indicate significant differences at $p = 0.05$. (**A**) Net Photosynthesis rate, (**B**) Transpiration rate, (**C**) Stomatal conductance, (**D**) intercellularCO_2 concentration.

2.4.2. Effect of La on Bahiagrass Chlorophyll Fluorescence Parameters

F_v/F_m is the maximum quantum yield of a completely open PSII reaction center, reflecting the potential maximum photosynthetic capacity of the plant; F_v'/F_m' is the excitation light capture efficiency of leaves; ΦPSII refers to the actual quantum yield of PSII with partial closure of the PSII reaction center under light, reflecting the current actual photosynthetic efficiency of the leaves; qP is the photochemical quenching coefficient, reflecting the level of photosynthetic activity of the plant; and ETR refers to the photosynthetic electron transport rate [15–17]. La treatment significantly improved the chlorophyll fluorescence parameters, indicating that La effectively promoted photosynthesis in bahiagrass ($p < 0.05$) (Figure 5).

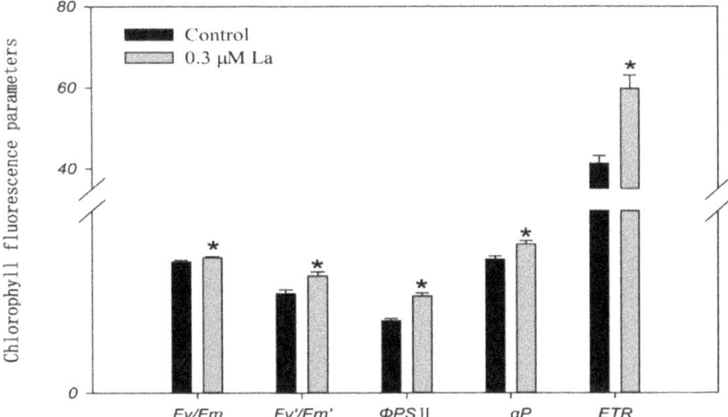

Figure 5. Effect of La on chlorophyll fluorescence parameters in bahiagrass Note: Bars within the same parameter with asterisks are significantly different at $p = 0.05$.

2.4.3. Effect of La on the Chlorophyll Content in Bahiagrass during Photosynthesis

Chlorophyll is important in the process of photosynthesis, including important roles in electron transport [18–20]. However, La treatment did not increase the chlorophyll content in bahiagrass leaves ($p > 0.05$) (Figure 6).

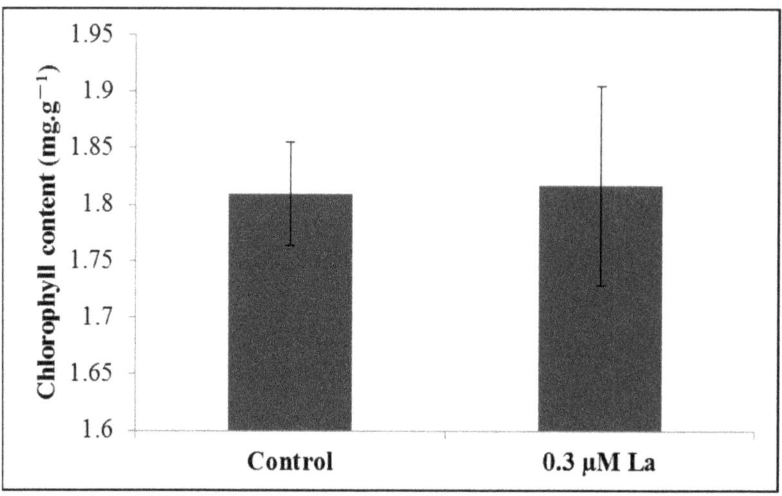

Figure 6. Effect of La on the chlorophyll content of bahiagrass.

2.5. Effect of La on Bahiagrass Respiration

2.5.1. Effect of La on the Respiration Rate of Bahiagrass Roots

Respiration provides a direct energy source for ATP in plant metabolism. The respiration rate is the most direct physiological indicator of respiration. La treatment significantly increased the respiration rate of bahiagrass roots ($p < 0.05$), providing sufficient energy for root growth (Figure 7).

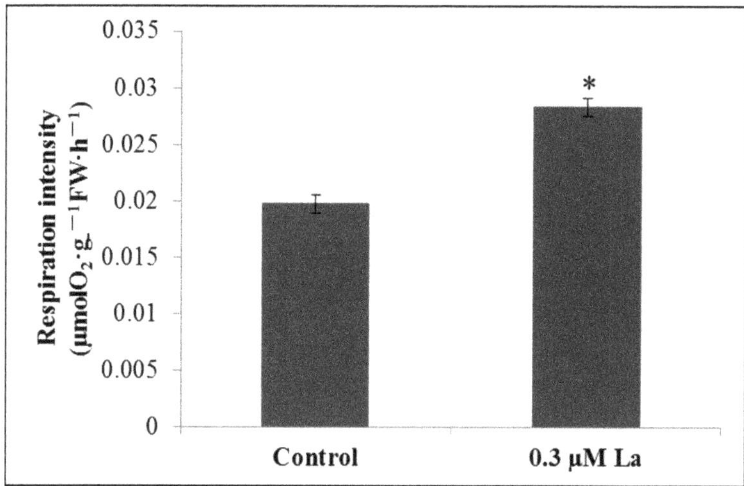

Figure 7. Effect of La on respiration intensity of bahiagrass roots. Note: Asterisks (*) indicate significant differences at $p = 0.05$.

2.5.2. Effect of La on Enzymes Related to Bahiagrass Root Respiration

Respiration requires a specific complex enzyme reaction system, wherein AAO and PPO enzyme activities are representative physiological indicators. In this study, La treatment significantly increased AAO activity ($p < 0.05$) but had no effect on PPO activity (Figure 8).

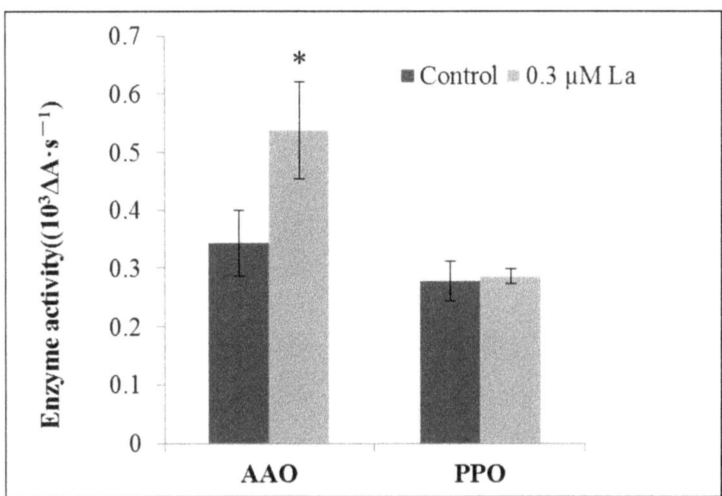

Figure 8. Effect of La on ascorbic acid oxidase (AAO) and polyphenol oxidase (PPO) activity in bahiagrass roots. Note: Bars within the same enzyme with asterisks are significantly different at $p = 0.05$.

2.6. Effect of La on Mature Mitochondria in Bahiagrass Roots

The respiration process in plants is completed in mitochondria. In this study, transmission electron microscopy was performed to evaluate mature root cells. The shape and number of mitochondria in nine cells were observed. As shown in Figure 9B,D, La treatment had no effect on mitochondrial morphology. As shown in Figure 9A,C, La treatment increased the number of mitochondria in mature cells. A quantitative analysis revealed that mature La-treated cells, on average, had 3.5 mitochondria, 1.75 more mitochondria than the corresponding number in the control (2 mitochondria).

2.7. Effect of La on Transcript Levels

In this study, the total RNA of bahiagrass roots was sequenced using high-throughput transcriptome sequencing technology. Approximately 130 million raw reads were obtained and approximately 110 million clean reads were retained for subsequent analyses after filtering. After assembling the clean reads, approximately 210,000 transcripts were obtained, containing approximately 120,000 potential genes. In total, 24,879 genes were significantly differentially expressed between the La treatment group and control group, of which 15,329 were up-regulated and 9550 were downregulated in the La treatment group.

A GO analysis of these differentially expressed genes (DEGs) showed that in the biological process category, the DEGs were mainly concentrated in cellular and metabolic processes. In the cellular component category, the DEGs were mainly concentrated in cells, cell parts, and organelles. In the molecular function category, the DEGs were mainly concentrated in binding and catalytic functions (Figure 10). In each major category, the two subcategories where DEGs were concentrated were further refined into subterm classifications. In the cellular and metabolic process category, most of the terms were related to energy metabolism, including enrichment for the electron transport chain, TCA cycle, glycolysis, and other growth-related terms. In the cell and organelle category, there was an enrichment of organelles associated with energy metabolism (mitochondria and chloroplasts). In the binding and catalytic function category, there was an enrichment of binding and enzymes predominantly related to nucleic acids (Table 1).

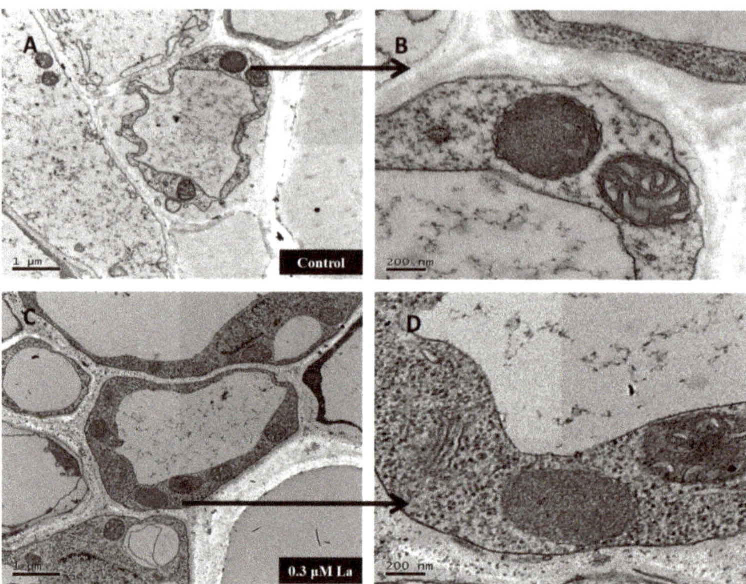

Figure 9. Effect of La on the subcellular distribution of mitochondria in bahiagrass mature root cells. (**A**) Mitochondrias in bahiagrass mature control root cells (Bar =1 μm), (**B**) Individual mitochondria in bahiagrass mature control root cells (Bar =200 nm), (**C**) Mitochondrias in bahiagrass mature treatment (0.3 μM La) root cells (Bar =1 μm), (**D**) Individual mitochondria in bahiagrass mature treatment (0.3 μM La) root cells (Bar =200 nm).

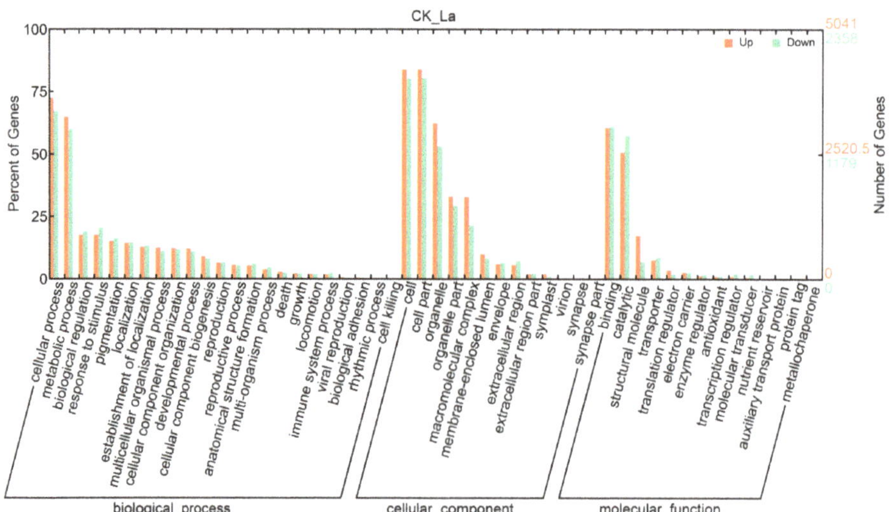

Figure 10. Gene ontology (GO) classification of differentially expressed genes. Note: The *x*-axis indicates the subterm within each main term. The left *y*-axis indicates the percentage of a specific term for differentially expressed genes in each main term. The right *y*-axis represents the number of differentially expressed genes.

Table 1. Gene ontology (GO) subterm classification of differentially expressed genes.

GO Term	GO Subterm	Up Count	Percent	Down Count	Percent
biological_process	electron transport chain	95	0.018845	42	0.017812
biological_process	ATP catabolic process	88	0.017457	43	0.018236
biological_process	GTP catabolic process	80	0.01587	19	0.008058
biological_process	tricarboxylic acid cycle	76	0.015076	42	0.017812
biological_process	growth	60	0.011902	10	0.004241
biological_process	glycolysis	57	0.011307	23	0.009754
biological_process	carbohydrate metabolic process	53	0.010514	30	0.012723
biological_process	cell division	50	0.009919	17	0.00721
biological_process	ATP biosynthetic process	44	0.008728	15	0.006361
biological_process	cell differentiation	38	0.007538	23	0.009754
cellular_component	cytoplasm	774	0.153541	330	0.139949
cellular_component	nucleus	635	0.125967	325	0.137829
cellular_component	ribosome	504	0.09998	90	0.038168
cellular_component	plasma membrane	467	0.09264	257	0.108991
cellular_component	cytosol	354	0.070224	188	0.079729
cellular_component	mitochondrion	301	0.05971	134	0.056828
cellular_component	nucleolus	248	0.049197	60	0.025445
cellular_component	extracellular region	145	0.028764	105	0.044529
cellular_component	chloroplast	134	0.026582	53	0.022477
cellular_component	mitochondrial inner membrane	130	0.025789	54	0.022901
molecular_function	ATP binding	1012	0.200754	518	0.219678
molecular_function	metal ion binding	582	0.115453	286	0.121289
molecular_function	GTP binding	217	0.043047	57	0.024173
molecular_function	nucleotide binding	212	0.042055	94	0.039864
molecular_function	GTPase	146	0.028963	40	0.016964
molecular_function	electron carrier	138	0.027376	62	0.026293
molecular_function	protein serine/threonine kinase	136	0.026979	125	0.053011
molecular_function	translation elongation factor	117	0.02321	17	0.00721
molecular_function	ATPase	89	0.017655	52	0.022053
molecular_function	oxidoreductase	74	0.01468	51	0.021628

A KEGG pathway enrichment analysis indicated that DEGs were involved in energy metabolism, oxidative phosphorylation, TCA cycle and sucrose metabolism (Table 2).

Table 2. Enriched KEGG metabolic pathways for differentially expressed unigenes.

Name	Gene_in_DE (Number)	Gene_in_Background (Number)	p	q	Result
Oxidative phosphorylation	181	602	0.000388	0.004069	yes
Starch and sucrose metabolism	59	326	0.010324	0.041963	yes
Citrate cycle (TCA cycle)	114	368	0.010918	0.04299	yes

Note: $q < 0.05$ means significant difference. DE means differential expression.

Six DEGs were randomly selected for qRT-PCR validation, and the results are summarized in Figure 11. The qRT-PCR results generally agreed with the high-throughput transcript sequencing results, with a high correlation coefficient of 0.92, indicating that the high-throughput transcript sequencing results were reliable.

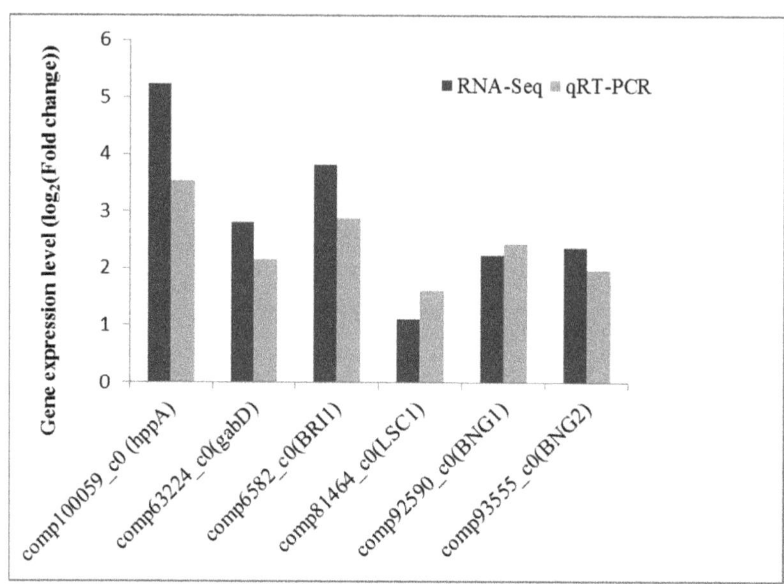

Figure 11. Verification of RNA-seq results by real-time quantitative PCR (qRT-PCR).

3. Discussion

3.1. Effect of Exogenous La on the La Content in Bahiagrass Leaves and Roots

La has a relatively high molecular weight. It mostly accumulates around cell walls after it is absorbed by plant roots [11]. However, our results showed that La could also be transported to aboveground parts, although the amount of La absorbed by stems and leaves was limited after a threshold level was reached. Liu et al. obtained similar results when studying cerium, another rare earth element, and pointed out that cerium accumulation is higher in the roots of rice than in the stems and leaves [3]. Our results for La accumulation in leaves provide indirect support for the theory that La treatment in the roots could affect plant physiological processes in aboveground parts.

3.2. Effect of La on the Root Activity and Cell Wall Polysaccharide Content of Roots

We have previously discovered that La could promote root growth in bahiagrass [10]. In this study, we found that La could also increase root activity. Shi et al. and Liu et al. have demonstrated that La promotes the root activity of red bean *(Ormosia microphylla)* [21] and wheat (*Triticum aestivum*) [22]. These results suggest that La enhances the biological function of bahiagrass roots.

Even though La binds to the root cell wall, many reports have pointed out that La could change the uronic acid content or rigidity of the cell wall [23,24]. We did not detect an effect of La on the polysaccharide content of root cell walls. Similarly, Yang et al. reported that La does not influence the polysaccharide content of rice apical cell walls (pectin, HCI, and HCII) [25].

3.3. Effect of La on Photosynthesis in Bahiagrass Leaves

We obtained a comprehensive overview of the effect of La on photosynthesis in bahiagrass. La improved the net photosynthetic rate, transpiration rate, and chlorophyll fluorescence parameters (F_v/F_m, F_v'/F_m', ΦPSII, qP, and ETR). The increase in the net photosynthetic rate indicated that La treatment could enhance the photosynthetic capacity of bahiagrass. Wen et al. and Wang et al. also reported that low-level La(III) treatment improved photosynthesis in soybean [26] and rice [27], respectively. The observation that La treatment did not increase stomatal conductance or the intracellular CO_2 concentration indicated that La promoted photosynthesis via the regulation of non-stomatal factors. Given the lack of an increase in stomatal conductance, the significant increase in transpiration rate can be explained by the effect of La on metabolic processes in the whole bahiagrass plant. This might involve the increase in water consumption by plant cells, which reduced the cell potential. The reduced cell potential accelerated water absorption by roots and upward transport, thereby increasing the transpiration rate of the plant.

Chlorophyll fluorescence is an indicator of photosynthesis. In this study, the enhancement of F_v/F_m, F_v'/F_m', and ΦPSII indicates that La significantly promoted PSII activity and the primary reaction of photosynthesis as well as the conversion of captured light to chemical energy in leaves. The increase in qP and ETR suggests that La increases the electron transport rate during photosynthesis. Hong et al. pointed out that La could enter spinach chloroplasts and significantly increase the formation of PSII and the electron transport rate [28]. Yan et al. further suggested that 90% of La is present in PSII inside spinach chloroplasts [29]. These results indicate that La could promote the function of PSII and the electron acceptor. Furthermore, in this study, the chlorophyll content in leaves was not increased by La. Shi et al. also showed that La had no significant effect on the chlorophyll content of wheat [21]. Our findings indicate that La raised the electron transport rate in bahiagrass without increasing the chlorophyll content, indicating that La improved the ability of photosynthetic pigments to capture and transport electrons. Hong et al. confirmed that La inside plant chloroplasts could also combine with chlorophyll [28], thereby affecting the electron transport rate during photosynthesis.

3.4. Effect of La on Respiration in Bahiagrass Roots

Respiration provides the most direct energy for plant growth and development, including root growth. This study provides a comprehensive overview of the impact of La on the respiration of bahiagrass roots based on the respiration rate, respiration-related enzyme activity, and mitochondrial ultrastructures in mature root cells. Our results showed that La significantly increased the respiration rate of bahiagrass roots, the activity of ascorbic acid oxidase (AAO), and the number of mitochondria in mature cells. This is the first report on the effect of La on the respiration of plant seedlings. Early in the last century, Palmer et al. observed that La could affect human mitochondria [30]. Brooks also reported that an appropriate concentration of La ions could improve the respiration rate of *Bacillus subtilis* [31]. In addition, Hong et al. and Fashui et al. reported that La could promote respiration in rice seeds [32,33]. Overall, our study shows that La clearly influences respiration processes of various taxa. Our research fills the gap regarding the effect of La on respiration in plant seedlings and clearly reveals that La promotes respiration in bahiagrass roots by improving root AAO activity and increasing the number of mitochondria in mature cells.

3.5. Effect of La on Transcript Levels in Bahiagrass Roots

In this study, the mechanism by which La promotes bahiagrass root growth was evaluated at the transcriptome level by high-throughput transcriptome sequencing. In the GO enrichment analysis of DEGs, significant enrichment for various functions related to the electron transport chain, chloroplast, and metal ions was observed, providing molecular-level explanations for the promotion of photosynthesis by the combination of La with chlorophyll, the increase in PSII activities, and the efficiency of the electron transport chain at the physiological level. Enrichment of the TCA cycle, glycolysis, mitochondria, inner mitochondrial membrane, and oxidoreductase as well as the significant enrichment of the oxidative phosphorylation, TCA cycle, and sucrose metabolic pathways identified by a KEGG pathway analysis provided further molecular-level support for the physiological-level effect of La, which promotes the respiration of bahiagrass roots by improving AAO activity and increasing the number of mitochondria in mature cells. In addition, the detection of enrichment terms related to cell growth, division, and differentiation in the GO enrichment analysis of DEGs further indicates that La can promote the growth of bahiagrass roots. Finally, quantitative fluorescence PCR verified the reliability of the high-throughput transcriptome sequencing results.

4. Materials and Methods

4.1. Plant Materials and Treatment

Bahiagrass seeds were obtained from Clover Group Corporation, Ltd. (Beijing, China). Bahiagrass seed germination and plant growth conditions were the same as those described by Liu et al. [10]. Healthy seeds were germinated on filter paper for 6 days. Seedlings with uniform root lengths were selected. The concentration of La^{3+} ($LaCl_3$) was 0.3 µM. The treatment duration was 6 days. The containers were placed in a growth chamber at 28 °C, 80% relative humidity, 650 µmol m^{-2} s^{-1} photosynthetic active radiation, and a photoperiod of 12 h. La treatment and the control were in the same condition.

4.2. Elemental Analysis

Fresh plants were collected and separated into roots and leaves. They were then washed thoroughly with deionized water, and a microwave-assisted digestion procedure was applied. Approximately 0.1 g of sample (FW) was weighed into Teflon bombs. Then, 5 mL of HNO_3 (70%) and 2 mL of H_2O_2 (30%) were added, followed by digestion. The digestion adopted the fractional stepwise temperature raising method. The temperature was heated to 120 °C for 5 min, then to 150 °C for 5 min, and then to 180 °C for 5 min. After digestion, the samples were transferred into polypropylene tubes and filled to 10.0 cm^3. Subsequently, the La content was investigated by inductively coupled plasma mass spectrometry (ICP-MS; Agilent, Hachioji, Japan).

4.3. Root Activity

Root activity was measured using the modified triphenyl tetrazolium chloride (TTC) method [34,35]. TTC is mainly succinate dehydrogenase, which is reduced by dehydrogenases. Dehydrogenase was expressed as the quantity of deoxidized TTC, which is an index of root activity. In brief, 50 fresh root tips (1 cm) were immersed in 10 mL of an equally mixed solution of TTC (0.4%) and phosphate buffer (0.1 mol·L^{-1}, pH 7.0) and kept in the dark for 3 h at 37 °C. Subsequently, 2 mL of H_2SO_4 (1 mol·L^{-1}) was added to stop the reaction. The immersed root tips were dried with filter paper and extracted with ethyl acetate. The extracted solution was transferred into a tube with ethyl acetate cleaning solution to a total volume of 10 mL, and absorbance was read at 485 nm. Root activity = amount of TTC reduction (µg)/100 fresh root tips (root) × time (h).

4.4. Cell Wall Polysaccharide Content

Cell walls were isolated from the upper 1 cm of root tips of 6-day-old seedlings of bahiagrass according to the procedure described by Zhong and Lauchli [36]. After freeze-drying, the cell wall materials were fractionated into three fractions: pectin, HCI, and HCII. The pectin fraction was extracted twice with 0.5% ammonium oxalate buffer containing 0.1% $NaBH_4$ (pH 4) in a boiling water bath for 1 h, and the supernatants were pooled. Pellets were subsequently subjected to triple extraction with 4% KOH containing 0.1% $NaBH_4$ at 25 °C for 24 h, followed by extraction with 24% KOH containing 0.1% NaBH4. The pooled supernatants from the 4% and 24% KOH extraction yielded the HCI and HCII fractions, respectively.

The uronic acid content was assayed according to the method of Blumenkrantz and Asboe-Hansen [37] for each cell wall fraction. Galacturonic acid (GalA) was used as a calibration standard, and the root pectin content is expressed as GalA equivalents.

4.5. Photosynthetic Characteristics

Photosynthetic properties (net photosynthetic rate, transpiration rate, stomatal conductance, and intercellular CO_2 concentration) were measured using a portable photosynthesis system (Li-6400; Li-Cor Inc., Lincoln, NE, USA). Measurements were carried out in a

2 cm^2 leaf area with an airflow rate of 500 μmol s^{-1}. The photon flux density (PPFD) was 1500 μmol m^{-2} s^{-1}, and the leaf temperature was controlled at 25 °C. The characteristic parameters evaluated were the net photosynthetic rate, transpiration rate, stomatal conductance, and intercellular CO$_2$ concentration.

4.6. Chlorophyll Fluorescence Parameters

After a seedling adaptation period of 30 min in the dark under a controlled temperature (25 °C) and treatment for 6 days with La, the minimum fluorescence (F_o), maximum fluorescence (F_m), constant fluorescence (F_s), maximum fluorescence with actinic light (F_m'), minimum fluorescence with far-red light (F_o'), effective photochemical quantum yield adapted to light (ΦPSII), photochemical quenching coefficient (qP), and electron transport rate (ETR) were measured with a portable chlorophyll fluorometer (PAM-2000; WALZ, Effeltrich, Germany) using a leaf-clip holder (2030-B; Walz, Germany). The photon flux density (PPFD) was 1500 μmol m^{-2} s^{-1}, and the actinic light treatment time was 30 min. Other parameters were calculated as $F_v/F_m = (F_m - F_o)/F_m$ and $F_v'/F_m' = (F_m' - F_o')/F_m'$.

4.7. Chlorophyll Content

Fresh leaves were weighed and 0.1 g was obtained after treatment for 6 days. Samples were placed into a 25 mL scale test tube after shearing. Then, 80% acetone was added and the leaves were carefully washed and adhered to the wall of the bottle. The bottle was covered and kept in a dark box overnight and shocked three times. After filtration or centrifugation, the chloroplast pigment extract was poured into a colorimetric cup with a light diameter of 1 cm. Absorbance was measured at wavelengths of 663 nm and 645 nm with 80% acetone as a blank control. The chlorophyll content was calculated using the following equations: chlorophyll (C) = 20.2D645 + 8.02D633 [38] and chlorophyll content (mg/g) = (chlorophyll concentration C × extracted liquid volume × dilution factor)/sample fresh weight.

4.8. Root Respiration Rate

After 6 days of treatment, bahiagrass root (approximately 0.2 g fresh weight) was taken and immediately placed into an oxygen suction tank containing 8 mL of reaction medium. The Clark oxygen electrode (YSI 5300, Yellow Springs, OH, USA) was used to measure root respiration rate, and the control and treatment were repeated three times each.

4.9. Respiration-Related Enzyme Activity

The enzyme activities of two respiration-related enzymes, ascorbic acid oxidase (AAO) and polyphenol oxidase (PPO), were measured. The method used to determine AAO enzyme activities was described by Oberbacher and Vines [39]. PPO enzyme activity was determined according to Jiang [40].

4.10. Observation of the Mitochondrial Ultrastructure in Mature Root Cells

The mitochondrial ultrastructure of mature cells in bahiagrass roots was observed by transmission electron microscopy (TEM; Tecnai 12, FEI, Eindhoven, The Netherlands). After treatment for 6 days, 10 plant roots were randomly selected for sample preparation for each treatment, three samples were randomly selected for slicing at the end of sample preparation, and three cells were randomly selected for observation. Root samples were prepared according to the method described by Xu et al. [41].

4.11. Transcriptome Sequencing

Total RNA was extracted from root tissues using TRIzol reagent (Life Technologies, Carlsbad, CA, USA) according to the manufacturer's protocol. RNA degradation and contamination were detected by 1% agarose gel electrophoresis. RNA purity was checked using a kaiaoK5500® spectrophotometer (Kaiao, Beijing, China). RNA integrity and concentration were assessed using the RNA Nano 6000 Assay Kit and the Bioanalyzer 2100 system

(Agilent Technologies, Santa Clara, CA, USA). A total of 3 μg of RNA per sample was used as the input material for RNA sample preparation. Sequencing libraries were generated using a NEBNext® Ultra™ RNA Library Prep Kit for Illumina (#E7530L; NEB, Ipswich, MA, USA) following the manufacturer's recommendations, and index codes were added to attribute sequences to each sample. After cluster generation, the libraries were sequenced on the Illumina HiSeq 2500 platform to generate 50 bp single-end reads. De novo transcriptome assembly was performed using Trinity (http://trinityrnaseq.sf.net/). RPKM values were used to normalize transcript levels [42]. Fold change values of >2 indicated significant differences in gene expression between treatments. Unigenes were identified using TransDecoder (version r2013_08_14) [43] and were functionally annotated using the Blast2GO Gene Ontology (GO) functional annotation suite (E-value < 10^{-5}) (http://www.blast2go.org/). Metabolic pathways were predicted using Kyoto Encyclopedia of Genes and Genomes (KEGG) mapping.

4.12. Quantitative Real-Time PCR Analysis

A total of 1–2 μg of RNA was reverse-transcribed using the TaKaRa PrimeScript II 1st Strand cDNA Synthesis Kit (D6210A) according to the manufacturer's instructions. qRT-PCR was performed using a Bio-Rad CFX96 Real-Time PCR System (Bio-Rad, Hercules, CA, USA). TaKaRa SYBR Premix Ex Taq II (Perfect Real Time) was used for amplification. The β-tubulin sequence was used as an internal control to measure the relative transcript levels. Information on the oligonucleotide sequences used for qRT-PCR is provided in Table 3.

Table 3. Primer sequences for qRT-PCR.

Gene Name	Forward Primer (5′-3′)	Reverse Primer (5′-3′)
comp100059_c0 (hppA)	TTCCTGACTGCTGAGGGAGT	AATCGAGACGAGAGCGAATC
comp63224_c0(gabD)	GCAAGATCAGTGCTGCTGAG	GGTTTCCCACCTCGTCATTA
comp6582_c0(BRI1)	CATCCCTTGGCATTCTCACT	CTTCAGCTCTGTGCAGCTTG
comp81464_c0(LSC1)	AGAGGAGGAGGGACGAAGAG	TCATCTACCAGGGCTTCACC
comp92590_c0(BNG1)	ACCGTCAGGAGCACAAAGAT	ACAAGGGTGACCGAGAAATG
comp93555_c0(BNG2)	GGGAGTTAATGCGTCGAGAA	CAGAGAGGCCGGCATATAAA
β-tubulin	GTGGAGTGGATCCCCAACAA	AAAGCCTTCCTCCTGAACATGG

4.13. Statistical Analysis

Variance analyses were performed using SPSS (version 13.0; Chicago, IL, USA). Differences between the means of the treatments for each parameter were assessed using the least significance difference test (LSD) at $p = 0.05$.

5. Conclusions

La promoted the growth of bahiagrass roots mainly by improving root activity, photosynthesis (i.e., improving net photosynthetic rate, transpiration rate, chlorophyll fluorescence, PSII activity, electron transport chain efficiency, and oxidative phosphorylation), and respiration (i.e., increasing respiration intensity and the number of mitochondria in mature cells, enhancing AAO activity, TCA cycle, and sucrose metabolism), thereby providing energy for root growth. These results explain the reason why lanthanum promoted bahiagrass root growth to some extent and could provide a theoretical basis for using lanthanum to rapidly establish bahiagrass artificial slope protection grassland and other ecological management projects.

Author Contributions: J.Z. conceived and designed the research; Y.L. performed the research and wrote the manuscript. All authors have read and agreed to the published version of the manuscript.

Funding: This research was supported by the National Natural Science Foundation of China (U21A20241) and the Scientific and Technological Achievements Transformation Project of Qinghai province, China (2021-SF-134-03).

Institutional Review Board Statement: Not applicable.

Informed Consent Statement: Not applicable.

Data Availability Statement: The data presented in this study are available upon request from the corresponding author. The data are not publicly available due to the ongoing downstream studies based on this study.

Acknowledgments: The authors would like to thank Huawei Song for the physiological and biochemical tests and Yong Chen for providing the growth chamber facility. The authors are grateful to all editors and anonymous reviewers for their valuable suggestions on the manuscript.

Conflicts of Interest: The authors declare no conflict of interest.

References

1. Guo, B.; Xu, L.; Guan, Z.; Wei, Y. Effect of lanthanum on rooting of in vitro regenerated shoots of saussurea involucrata Kar. et Kir. *Biol. Trace Elem. Res.* **2012**, *147*, 334–340. [CrossRef] [PubMed]
2. Xie, Y.; Cai, X.; Liu, W.; Tao, G.; Chen, Q.; Zhang, Q. Effects of lanthanum nitrate on growth and chlorophyll fluorescence characteristics of Alternanthera philoxeroides under perchlorate stress. *J. Rare Earth.* **2013**, *31*, 823–829. [CrossRef]
3. Liu, D.; Wang, X.; Lin, Y.; Chen, Z.; Xu, H.; Wang, L. The effects of cerium on the growth and some antioxidant metabolisms in rice seedlings. *Environ. Sci. Pollut. Res.* **2012**, *19*, 3282–3291. [CrossRef] [PubMed]
4. He, Y.; Loh, C. Cerium and lanthanum promote floral initiation and reproductive growth of Arabidopsis thaliana. *Plant Sci.* **2000**, *159*, 117–124. [CrossRef]
5. Zheng, H.; Zhao, Z.; Zhang, C.; Feng, J.; Ke, Z.; Su, M. Changes in lipid peroxidation, the redox system and ATPase activities in plasma membranes of rice seedling roots caused by lanthanum chloride. *Biometals* **2000**, *13*, 157–163. [CrossRef]
6. Liu, M.; Hasenstein, K.H. La3+ uptake and its effect on the cytoskeleton in root protoplasts of *Zea mays* L. *Planta* **2005**, *220*, 658–666. [CrossRef]
7. He, J.Y.; Ren, Y.F.; Wang, F.J.; Pan, X.B.; Zhu, C.; Jiang, D.A. Characterization of cadmium uptake and translocation in a cadmium-sensitive mutant of rice (*Oryza sativa* L. Ssp japonica). *Arch. Environ. Con. Tox.* **2009**, *57*, 299–306. [CrossRef]
8. Liu, Y.; Wang, Y.; Wang, F.; Liu, Y.; Cui, J.; Hu, L.; Mu, K. Control effect of lanthanum against plant disease. *J. Rare Earth.* **2008**, *26*, 115–120. [CrossRef]
9. Lynch, J.P.; Brown, K.M. Root strategies for phosphorus acquisition. In *The Ecophysiology of Plant-Phosphorus Interactions*, 1st ed.; White, P.J., Hammond, J.P., Eds.; Springer: Dordrecht, The Netherlands, 2008; Volume 7, pp. 83–116.
10. Liu, Y.; Song, H.; Zhang, J.; Richardson, M.D. Growth characteristics of bahiagrass roots treated with micronutrients, rare earth elements, and plant hormones. *Horttechnology* **2016**, *26*, 176–184. [CrossRef]
11. Liu, D.; Wang, X.; Zhang, X.; Gao, Z. Effects of lanthanum on growth and accumulation in roots of rice seedlings. *Plant Soil Environ.* **2013**, *59*, 196–200. [CrossRef]
12. Berry, J.A.; Dowton, W.J.S. *Photosynthesis(Vol II): Development, Carbon Metabolism and Plant Productivity*; Academic Press: New York, NY, USA, 1982.
13. Chen, S.P.; Bai, Y.F.; Lin, G.H.; Liang, Y.; Han, X.G. Effects of grazing on photosynthetic characteristics of major steppe species in the Xilin River Basin, Inner Mongolia, China. *Photosynthetica* **2005**, *43*, 559–565. [CrossRef]
14. Peng, Y.; Jiang, G.M.; Liu, X.H.; Niu, S. Photosynthesis, transpiration and water use efficiency of four plant species with grazing intensities in Hunshandak Sandland, China. *J. Arid. Environ.* **2007**, *70*, 304–315. [CrossRef]
15. Kate, M.; Johnson, G.N. Chlorophyll fluorescence—A practical guide. *J. Exp. Bot.* **2000**, *51*, 659–668.
16. Qiu, Z.Y.; Wang, L.H.; Zhou, Q. Effects of bisphenol A on growth, photosynthesis and chlorophyll fluorescence in above-ground organs of soybean seedlings. *Chemosphere* **2013**, *90*, 1274–1280. [CrossRef] [PubMed]
17. Weng, J.H.; Lai, M.F. Estimating heat tolerance among plant species by two chlorophyll fluorescence parameters. *Photosynthetica* **2005**, *43*, 439–444. [CrossRef]
18. Ren, H.; Han, G.; Lan, Z.; Wan, H.; Philipp, S.; Martin, G.; Friedhelm, T. Grazing effects on herbage nutritive values depend on precipitation and growing season in Inner Mongolian grassland. *J. Plant Ecol.* **2016**, *9*, 712–723. [CrossRef]
19. Limantara, L.; Dettling, M.; Indrawati, R.; Tatas, I.; Panintingjati, H. Analysis on the chlorophyll content of commercial green leafy vegetables. *Procedia Chem.* **2015**, *14*, 225–231. [CrossRef]
20. Croft, H.; Chen, J.M.; Luo, X.; Rtlett, P.; Chen, B.; Staebler, R.M. Leaf chlorophyll content as a proxy for leaf photosynthetic capacity. *Glob. Chang. Biol.* **2017**, *23*, 3513–3524. [CrossRef]
21. Shi, D.; Yang, X.; Zhang, Y. Effects of lanthanum on the growth and physiological characteristics of wheat seedling under salt stress. *Xibei Zhiwu Xuebao* **2008**, *28*, 730–736.
22. Liu, F.; Zhang, Y.; Jia, R.; Ren, X. Effect of Seed Soaking with La(NO3)(3) on Root and Seedling Growth of Adzuki Bean under Different Concentrations NaCl Stresses. *Wuhan Zhiwuxue Yanjiu* **2009**, *27*, 397–402.
23. Wehr, J.B.; Menzies, N.W.; Blamey, F.P.C. Inhibition of cell-wall autolysis and pectin degradation by cations. *Plant Physiol. Bioch.* **2004**, *42*, 485–492. [CrossRef] [PubMed]

24. Kopittke, P.M.; Blamey, F.P.C.; Menzies, N.W. Toxicities of soluble Al, Cu, and La include ruptures to rhizodermal and root cortical cells of cowpea. *Plant Soil* **2008**, *303*, 217–227. [CrossRef]
25. Yang, J.L.; Li, Y.Y.; Zhang, Y.J.; Zhang, S.S.; Wu, Y.R.; Wu, P.; Zheng, S.J. Cell wall polysaccharides are specifically involved in the exclusion of aluminum from the rice root apex. *Plant Physiol.* **2008**, *146*, 602–611. [CrossRef]
26. Wen, K.; Liang, C.; Wang, L.; Gang, H.; Zhou, Q. Combined effects of lanthanumion and acid rain on growth, photosynthesis and chloroplast ultrastructure in soybean seedlings. *Chemosphere* **2011**, *84*, 601–608. [CrossRef] [PubMed]
27. Wang, L.; Wang, W.; Zhou, Q.; Huang, X. Combined effects of lanthanum (III) chloride and acid rain on photosynthetic parameters in rice. *Chemosphere* **2014**, *112*, 355–361. [CrossRef]
28. Hong, F.; Wei, Z.; Zhao, G. Mechanism of lanthanum effect on chlorophyll of spinach. *Sci. China Ser. C Life Sci.* **2002**, *45*, 166–176. [CrossRef]
29. Yan, W.; Yang, L.; Wang, Q. Distribution of lanthanum among the chloroplast subcomponents of spinach and its biological effects on photosynthesis: Location of the lanthanum binding sites in photosystem II. *Chinese Sci. Bull.* **2005**, *50*, 1714–1720. [CrossRef]
30. Palmer, R.J.; Butenhoff, J.L.; Stevens, J.B. Cytotoxicity of the rare earth metals cerium, lanthanum, and neodymium in vitro: Comparisons with cadmium in a pulmonary macrophage primary culture system. *Environ. Res.* **1987**, *43*, 142–156. [CrossRef]
31. Brooks, M.M. Comparative studies on respiration xiv. Antagonistic action of lanthanum as related to respiration. *J. Gen. Physiol.* **1921**, *3*, 337–342. [CrossRef]
32. Hong, F.; Wei, Z.; Zhao, G. Effect of lanthanum on aged seed germination of rice. *Biol. Trace Elem. Res.* **2000**, *75*, 205–213.
33. Fashui, H.; Ling, W.; Chao, L. Study of lanthanum on seed germination and growth of rice. *Biol. Trace Elem. Res.* **2003**, *94*, 273–286. [CrossRef]
34. Wang, H.F.; Zhu, Y.H.; Sun, H.J. Determination of drought tolerance using root activities in Robinia pseudoacacia 'Idaho' transformed with mtl-D gene. *For. Ecosyst.* **2006**, *8*, 75–81.
35. Richter, A.K.; Frossard, E.; Brunner, I. Polyphenols in the woody roots of Norway spruce and European beech reduce TTC. *Tree Physiol.* **2007**, *27*, 155–160. [CrossRef] [PubMed]
36. Zhong, H.; Lauchli, A. Changes of cell wall composition and polymer size in primary roots of cotton seedlings under high salinity. *J. Exp. Bot.* **1993**, *44*, 773–778. [CrossRef]
37. Blumenkrantz, N.; Asboe-Hansen, G. An improved method for the assay of hydroxylysine. *Anal. Biochem.* **1973**, *56*, 10–15. [CrossRef]
38. Arnon, D.I. Copper enzymes in isolated chloroplasts. Polyphenoloxidase in beta vulgaris. *Plant Physiol.* **1949**, *24*, 1–15. [CrossRef]
39. Oberbacher, M.F.; Vines, H.M. Spectrophotometric assay of ascorbic acid oxidase. *Nature* **1963**, *197*, 1203–1204. [CrossRef]
40. Jiang, Y. Role of anthocyanins, polyphenol oxidase and phenols in lychee pericarp browning. *J. Sci. Food Agric.* **2000**, *80*, 305–310. [CrossRef]
41. Xu, Q.S.; Hu, J.Z.; Xie, K.B.; Yang, H.Y.; Du, K.H.; Shi, G.X. Accumulation and acute toxicity of silver in *Potamogeton crispus* L. *J. Hazard. Mater.* **2010**, *173*, 186–193. [CrossRef]
42. Mortazavi, A.; Williams, B.A.; Mccue, K.; Schaeffer, L.; Wold, B. Mapping and quantifying mammalian transcriptomes by RNA-Seq. *Nat. Methods* **2008**, *5*, 621–628. [CrossRef]
43. Grabherr, M.G.; Haas, B.J.; Yassour, M.; Levin, J.Z.; Thompson, D.A.; Amit, I.; Adiconis, X.; Fan, L.; Raychowdhury, R.; Zeng, Q. Full-length transcriptome assembly from RNA-Seq data without a reference genome. *Nat. Biotechnol.* **2011**, *29*, 644–652. [CrossRef] [PubMed]

Article

Expression of *ZjPSY*, a Phytoene Synthase Gene from *Zoysia japonica* Affects Plant Height and Photosynthetic Pigment Contents

Di Dong [1,†], Yuhong Zhao [2,†], Ke Teng [1,3], Penghui Tan [4], Zhuocheng Liu [1], Zhuoxiong Yang [1], Liebao Han [1,*] and Yuehui Chao [1,*]

[1] School of Grassland Science, Beijing Forestry University, Beijing 100083, China; didoscori@163.com (D.D.); tengke@grass-env.com (K.T.); liuzhuocheng@bjfu.edu.cn (Z.L.); yangzhuoxiong@bjfu.edu.cn (Z.Y.)
[2] Animal Science College, Tibet Agriculture & Animal Husbandry University, Nyingchi 860000, China; zhaoyh@xza.edu.cn
[3] Beijing Academy of Agriculture and Forestry Sciences, Beijing 100083, China
[4] Beijing Chaoyang Foreign Language School, Beijing 100101, China; penghuitan@bjfu.edu.cn
[*] Correspondence: hanliebao@bjfu.edu.cn (L.H.); chaoyuehui@bjfu.edu.cn (Y.C.); Tel.: +86-10-6233-6399 (L.H. & Y.C)
[†] These authors contributed equally to this work.

Citation: Dong, D.; Zhao, Y.; Teng, K.; Tan, P.; Liu, Z.; Yang, Z.; Han, L.; Chao, Y. Expression of *ZjPSY*, a Phytoene Synthase Gene from *Zoysia japonica* Affects Plant Height and Photosynthetic Pigment Contents. *Plants* **2022**, *11*, 395. https://doi.org/10.3390/plants11030395

Academic Editor: Tomoko Shinomura

Received: 29 November 2021
Accepted: 28 January 2022
Published: 31 January 2022

Abstract: Phytoene synthase (PSY) is a key limiting enzyme in the carotenoid biosynthesis pathway for regulating phytoene synthesis. In this study, *ZjPSY* was isolated and identified from *Zoysia japonica*, an important lawn grass species. *ZjPSY* cDNA was 1230 bp in length, corresponding to 409 amino acids. *ZjPSY* showed higher expression in young leaves and was downregulated after GA$_3$, ABA, SA, and MeJA treatments, exhibiting a sensitivity to plant hormones. Regulatory elements of light and plant hormone were found in the upstream of *ZjPSY* CDS. Expression of *ZjPSY* in *Arabidopsis thaliana* protein led to carotenoid accumulation and altered expression of genes involved in the carotenoid pathway. Under no-treatment condition, salt treatment, and drought treatment, transgenic plants exhibited yellowing, dwarfing phenotypes. The carotenoid content of transgenic plants was significantly higher than that of wild-type under salt stress and no-treatment condition. Yeast two-hybrid screening identified a novel interacting partner *ZjJ2* (*DNAJ homologue 2*), which encodes heat-shock protein 40 (HSP40). Taken together, this study suggested that ZjPSY may affect plant height and play an important role in carotenoid synthesis. These results broadened the understanding of carotenoid synthesis pathways and laid a foundation for the exploration and utilization of the *PSY* gene.

Keywords: carotenoids; dwarfing; *Zoysia japonica*

1. Introduction

In plants, carotenoids are important isoprenoid compounds, which play classical roles in photosynthetic biological processes, including photomorphogenesis and photoprotection. Carotenoids also participate in plant color formation as pigments, which affect plant color through different levels of aggregation in the chromoplasts [1]. Carotenoids range in color from colorless to yellow and red and are reflected in the fruits and leaves of many plants [2]. In addition, carotenoids and their oxidative and enzymatic lysates are considered to be signaling molecules for interactions of plants with environment, so carotenoids are proposed to play an important regulatory role in plant growth and development [3].

Phytoene is a precursor for the synthesis of all carotenoid substances. *Phytoene synthase* (*PSY*), a key enzyme regulating carotenoid synthesis, can catalyze the conversion of geranylgeranyl pyrophosphate (GGPP) to phytoene, which is the first rate-limiting reaction in the carotenoid synthesis pathway (Figure 1) [4–7]. GGPP also acts as a precursor for gibberellin, abscisic acid (ABA), and chlorophylls [4].

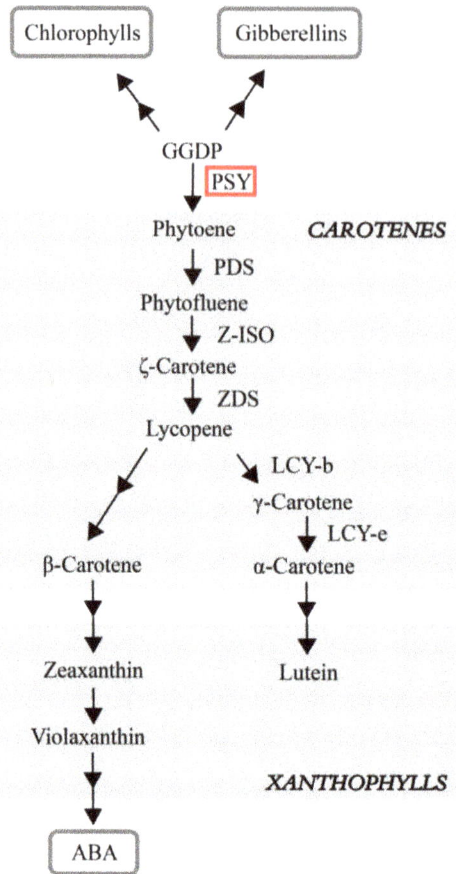

Figure 1. *PSY* expression regulates the synthesis of carotenoids and is involved in the synthesis of hormones [4–7]. PDS, phytoene desaturase. Z-ISO, 15-cis-ζ-carotene isomerase. ZDS, ζ-carotene desaturase. LCY-b, Lycopene β-cyclase. LCY-e, lycopene ε-cyclase.

PSY is a preferred candidate gene for understanding the molecular regulation of carotenoid accumulation in most species [8]. Constitutive PSY overexpression could increase total carotenoid and β-carotene contents [2,9]. Transgenic tobacco (*Nicotiana benthamiana*) with overexpression of *PSY* significantly increased carotenoid content, and the *PSY* gene showed a key role in lycopene accumulation in transgenic autumn olive fruits (*Elaeagnus umbellata*) [9,10]. The *PSY* genes contain several isoforms that show tissue-specific expression. There is only one *PSY* gene in *Arabidopsis thaliana*, and overexpression of endogenous *PSY* leads to delayed seed germination and increased carotenoid and chlorophyll levels [11]. There are two PSY genes reported in carrots (*Daucus carota*) and three in tomato (*Solanum lycopersicum*) and rice (*Oryza sativa*) [12–14]. *SlPSY1* is required for carotenoid synthesis in tomato fruits, and *SlPSY2* is necessary for carotenoid synthesis in leaf tissues, whereas *SlPSY3* is related to root resistance under stress [15]. The *PSY* gene may also be involved in plant dwarfing; expression of *PSY* from *Oncidium* Gower Ramsey in Tobacco showed dwarfing and reduced leaf area phenotypes [16]. The study of the *PSY* gene is beneficial to deepening the understanding of basic metabolism and regulation of the carotenoid synthesis pathway.

Zoysia japonica is a common type of warm-season turfgrass. Due to its resistance to drought and salt, it has been widely used in golf courses, sports grounds, and urban greening. Although the genes involved in the carotenoid biosynthesis pathway have been identified, the regulation of these genes in plant growth, development, and stress tolerance is not fully understood. Although the carotenoid synthesis pathway has been confirmed to be critical in plant growth regulation [1–3], the regulation mode of carotenoid accumulation and the regulation mechanism of the *PSY* gene in *Zoysia japonica* have not been studied. In addition, dwarfing is an important index in the production and application of turfgrass. Dwarfing can improve planting density, enhance photosynthetic efficiency and reduce pruning rate, thus improving the quality of turfgrass [17]. In this study, the *PSY* gene of *Zoysia japonica* was isolated and identified for a biological and functional study, which could lead to further insight into the function of the *PSY* gene and lay a foundation for studying the role of carotene in Zoysia.

2. Results

2.1. Identification and Bioinformatic Analysis of the ZjPSY Gene in Zoysia japonica

According to the Zoysia Genome Database [18], the DNA fragment containing *ZjPSY* coding domain sequences (CDSs) was cloned and ligated into cloning vector pMD19-T. The ZjPSY cDNA sequences were deposited in the NCBI database with accession numbers KY264127.1. The full-length *ZjPSY* cDNA contains a 1230 bp ORF encoding a protein of 409 amino acids which belongs to the Isoprenoid_Biosyn_C1 superfamily. According to the Compute pI/MW tool of Expasy, the molecular weight of ZjPSY was 46.39 kD, and the theoretical isoelectric point was 9.20. Intron/exon organization analysis showed that *ZjPSY* had five exons (Figure S1A). Secondary structure analysis indicates that ZjPSY has 63% alpha helix, 1% beta strand, and 25% disorder proteins (Figure S1B). A 3D Tertiary structure model of ZjPSY provides insights into its molecular geometry (Figure S1C). The stereochemical structure supported the α-helical nature of the enzyme.

Phylogenetic analysis of PSY proteins was performed based on the protein sequences from different species collected from NCBI database, with a neighbor-joining method (Figure 2). Phylogenetic analysis of the PSY protein revealed that ZjPSY has the highest homology with PSY of *Zoysia matrella*, though it was most closely related to OsPSY1 of *Oryza sativa*. To further dissect the homology of PSY, the proteins in 22 different species were further analyzed for the occurrence of conserved motifs. Motif sequence scan analysis showed the top three scored motifs with presence of amino-acid rich profiles (Figure 3A). There were distinct patterns indicating that the three motifs were found in all PSY protein sequences (Figure 3B). The three motifs were found to be highly conserved. In this respect, the phylogenetic analysis and motif analysis revealed that ZjPSY had a low similarity to PSY3 from *Oryza sativa* and *Zea mays* and PSY from *Rubus idaeus*.

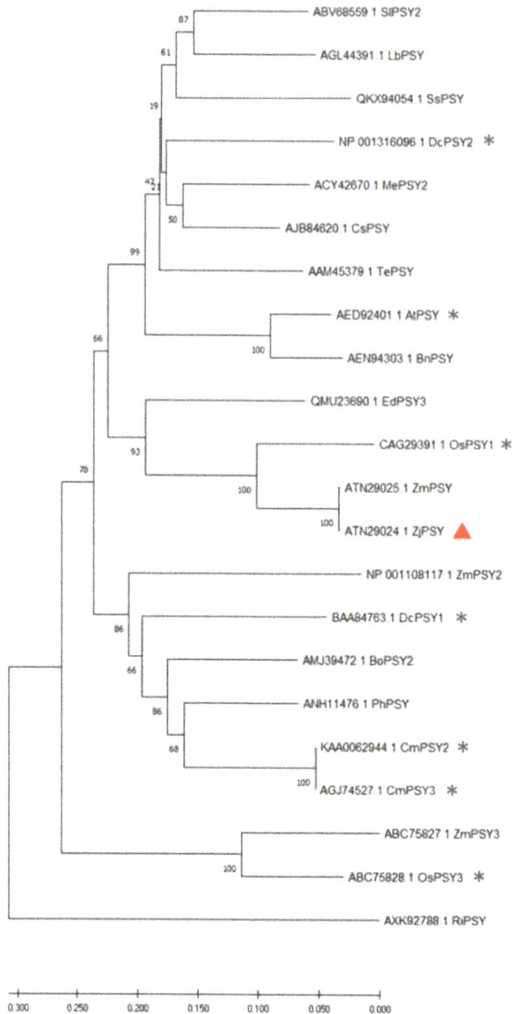

Figure 2. Phylogenetic tree of PSY. Phylogenetic tree analysis among 22 PSY proteins. Protein sequence accession numbers are as follows: SlPSY2 (ABV68559.1), *Solanum lycopersicum*; IbPSY (AGL44391.1) [19], *Ipomoea batatas*; SsPSY (QKX94054.1), *Salvia splendens*; MePSY2 (ACY42670.1), *Manihot esculenta*; CsPSY (AJB84620.1) [20], *Camellia sinensis*; DcPSY2 (NP_001316096.1) [21], *Daucus carota*; TePSY (AAM45379.1), *Tagetes erecta*; AtPSY (AED92401.1) [22], *Arabidopsis thaliana*; BnPSY (AEN94303.1), *Brassica napus*; EdPSY3 (QMU23690.1), *Eleocharis dulcis*; OsPYS1 (CAG29391.1) [23], *Oryza sativa*; ZmPSY (ATN29025.1), *Zoysia matrella*; ZjPSY (ATN29024.1), *Zoysia japonica*; ZmPSY2 (NP_001108117.1), *Zea mays*; DcPSY1 (BAA84763.1) [21], *Daucus carota*; CmPSY2 (KAA0062944.1) [24], *Cucumis melo*; CmPSY3 (AGJ74527.1) [24], *Cucumis melo*; BoPSY2 (AMJ39472.1), *Bixa orellana*; PhPSY (ANH11476.1), *Prunus humilis*; RiPSY (AXK92788.1), *Rubus idaeus*; ZmPSY3 (ABC75827.1), *Zea mays*; OsPSY3 (ABC75828.1) [25], *Oryza sativa*. ▲ Mark represented the ZjPSY. * Represents the PSY protein was analyzed.

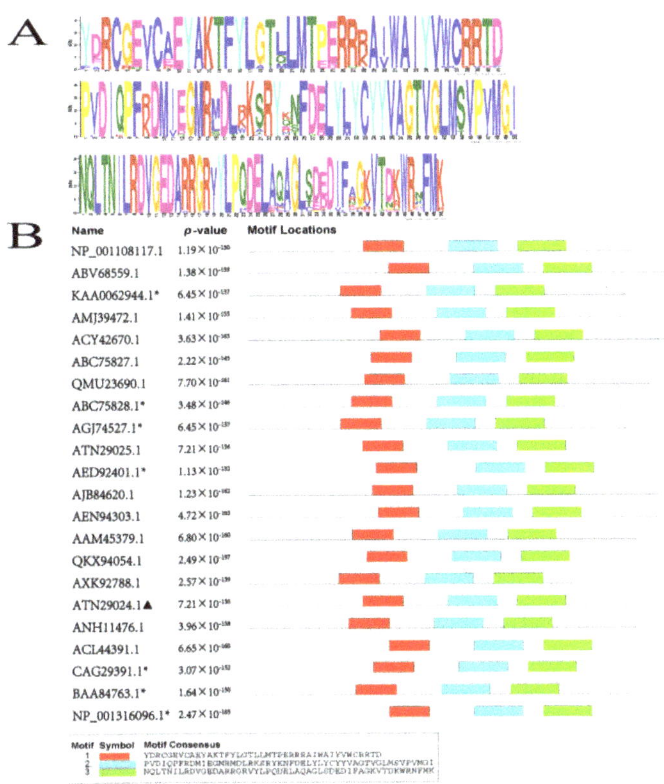

Figure 3. Conserved structure analysis of PSYs. (**A**) The conserved motifs in 22 PSY proteins of different species. Top scoring three motifs consisting of amino acids composition are listed. (**B**) Protein motif analyses of PSY protein sequence. Each color shows the different conserved motif structure identified in the PSY protein. ▲ Mark represented the ZjPSY. * Represents the PSY protein has been analyzed.

2.2. Prediction of Cis-Regulatory Elements of ZjPSY Promoter

To analyze the regulatory pathway and function of ZjPSY, a 2000 bp upstream sequence from the translation start site (ATG) was examined using the PlantCARE database (Figure 4) [26]. Multiple light responsive motifs, including I-box and G-box, were identified in the frequency of occurrence of cis-elements. In addition, hormone response elements, such as gibberellin responsiveness P-box, salicylic acid responsiveness TCA-element, MeJA-responsiveness TGACG-motif, and CGTCA-motif were also identified.

In addition, several cis-acting elements involved growth and development were found, such as endosperm expression element GCN4_motif and zein metabolism regulation element O2-site.

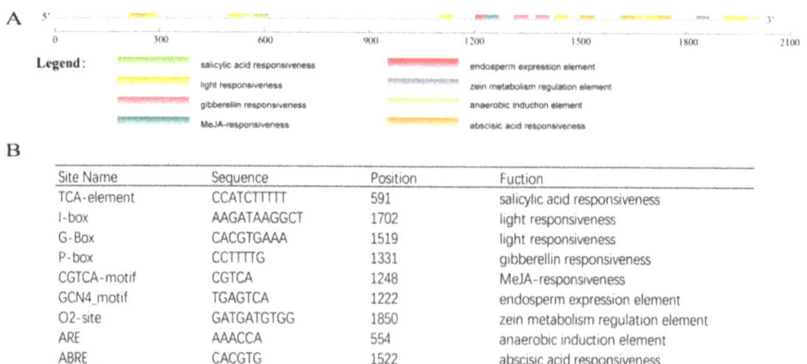

Figure 4. Analysis of up-stream sequence of *ZjPSY* gene. (**A**) Cis-acting elements upstream of *ZjPSY* gene. (**B**) List of predicted binding sites for the transcription factors upstream of *ZjPSY* gene.

2.3. ZjPSY Was Localized in the Chloroplast and Cytoplasm

We next sought to examine the subcellular localization of ZjPSY proteins in tobacco leaf via transient transformation. The untargeted YFP was transfected as a control and the fluorescence signal localized in the nucleus and cytoplasm (Figure 5A–D). In the leaf tissue which transiently expressed YFP fusion proteins (35S::*ZjPSY*:YFP), YFP fluorescence signal colocalized with the autofluorescence of chlorophyll in the chloroplast. The fluorescence accumulated in the chloroplast and cytoplasm (Figure 5E–H), suggesting that ZjPSY mainly localized in chloroplast and cytoplasm.

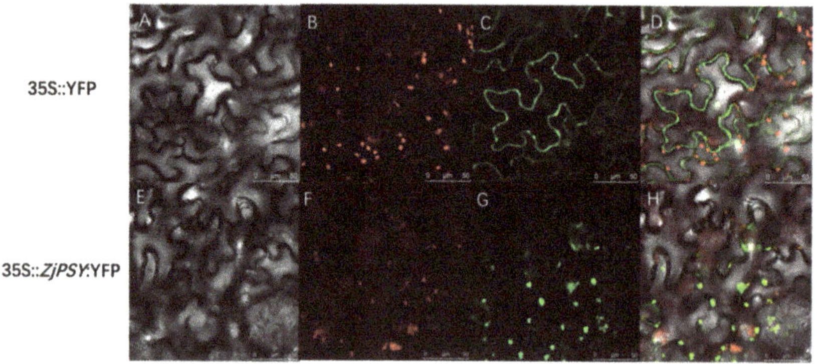

Figure 5. Subcellular localization of ZjPSY in plant cells. (**A–D**) 35S::YFP fluorescent detection. (**E–H**) 35S::ZjPSY:YFP fluorescent detection. (**A,E**): bright light; (**B,F**): chlorophyll autofluorescence; (**C**) and (**G**): YFP fluorescence; (**D,H**): merged signal.

2.4. ZjPSY Interacts with DNAJ Homologue 2

To analyze the protein interaction of ZjPSY, yeast two-hybrid assays and Bimolecular Fluorescence Complementation (BiFC) assays were performed. The *J2* gene in Zoysia was identified, and it had high homology with *J2* in Arabidopsis. In Arabidopsis, *J2* encodes heat-shock protein 40 (HSP40) [27,28]. The results showed yeast cells with pGADT7-*ZjJ2* and pGBKT7-*ZjPSY* grew well and turned blue on QDO/X/A plates, similar to the positive control (Figure 6). These results indicate that ZjJ2 protein interact with ZjPSY protein.

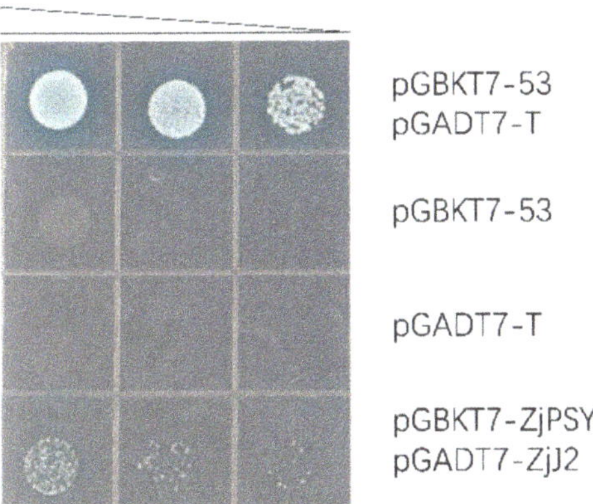

Figure 6. Screening for ZjPSY interacting proteins with yeast two-hybrid assay. Interaction between *ZjJ2* and *ZjPSY*. Yeast cells were spotted on higher stringency QDO/X/A agar plates by 10-fold serial dilutions. Blue clones were positive and white or absent clones were negative. pGBKT7-53 or pGADT7-T was used as the negative control, and pGBKT7-53 and pGADT7-T were co-transformed as positive control.

BiFC was used to confirm the interaction of ZjPSY and ZjJ2 in plant cells. ZjPSY and ZjJ2 were fused with the C and N terminals of YFP, respectively, and transiently expressed in tobacco cells. When YNE-ZjPSY and YCE-ZjJ2 were co-expressed, the YFP fluorescence was captured (Figure 7A–D), while no YFP signal was found in negative samples (Figure 7E–L). Those results demonstrated and confirmed the interaction of ZjPSY with ZjJ2.

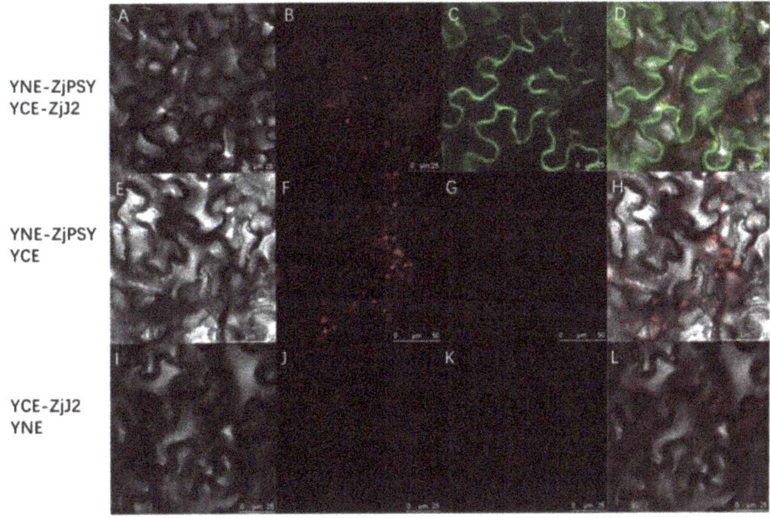

Figure 7. The interaction between ZjPSY and ZjJ2 in living cells shown by BiFC. (**A–D**) YNE-ZjPSY and YCE-ZjJ2 fluorescent detection. (**E–H**) YNE-ZjPSY and YCE fluorescent detection. (**I–L**) YCE-ZjJ2 and YNE fluorescent detection. (**A,E,I**): Bright light; (**B,F,J**): chlorophyll autofluorescence; (**C,G,K**): YFP fluorescence; (**D,H,L**): merged signal. Co-expression of YNE-ZjPSY with YCE or YCE-ZjJ2 with YNE was used as negative control.

2.5. Expression Pattern of ZjPSY

Real-time PCR was conducted to examine the expression pattern of *ZjPSY* in different tissues and development stages. The expression profile results showed that *ZjPSY* was detected in root, stem, and leaf tissues, and the expression was significantly abundant in leaf (Figure 8A). *ZjPSY* expression was relatively high in early leaf growth stage and slightly decreased in fast-growing and mature leaves (Figure 8B). The expression levels of *ZjPSY* were downregulated by multiple hormone treatments, including GA_3, ABA, SA, and MeJA (Figure 8C–F). After leaves were treated with these hormones for 3 h, the expression of ZjPSY reached the lowest level at 3 h. The expression patterns suggested that *ZjPSY* may be involved in leaf development and hormone responses.

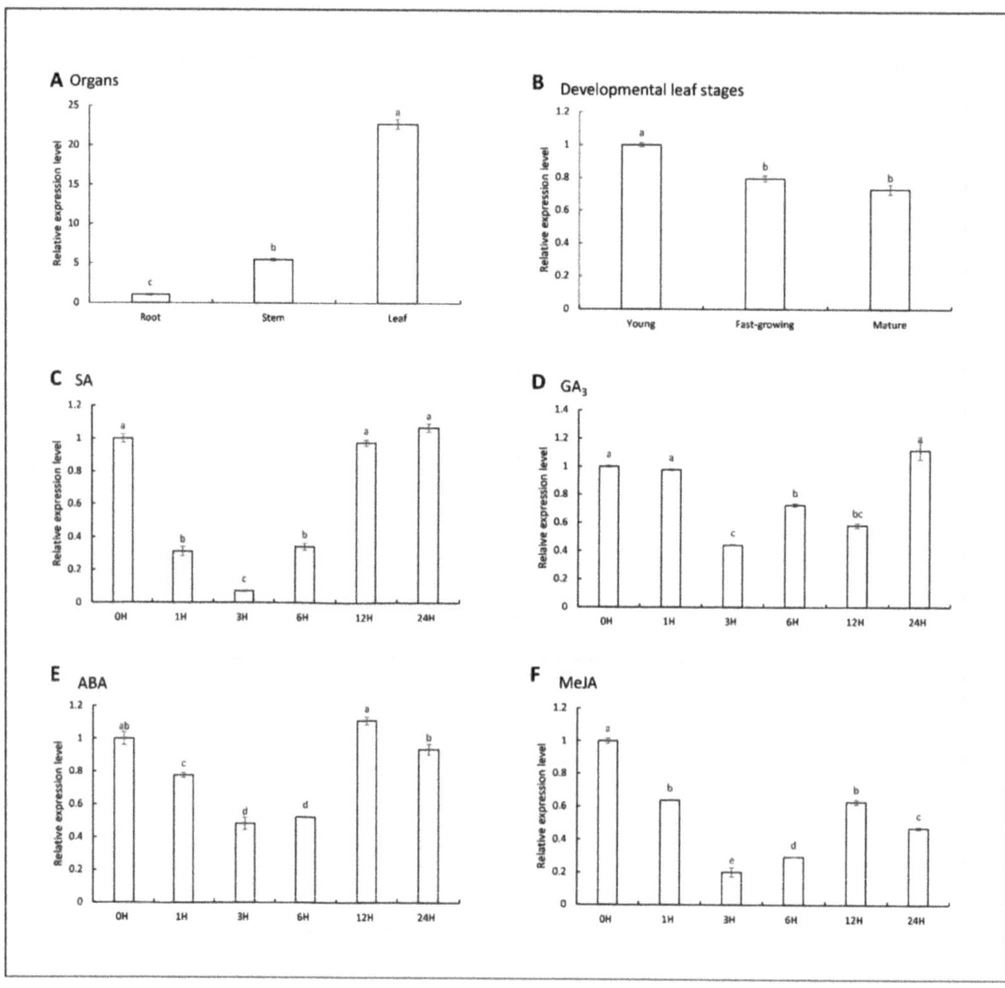

Figure 8. Expression profiles of *ZjPSY*. (**A**) *ZjPSY* expression pattern in root, stem, and leaf. (**B**) *ZjPSY* expression pattern in leaves at different developmental stages. (**C–F**) Expression pattern in leaf under 10 µM GA_3 treatment (**C**), 10 µM ABA treatment (**D**), 0.5 mM SA treatment (**E**), and 10 µM MeJA treatment (**F**). Different letters above the columns indicate significant differences ($p \leq 0.05$, $n = 3$). *ZjPSY* expression in root (**A**), young stage (**B**), and 0 h (**C–F**) was set as calibrator sample with a relative expression equal to 1.

2.6. Chlorophyll and Carotenoid Contents in Transgenic Plants

To investigate whether *ZjPSY* expression affects chlorophyll and carotenoid contents in the plant, genetic transformation was conducted using the Agrobacterium-mediated method. Expression levels in different lines were measured by qRT-PCR, and most transgenic lines had much higher expression levels, indicating the *ZjPSY* gene was successfully expressed in transgenic Arabidopsis (Figure 9A). To explore the effect of *ZjPSY*, chlorophyll and carotenoid contents were detected in transgenic lines and wild-type plants (Figure 9B,C). The lines with the high *ZjPSY* expression levels enhanced carotenoid levels and decreased chlorophyll levels, which indicates that expression of *ZjPSY* affects carotenoid and chlorophyll contents.

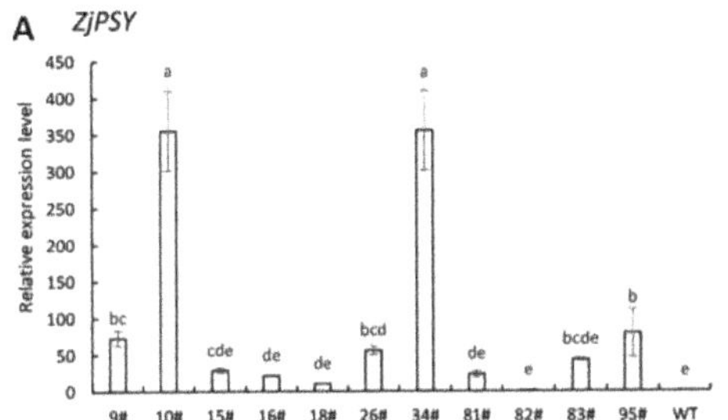
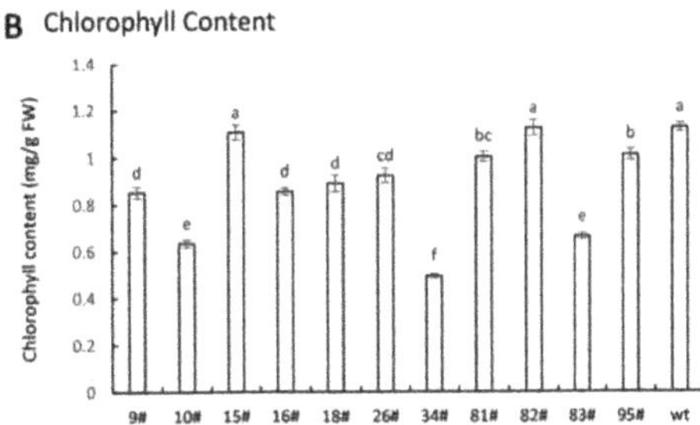
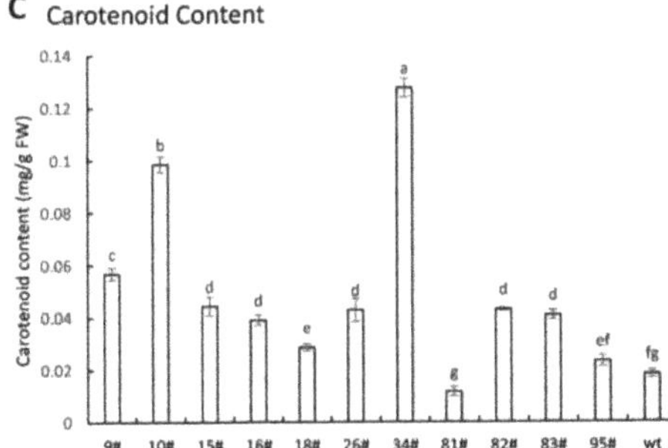

Figure 9. (**A**) The qRT-PCR analysis of ZjPSY expression in wild-type plants and the transgenic lines. Transgenic line designated #82 was set as calibrator sample with a relative expression equal to 1. (**B**) Chlorophyll contents of wild-type plants and the transgenic lines. (**C**) Carotenoid contents of wild-type plants and the transgenic lines. Different letters above the columns indicate significant differences ($p \leq 0.05$).

2.7. ZjPSY Expression in Arabidopsis Thaliana Resulted in Leaf Yellowing and Plant Dwarfing

Based on the expression levels and carotenoid and chlorophyll contents, three positive transgenic lines designated #9, #10, #34 were selected for further study of the *ZjPSY* gene. Grown on MS plates, transgenic seedlings had etiolated cotyledons, showing *ZjPSY* decreased chlorophyll contents in early stages of transgenic plants (Figure 10A). Grown in soil, transgenic plants displayed dwarf phenotype and senescent leaves compared with the WT (Figure 10B,C). These results suggest that *ZjPSY* may be involved in leaf pigment accumulation and plant senescence.

Figure 10. Leaf yellowing and dwarfing phenotype of transgenic Arabidopsis expressing the *ZjPSY* gene and wild-type Arabidopsis. (**A**) Transgenic Arabidopsis #34 expressing *ZjPSY* gene and wild-type Arabidopsis of 5 days. (**B**) Comparison of the phenotypes of transgenic Arabidopsis and wild-type Arabidopsis at 30 days. (**C**) Typical leaves in WT and transgenic Arabidopsis at 45 days.

2.8. Regulation of Plant Growth, Plant Pigment, and Pigment-Related Gene Expression by ZjPSY

To further detect the transgenic plants, the three selected transgenic lines and the wild-type were transplanted into MS medium (Figure S2). The rosette leaf lengths and taproot lengths of the three transgenic lines were significantly smaller than WT plants, especially in the transgenic line (#34) with higher *ZjPSY* expression levels (Figures S2 and 11A,B). According to statistical analysis, the transgenic plants in the line #34 had 39.01% smaller rosette leaf length and 46.03% smaller taproot length, compared with wild-type (Figure 11A,B). The chlorophyll content of transgenic lines was significantly lower than that of WT plants, while the carotenoid content of transgenic lines was significantly higher than that of WT plants (Figure 11C,D). These results indicated that expression of *ZjPSY* in *Arabidopsis thaliana* significantly increased carotenoid contents and decreased chlorophyll contents.

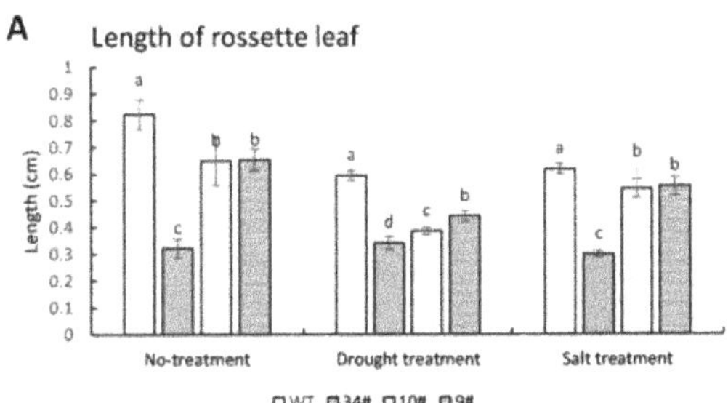

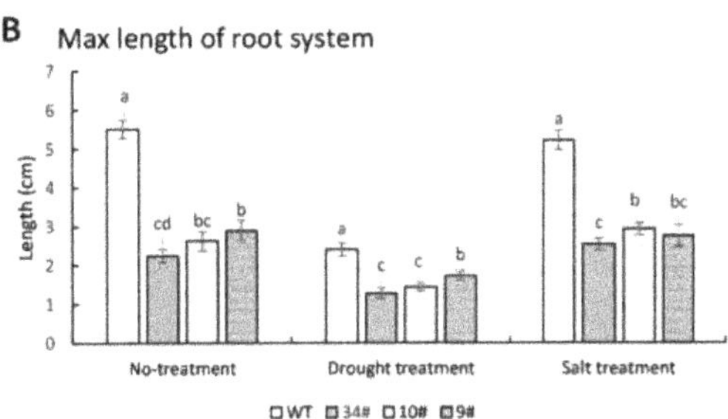

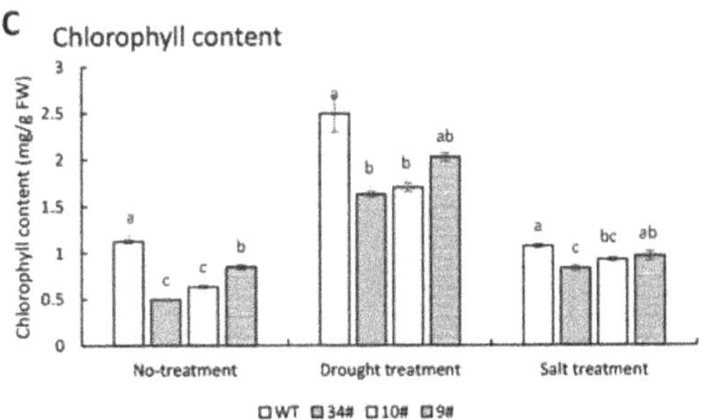

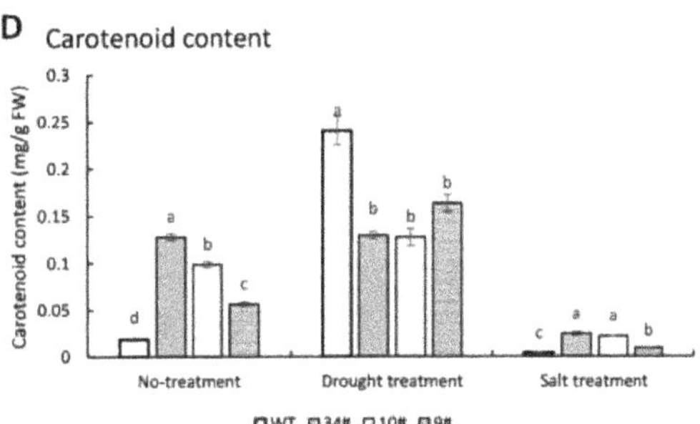

Figure 11. Statistical analysis of plant height and chlorophyll and carotenoid contents in transgenic and WT Arabidopsis. The transgenic and wild-type Arabidopsis grown in MS agar plates for 14 days: 200 mM Nacl and 500 mM mannitol were added to MS medium in drought treatment and salt treatment groups, respectively. (**A,B**) Length of rosette leaf (**A**) and maximum length of root system (**B**) in transgenic lines and wild-type. Different letters above the columns indicate significant differences ($p \leq 0.05$, $n = 20$). (**C,D**) Chlorophyll (**C**) and carotenoid (**D**) contents in transgenic lines and wild-type Arabidopsis. Different letters above the columns indicate significant differences ($p \leq 0.05$, $n = 3$).

Stresses can affect plant pigment contents. Different lines were placed on MS agar plates containing 200 mM mannitol for drought treatment or 100 mM NaCl for salt stress (Figure S2B,C). Similar to no-treatment conditions, the plant height of MS transgenic lines was significantly lower than WT lines after mannitol or NaCl was added (Figures S2B,C and 11A,B). Under drought treatment and salt treatment for 14 days, transgenic lines showed an obvious dwarfing phenotype with root lengths and rosette leaf lengths significantly lower than those of wild-type. Under drought treatment, the carotenoid content of the transgenic plants did not show similar performance to that of no treatment; the increase in carotenoid content of transgenic plants was much lower than that of wild-type (Figure 11D). The chlorophyll content of transgenic plants #34 and #10 was lower than that of wild-type, but the chlorophyll content of transgenic plant #9 was not significantly different from that of wild-type (Figure 11C). Under salt treatment, the chlorophyll contents of transgenic plants #34 and #10 were lower than those in wild-type (Figure 11C), and the carotenoid level of the transgenic plants was higher than that of the wild-type (Figure 11D). The results suggests that expression of *ZjPSY* may be involved in inhibiting carotenoid accumulation or accelerating its degradation under drought treatment. In plants subjected to salt stress, transgenic plants with *ZjPSY* gene overexpression showed higher levels of carotenoids than WT plants.

To gain more insight into the role of *ZjPSY* in plant growth and development, the expression profiles of six pigment-genes, *AtCLH (Chlorophyllase)*, *AtNYC (NON-YELLOWCOLORING)*, *AtNOL (NYC1-LIKE)*, *AtPAL (Phenylalanine ammonia-lyase)*, *AtNCED (9-cis-epoxycarotenoid dioxygenase)*, and *AtZDS (Z-carotene desaturase)*, were determined using qRT-PCR (Figure 12). *AtCLH* catalyzes the hydrolysis of ester bonds to chlorophyllide and phytol, which is the first step of chlorophyll degradation [29]. *AtNYC* and *AtNOL* are involved in chlorophyll band light-harvesting complex II degradation [30–32]. *AtPAL*, a key enzyme in the phenylalanine pathway, catalyzes the deamination of phenylalanine into trans-cinnamic acid, which is related to anthocyanin synthesis [33]. *AtNCED*, a kind of carotenoid cleavage enzyme, is a key enzyme that regulates ABA biosynthesis under stress [34]. *AtZDS* plays a regulatory role in the catalysis of Z-carotene to tetra-cis lycopene and is also an important gene in the production of carotenoids [35]. Under no-treatment conditions, the relative expression levels of several pigment-related genes measured in transgenic plants were lower than those in wild-type. Expect for *AtCLH* and *AtPAL*, there was no significant difference in the relative expression levels of other measured genes between wild-type and transgenic plants under drought treatment. In salt stress, *AtCLH* was significantly increased comparing with WT, and the relative expression of *AtZDS* was also slightly increased. Salt treatment may not influence the expression of *AtNYC*, *AtNOL*, *AtPAL*, or *AtNCED*.

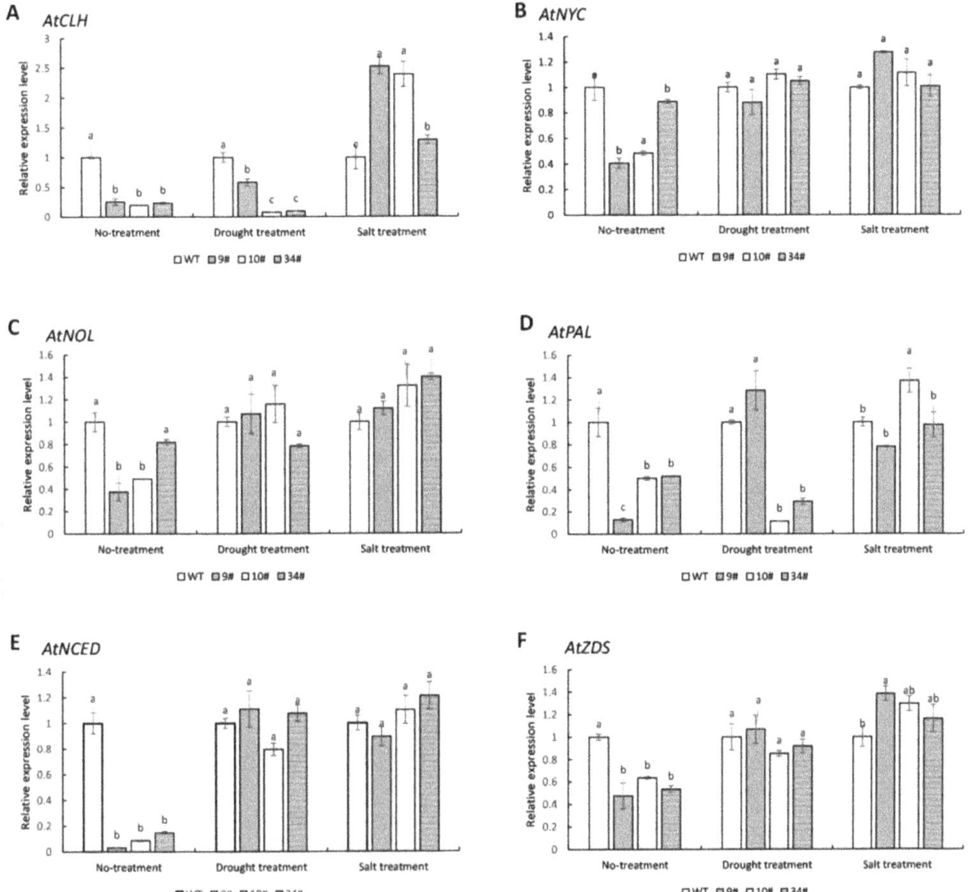

Figure 12. Pigment-related gene expression level of wild-type and transgenic plants subjected to no-treatment, drought, and salt stress conditions. Pigment-related genes include *AtCLH* (**A**), *AtNYC* (**B**), *AtNOL* (**C**), *AtPAL* (**D**), *AtNCED* (**E**), and *AtZDS* (**F**). WT in no-treatment, drought treatment, or salt treatment was set as calibrator sample with a relative expression equal to 1.

3. Discussion

The *PSY* gene regulates the first committed step in the process of carotenoid biosynthesis. In this study, the *ZjPSY* gene was cloned from *Zoysia japonica*. Bioinformatics analysis showed that the open reading frame of the *ZjPSY* gene was 1230 bp, encoding 409 amino acids. The protein belongs to the isoprenoid_biosyn_C1 superfamily. The large central cavity formed by the antiparallel α helixes and the two aspartic acid rich regions in the opposite wall constitute its catalytic sites. Structural analysis shows that ZjPSY contains many α helixes, and its three-dimensional analog structure contains a large central cavity (Figure S1C). Evolutionarily conserved motifs in different species revealed similar functions [36], and three motifs were found in 22 PSY expressed species. Phylogenetic analysis of PSY proteins with a neighbor-joining method revealed that ZjPSY is also more homologous with OsPSY1 and OsPSY3 in Oryza sativa. In addition, ZjPSY had a low similarity to PSY in model plant *Arabidopsis thaliana*.

According to the prediction of regulatory elements, the ZjPSY promoter is expected to mainly include the following cis-acting sequences: TCA-element, I-box, G-box, P-box,

CGTCA-motif, GCN4-motif, O2-site, ARE, and ABRE. The regulatory element analysis of the *ZjPSY* promoter contains of a variety of hormone responsive elements. TCA Element, P-box, TGACG Motif, and ABRE element are involved in responsiveness to salicylic acid, gibberellin, MeJA, and ABA, respectively [37–40]. The G-Box is involved in light responsiveness. Transcription factor *HY5* (*LONG HYPOCOTYL 5*) and *PIFs* (*PHYTOCHROME INTERACTING FACTORS*), which are regulators of chlorophyll and carotenoid biosynthesis, impart regulation to PSY through direct binding to the G-box cis-element in *Arabidopsis thaliana* [41]. Based on the cis-element analysis of the *ZjPSY* promoter, we further explored the expression pattern of *ZjPSY* using qRT-PCR. The expression pattern results suggest that *ZjPSY* was abundant in developing leaves. *ZjPSY* might be highly sensitive to hormone signaling; it was significantly downregulated under GA$_3$, ABA, SA, and MeJA treatment. These results indicated it was involved in pathways of different hormone signal response.

To investigate the functions of *ZjPSY* in plants, the 35S::*ZjPSY*:YFP was constructed and transformed to *Arabidopsis thaliana* through the floral dip method. The transgenic plants showed yellowing and dwarfing phenotypes. *OncPSY* expression has also contributed to the dwarfing of tobacco [16]. The dwarfed phenotype in transgenic plants may be caused by the conversion of large amounts of GGPP to phytoene due to the increase in active PSY protein, which leads to the suppression of the gibberellin synthesis pathway (Figure 1) [42].

The color of carotenoids in photosynthetic tissues is usually masked by chlorophyll, and carotenoids could provide bright coloration characteristics in tissue with low chlorophyll content [43]. The content of chlorophyll and carotenoids in wild-type and transgenic plants were determined. Compared with wild-type, carotenoid content of transgenic plants increased significantly, and chlorophyll content decreased significantly under no treatment, which indicated *ZjPSY* not only changed leaf color by increasing the accumulation of carotenoids but also affected chlorophyll. Studies have shown that plants adapt to photoprotective pigments by regulating the development of plastid structures. *PSY* is involved in the coordination process of carotenoid biosynthesis and storage with the molecular factors of photosynthetic development by participating in carotenogenesis (Figure 13) [44]. The ZjPSY was localized specifically to the chloroplast and cytoplasm consistent with the function of carotenogenesis and involvement of chloroplast development.

In order to further explore the effect of *ZjPSY* on plant leaf color, the expression levels of six pigment-related genes, *AtCLH*, *AtNYC*, *AtNOL*, *AtPAL*, *AtNCED*, and *AtZDS*, were assessed. Under no-treatment conditions, the expression levels of six genes were lower than those of wild-type plants. Under drought treatment, the relative expression levels of most genes had no significant difference except *AtCLH* and *AtPAL*. However, under salt treatment, the expression pattern of *AtCLH* was completely opposite to the no-treatment condition. Correspondingly, the amount of carotenoid and chlorophyll synthesis was different under different treatments. Expression of *ZjPSY* could increase carotenoid content under no-treatment conditions and salt treatment but decreased under drought treatment. Expression of *LbPSY* in *Escherichia coli* and yeast cells could improve tolerance in high salinity and drought environment [44]. Studies on *Daucus carota* have confirmed that *DcPSY* participates in ABA-mediated salt stress tolerance through the binding of promoters and ABRE transcription factors [45]. In this experiment, the transgenic plants showed no significant enhanced stress tolerance and even showed premature senescence under drought treatment. Due to the low homology of PSY between model plants and *Zoysia japonica*, it may be caused by differences in PSY between species. Copy volume and expression level of *PSY* in different transgenic lines also affects the phenotype of the transgenic plants [44].

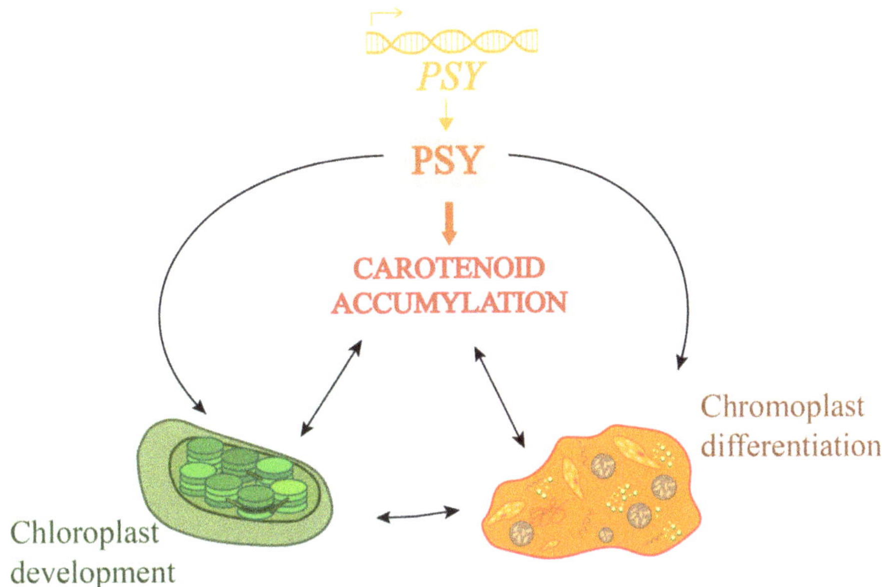

Figure 13. Molecular schematic of PSY's involvement in carotene accumulation and chloroplast development in plants [1]. The development of chloroplasts and chromoplast differentiation are related to PSY activity and sink capacity for carotenoids.

In this study, yeast two-hybrid screening identified *J2* as a novel interacting partner of *ZjPSY*. The *J2* gene encodes heat-shock protein 40 (HSP40) isoforms J2, a molecular chaperone that helps to prevent proteins from misfolding [27]. DnaJ-like chaperone has been shown to be involved in carotenoid synthesis in several species. Expression of Orange protein that belongs to DnaJ-Like chaperones in *Arabidopsis thaliana* can increase the abundance of active PSY protein, but the transcription level of the *PSY* gene does not change. It was inferred to be the chief posttranscriptional regulator of PSY in carotenoid biosynthesis [46]. Expression of gene for orange protein increased lutein and β-carotene in *Chlamydomonas reinhardtii* [47]. DnaJ-like chaperone is considered to support phytoene synthase [43]. In addition, experiments have shown that defective farnesylation of HSP40 is sufficient to induce ABA hypersensitivity [5]. It is possible that *PSY* and *J2* work together by affecting ABA activity in plants. The pathway of carotenoid synthesis greatly affects the endogenous ABA levels [5]. In this study, *ZjJ2* was confirmed as a novel interactive partner of *ZjPSY*, but its interaction mode needs further experimental verification.

In summary, ZjPSY may be involved in plant development and pigment synthesis. In response to abiotic stress, overexpression of ZjPSY may have different regulatory effects under no-treatment condition, salt stress, and under drought stress. However, the molecular mechanisms need to be further studied.

4. Materials and Methods

4.1. Plant Materials and Growth Conditions

Zoysia japonica cultivar 'Compadre' seeds were purchased from the Hancock seed company (Hancock, FL, USA). *Zoysia japonica* was cultivated in pots, kept at 28/23 °C (day/night) with 16 h (at 400 mmol/m^2/s)/8 h photoperiod and 50% humidity. All the Arabidopsis materials used in this study were of a Col ecotype background. The Arabidopsis seeds were sown on Murashige and Skoog plates (4.43 g/L Murashige and

Skoog powder, 8 g/L agar, pH 5.8), with a 16 h white light (at 90 mmol/m2/s)/8 h dark cycle and 50% humidity at 25 °C. After the fourth leaves appeared on the seedlings, the seedlings were transferred to sterilized soil or culture medium to continue growth under the same growth conditions.

4.2. Identification and Cloning of ZjPSY

Total RNA was isolated from 'Compadre' leaves of 3-month-old samples using Plant RNA Kit (Omega Bio-tek, Atlanta, USA), and the first-strand cDNA was synthesized using the PrimeScriptTM II 1st Strand cDNA Synthesis Kit (TAKARA, Dalian, China). Based on the sequence of *ZjPSY* (Zjn_sc00008.1.g00200.1.sm.mkhc) in the Zoysia Genome Database [18], primer sequences for cloning were designed, and PCR products were ligated into the cloning vector pDM19-T for storage and further experiments. All primers used in this study are listed in Table S1.

4.3. Bioinformatics Analysis

BLAST analysis of the NCBI database was used to identify homologs, and 22 PSY protein sequences from other species were obtained. The phylogenetic tree and ZjPSY gDNA structure were constructed using MEGA version 6.0 with the neighbor-joining method and GSDS 2.0 (http://gsds.gao-lab.org/Gsds_about.php (accessed on 6 October 2021)) [48]. The molecular weights and theoretical isoelectric points were analyzed by Compute pI/MW tool (http://web.expasy.org/compute_pi/ (accessed on 15 September 2021)). The motif analysis of PSY amino acids and cis-regulatory elements in the promoter involved the use of MEME (http://meme-suite.org/tools/meme (accessed on 15 September 2021)), PlantCARE database (http://bioinformatics.psb.ugent.be/webtools/plantcare/html/ (accessed on 15 September 2021)) and TBtools [49,50]. The ZjPSY protein model was generated by using the Phyre v2.0 tool (www.sbg.bio.ic.ac.uk/phyre2/ (accessed on 6 October 2021)).

4.4. Vector Construction and Generation of Transgenic Plants

The completed coding region of *ZjPSY* was amplified by PCR with primers ZjPSY-F and ZjPSY-R (Table S1). The fragment containing the complete CDS was ligated to the vector plasmid pMD19-T and stored in *E. coli* for long-term preservation. The 35S::*ZjPSY*:YFP was constructed by fusing completed *ZjPSY* CDS followed by YFP with primers, 3302Y-ZjPSY-F and 3302Y-ZjPSY-R. The recombination was introduced into *Agrobacterium tumefaciens* GV3101, which was then used for the transformation of *Arabidopsis thaliana* through the floral dip method [51]. Transgenic plants were selected using 2 mg·L^{-1} glufosinate and verified by PCR. T$_3$ transgenic lines were generated by self-pollination for subsequent experiments.

4.5. Stress Treatments and RNA Extraction

Three different transgenic lines with high expression levels of *PSY* were selected. The wild-type and transgenic seedlings were cultivated on 1/2 MS medium containing 200 mM mannitol or 100 mM NaCl for 14 days under conditions of 26/20 °C (day/night), 16/8 h (light/dark), and humidity of 70%. Length of rosette leaf and maximum length of root system of transgenic and WT plants were measured, and 20 biological replicates were performed. The concentration of chlorophyll and carotenoids was extracted in 95% alcohol using the soaking method [52], and three biological replicates were performed. All comparisons in the research were analyzed using one-way ANOVA with Tukey's test. Significance was defined as $p < 0.05$. Total RNA was isolated from each sample as described above. The first-strand cDNA was used for real-time quantitative RT-PCR (qRT-PCR) analysis.

4.6. Expression Levels of the PSY Gene and Pigment-Related Genes

Quantitative real-time PCR was carried out to measure the expression levels of six pigment-related genes from different samples with the $2^{-\triangle\triangle Ct}$ method, following the manufacturer's procedures of TB Green Premix Ex Taq II (Takara). The AtUBQ10 (NM_116771) gene was used as an internal reference gene for evaluating transcriptional abundance in Arabidopsis [53]. Pigment-related genes include *AtCLH* (AT1G19670.1), *AtNYC* (AT4G13250.1), *AtNOL* (AT5G04900.1), *AtPAL* (AT5G04230), *AtNCED* (AT1G30100), and *AtZDS* (AT3G04870). Relevant primer sequences used are shown in Table S1. All experiments in the research contained three biological replicates.

Quantitative real-time PCR was carried out to measure the expression levels of ZjPSY in different developmental stages and tissues of zoysiagrass and leaves subjected to no-treatment, hormone treatment, drought, or salt stress conditions. The 2-month-old zoysiagrass was inducted with 10 μM GA$_3$, 10 μM ABA, 10 μM MeJA, or 0.5 mM SA. Leaves at different growth stages (young, fast-growing, and mature), root tissues and stem tissues were also collected. RNA isolation and cDNA synthesis were performed as above. The *Z. japonica* beta-actin was used as the internal reference gene (GenBank accession no. GU290546) [54]. Relevant primer sequences used are shown in Table S1. All quantitative real-time PCR experiments in the research contained three biological replicates.

4.7. Yeast Two-Hybrid Assay

The ZjPSY ORFs were cloned using the primers pGBKT7-ZjPSY-F/R and inserted into the pGBKT7 vector. The recombinant bait plasmid pGBKT7-ZjPSY was co-transformed into yeast strain Y2H as described in the Yeastmaker™ Yeast Transformation System (Clontech, CA, USA) manufacturer's instructors. The MATCHMAKER GAL4 yeast two-hybrid system was utilized according to the Match7maker™ Gold Yeast Two-Hybrid System manufacturer's instructors (Clontech, CA, USA). Matchmaker Insert Check PCR Mix 2 (Clontech, CA, USA) is designed to be used to analyze cDNA inserts (Table S2) in library prey vectors.

4.8. Subcellular Localization Determination

The YFP plasmid and recombinant plasmid 35S::*ZjPSY*:YFP were introduced into *Agrobacterium tumefaciens* strain EHA105 by direct transformation. The young leaves of tobacco were infiltrated with recombinant Agrobacterium strains as described previously [55]. The visualization of YFP and YFP fusion proteins in leaves was performed using Leica TCS SP 8 confocal microscope.

4.9. Bimolecular Fluorescence Complementation (BiFC) Analysis

BiFC assays were conducted to verify the interaction between ZjPSY and ZjJ2. Coding sequences of *ZjPSY* and *ZjJ2* were cloned into YNE and YCE vectors, respectively. The primers YNE-ZjPSY-F/R and YCE-ZjJ2-F/R were used. YNE and YCE were transiently expressed in tobacco leaves as described previously [56]. YFP fluorescence signals were analyzed with Leica TCS SP 8 confocal microscope.

Supplementary Materials: The following supporting information can be downloaded at: https://www.mdpi.com/article/10.3390/plants11030395/s1, Figure S1: Sequence analysis of ZjPSY. (A) Intron/Exon organization in the ZjPSY genes. Introns and exons are shown as black lines and yellow boxes, respectively. (B) Secondary structure analysis of ZjPSY protein. Green helix, alpha helix; Blue arrow, beta strand; Faint lines, coil; SS confidence line, the prediction confidence. (C) Protein structure simulation of ZjPSY.; Figure S2: Performance of wild-type and transgenic plants. The transgenic and wild-type Arabidopsis grown in MS agar plates for 14 days (A). 500 mM mannitol (B) and 200 mM Nacl (C) were added to MS medium in drought treatment and salt treatment groups, respectively; Table S1: Primers used in the study; Table S2: cDNA inserts in library prey vectors.

Author Contributions: Conceptualization, Y.C. and L.H.; methodology, D.D., K.T., Z.Y. and P.T.; software, Z.L.; resources, L.H. and Y.Z.; writing—original draft preparation, D.D.; writing—review and editing, D.D. and Y.C.; project administration, L.H. All authors have read and agree to the published version of the manuscript.

Funding: The research was funded by the National Natural Science Foundation of China (no. 31971770 and 31860151) and Beijing Natural Science Foundation (no. 6204039).

Data Availability Statement: The data presented in this study are available in article or Supplementary Materials.

Conflicts of Interest: The authors declare no conflict of interest.

References

1. Llorente, B.; Martinez-Garcia, J.F.; Stange, C.; Rodriguez-Concepcion, M. Illuminating colors: Regulation of carotenoid biosynthesis and accumulation by light Briardo Llorente1, Jaime F Martinez-Garcia1,2, Claudia Stange3 and Manuel Rodriguez-Concepcion1. *Curr. Opin. Plant Biol.* **2017**, *37*, 49–55. [CrossRef] [PubMed]
2. Nisar, N.; Li, L.; Lu, S.; Khin, N.C.; Pogson, B.J. Carotenoid Metabolism in Plants. *Mol. Plant* **2015**, *8*, 68–82. [CrossRef] [PubMed]
3. Walter, M.H.; Strack, D. Carotenoids and their cleavage products: Biosynthesis and functions. *Nat. Prod. Rep.* **2011**, *28*, 663–692. [CrossRef] [PubMed]
4. Colasuonno, P.; Lozito, M.L.; Marcotuli, I.; Nigro, D.; Giancaspro, A.; Mangini, G.; De Vita, P.; Mastrangelo, A.M.; Pecchioni, N.; Houston, K.; et al. The carotenoid biosynthetic and catabolic genes in wheat and their association with yellow pigments. *BMC Genom.* **2017**, *18*, 1–18. [CrossRef]
5. Bradford, K.J.; Nonogaki, H. Regulation of ABA and GA levels during seed development and germination in Arabidopsis. *Annu. Plant Rev.: Seed Dev. Dormancy Germination* **2007**, *27*, 224–247.
6. Fraser, P.D.; Romer, S.; Shipton, C.A.; Mills, P.B.; Kiano, J.W.; Misawa, N.; Drake, R.G.; Schuch, W.; Bramley, P.M. Evaluation of Transgenic Tomato Plants Expressing an Additional Phytoene Synthase in a Fruit-Specific Manner. *Proc. Natl. Acad. Sci. USA* **2002**, *99*, 1092–1097. [CrossRef]
7. Ruiz-Sola, M.A.; Rodriguez-Concepcion, M. Carotenoid biosynthesis in Arabidopsis: A colorful pathway. *Arab. Book* **2012**, *10*, e0158. [CrossRef]
8. Obrero, A.; Gonzalez-Verdejo, C.I.; Roman, B.; Gomez, P.; Die, J.V.; Ampomah-Dwamena, C. Identification, cloning, and expression analysis of three phytoene synthase genes from *Cucurbita pepo*. *Biol. Plant.* **2015**, *59*, 201–210. [CrossRef]
9. Wang, T.; Hou, Y.; Hu, H.; Wang, C.; Zhang, W.; Li, H.; Cheng, Z.; Yang, L. Functional Validation of Phytoene Synthase and Lycopene ε-Cyclase Genes for High Lycopene Content in Autumn Olive Fruit (*Elaeagnus umbellata*). *J. Agric. Food Chem.* **2020**, *68*, 11503–11511. [CrossRef]
10. Busch, M.; Seuter, A.; Hain, R. Functional Analysis of the Early Steps of carotenoid Biosynthesis in Tobacco. *Plant Physiol.* **2002**, *128*, 439–453. [CrossRef]
11. Lindgren, L.O.; Stalberg, K.G.; Höglund, A. Seed-Specific Overexpression of an Endogenous Arabidopsis Phytoene Synthase Gene Results in Delayed Germination and Increased Levels of Carotenoids, Chlorophyll, and Abscisic Acid. *Plant Physiol.* **2003**, *132*, 779–785. [CrossRef] [PubMed]
12. Just, B.J.; Santos, C.A.; Fonseca, M.E.; Boiteux, L.S.; Oloizia, B.B.; Simon, P.W. Carotenoid biosynthesis structural genes in carrot (*Daucus carota*): Isolation, sequence-characterization, single nucleotide polymorphism (SNP) markers and genome mapping. *Theor. Appl. Genet.* **2007**, *114*, 693–704. [CrossRef] [PubMed]
13. Chen, L.; Li, W.; Li, Y.; Feng, X.; Du, K.; Wang, G.; Zhao, L. Identified trans-splicing of *Yellow-Fruited Tomato 2* encoding the Phytoene Synthase 1 protein alters fruit color by map-based cloning, functional complementation and Race. *Plant Mol. Biol.* **2019**, *100*, 647–658. [CrossRef] [PubMed]
14. Welsch, R.; Arango, J.; Bär, C.; Salazar, B.; Al-Babili, S.; Beltrán, J.; Chavarriaga, P.; Ceballos, H.; Tohme, J.; Beyer, P. Provitamin A Accumulation in Cassava (*Manihot esculenta*) Roots Driven by a Single Nucleotide Polymorphism in a Phytoene Synthase Gene. *Plant Cell* **2010**, *22*, 3348–3356. [CrossRef]
15. Kachanovsky, D.E.; Filler, S.; Isaacson, T.; Hirschberg, J. Epistasis in tomato color mutations involves regulation of phytoene synthase 1 expression by cis-carotenoids. *Proc. Natl. Acad. Sci. USA* **2012**, *109*, 19021–19026. [CrossRef]
16. Lee, W.; Huang, J.; Chen, F. Ectopic Expression of Oncidium Gower Ramsey Phytoene Synthase Gene Caused Pale Flower and Reduced Leaf Size in Transgenic Tobacco. *J. Int. Coop.* **2012**, *7*, 133–150.
17. Matsuyama, T.; Idota, S.; Horimoto, N.; Otabara, T.; Nomura, Y.; Sasamoto, H.; Akashi, R. Effect of growth rate and sprout promotion on the bivalent iron ions in japanese lawn grass (*Zoysia japonica* steud.). *J. Jpn. Soc. Turfgrass Sci.* **2010**, *38*, 162–166.
18. Tanaka, H.; Hirakawa, H.; Kosugi, S.; Nakayama, S.; Ono, A.; Watanabe, A.; Hashiguchi, M.; Gondo, T.; Ishigaki, G.; Muguerza, M.; et al. Sequencing and comparative analyses of the genomes of zoysiagrasses. *DNA Res.* **2016**, *23*, 171–180. [CrossRef]
19. Park, S.; Kim, H.S.; Jung, Y.J.; Kim, S.H.; Ji, C.Y.; Wang, Z.; Jeong, J.C.; Lee, H.S.; Lee, S.Y.; Kwak, S.S. Orange protein has a role in phytoene synthase stabilization in sweetpotato. *Sci. Rep.* **2016**, *6*, 33563. [CrossRef]
20. Bhat, A.; Mishra, S.; Kaul, S.; Dhar, M.K. Elucidation and functional characterization of CsPSY and CsUGT promoters in *Crocus sativus* L. *PLoS ONE* **2018**, *13*, e0195348. [CrossRef]

21. Wang, H.; Ou, C.G.; Zhuang, F.Y.; Ma, Z.G. The dual role of phytoene synthase genes in carotenogenesis in carrot roots and leaves. *Mol. Breed.* **2014**, *34*, 2065–2079. [CrossRef] [PubMed]

22. Maass, D.; Arango, J.; Wust, F.; Beyer, P.; Welsch, R. Carotenoid crystal formation in Arabidopsis and carrot roots caused by increased phytoene synthase protein levels. *PLoS ONE* **2009**, *4*, e6373. [CrossRef] [PubMed]

23. Hong, Y.; Zhang, H.; Huang, L.; Li, D.; Song, F. Overexpression of a Stress-Responsive NAC Transcription Factor Gene ONAC022 Improves Drought and Salt Tolerance in Rice. *Front. Plant Sci.* **2016**, *7*, 4. [CrossRef] [PubMed]

24. Qin, X.; Coku, A.; Inoue, K.; Tian, L. Expression, subcellular localization, and cis-regulatory structure of duplicated phytoene synthase genes in melon (Cucumis melo L.). *Planta* **2011**, *234*, 737–748. [CrossRef]

25. Welsch, R.; Wust, F.; Bar, C.; Al-Babili, S.; Beyer, P. A third phytoene synthase is devoted to abiotic stress-induced abscisic acid formation in rice and defines functional diversification of phytoene synthase genes. *Plant Physiol.* **2008**, *147*, 367–380. [CrossRef]

26. Lescot, M.; Dehais, P.; Thijs, G.; Marchal, K.; Moreau, Y.; Van de Peer, Y.; Rouze, P.; Rombauts, S. PlantCARE, a database of plant cis-acting regulatory elements and a portal to tools for in silico analysis of promoter sequences. *Nucleic Acids Res.* **2002**, *30*, 325–327. [CrossRef]

27. Zhou, R.; Kroczynska, B.; Hayman, G.T.; Miernyk, J.A. AtJ2, an Arabidopsis homolog of Escherichia coli dnaJ. *Plant Physiol.* **1995**, *108*, 821–822. [CrossRef]

28. Barghetti, A.; Sjögren, L.; Floris, M.; Paredes, E.B.; Wenkel, S.; Brodersen, P. Heat-shock protein 40 is the key farnesylation target in meristem size control, abscisic acid signaling, and drought resistance. *Gene Dev.* **2017**, *31*, 2282–2295. [CrossRef]

29. Tsuchiya, T.; Ohta, H.; Okawa, K.; Iwamatsu, A.; Shimada, H.; Masuda, T.; Takamiya, K. Cloning of chlorophyllase, the key enzyme in chlorophyll degradation: Finding of a lipase motif and the induction by methyl jasmonate. *Proc. Natl. Acad. Sci. USA* **1999**, *96*, 15362–15367. [CrossRef]

30. Sato, Y.; Morita, R.; Katsuma, S.; Nishimura, M.; Tanaka, A.; Kusaba, M. Two short-chain dehydrogenase/reductases, NON-YELLOW COLORING 1 and NYC1-LIKE, are required for chlorophyll b and light-harvesting complex II degradation during senescence in rice. *Plant J.* **2009**, *57*, 120–131. [CrossRef]

31. Morita, R.; Sato, Y.; Masuda, Y.; Nishimura, M.; Kusaba, M. Defect in non-yellow coloring 3, an α/β hydrolase-fold family protein, causes a stay-green phenotype during leaf senescence in rice. *Plant J.* **2009**, *59*, 940–952. [CrossRef] [PubMed]

32. Jia, T.; Ito, H.; Hu, X.; Tanaka, A. Accumulation of the NON-YELLOW COLORING 1 protein of the chlorophyll cycle requires chlorophyll b in *Arabidopsis thaliana*. *Plant J.* **2015**, *81*, 586–596. [CrossRef] [PubMed]

33. Yu, M.; Chen, J.; Qu, J.; Liu, F.; Zhou, M.; Ma, Y.; Xiang, S.; Pan, X.; Zhang, H.; Yang, M. Exposure to endophytic fungi quantitatively and compositionally alters anthocyanins in grape cells. *Plant Physiol. Bioch.* **2020**, *149*, 144–152. [CrossRef] [PubMed]

34. Espasandin, F.D.; Maiale, S.J.; Calzadilla, P.; Ruiz, O.A.; Sansberro, P.A. Transcriptional regulation of 9-cis-epoxycarotenoid dioxygenase (*NCED*) gene by putrescine accumulation positively modulates ABA synthesis and drought tolerance in Lotus tenuis plants. *Plant Physiol. Bioch.* **2014**, *76*, 29–35. [CrossRef]

35. Cazzonelli, C.I. Carotenoids in nature: Insights from plants and beyond. *Funct. Plant Biol.* **2011**, *38*, 833. [CrossRef] [PubMed]

36. Andolfo, G.; Iovieno, P.; Ricciardi, L.; Lotti, C.; Filippone, E.; Pavan, S.; Ercolano, M.R. Evolutionary conservation of mlo gene promoter signatures. *BMC Plant Biol.* **2019**, *19*, 1–11. [CrossRef]

37. Shah, J.; Klessig, D.F. Identification of a salicylic acid-responsive element in the promoter of the tobacco pathogenesis-related beta-1,3-glucanase gene, *PR-2d*. *Plant J.* **2010**, *10*, 1089–1101. [CrossRef]

38. Li, H.; Wang, Y.; Li, X.; Gao, Y.; Wang, M. A GA-insensitive dwarf mutant of *Brassica napus* L. correlated with mutation in pyrimidine box in the promoter of *GID1*. *Mol. Biol. Rep.* **2010**, *38*, 191–197. [CrossRef]

39. Rouster, J.; Leah, R.; Mundy, J.; Cameron-Mills, V. Identification of a methyl jasmonate-responsive region in the promoter of a lipoxygenase 1 gene expressed in barley grain. *Plant J.* **2010**, *11*, 513–523. [CrossRef]

40. Finkelstein, R.; Gampala, S.; Lynch, T.J.; Thomas, T.L.; Rock, C.D. Redundant and distinct functions of the ABA Response Loci *ABA-INSENSITIVE (ABI)* 5 and *ABRE-BINDING FACTOR (ABF)* 3. *Plant Mol. Biol.* **2005**, *59*, 253–267. [CrossRef]

41. Toledo-Ortiz, G.; Johansson, H.; Lee, K.P.; Bou-Torrent, J.; Stewart, K.; Steel, G.; Concepcion, M.R.; Halliday, K.J. The HY5-PIF regulatory module coordinates light and temperature control of photosynthetic gene transcription. *PLoS Genet.* **2014**, *10*, e1004416. [CrossRef] [PubMed]

42. Zhou, E.F. More is better: The diversity of terpene metabolism in plants Fei Zhou and Eran Pichersky. *Curr. Opin. Plant Biol.* **2020**, *55*, 1–10. [CrossRef] [PubMed]

43. Bartley, G.E.; Scolnik, P.A. Plant carotenoids: Pigments for photoprotection, visual attraction, and human health. *Plant Cell* **1995**, *7*, 1027–1038. [PubMed]

44. Cidade, L.C.; de Oliveira, T.M.; Mendes, A.F.S.; Macedo, A.F.; Floh, E.I.S.; Gesteira, A.S.; Soares-Filho, W.S.; Costa, M.G.C. Ectopic expression of a fruit phytoene synthase from Citrus paradisi Macf. promotes abiotic stress tolerance in transgenic tobacco. *Mol. Biol. Rep.* **2012**, *39*, 10201–10209. [CrossRef]

45. Simpson, K.; Fuentes, P.; Quiroz-Iturra, L.F.; Flores-Ortiz, C.; Contreras, R.; Handford, M.; Stange, C. Unraveling the induction of phytoene synthase 2 expression by salt stress and abscisic acid in Daucus carota. *J. Exp. Bot.* **2018**, *69*, 4113–4126. [CrossRef]

46. Zhou, X.; Welsch, R.; Yang, Y.; Álvarez, D.; Riediger, M.; Yuan, H.; Fish, T.; Liu, J.; Thannhauser, T.W.; Li, L. Arabidopsis OR proteins are the major posttranscriptional regulators of phytoene synthase in controlling carotenoid biosynthesis. *Proc. Natl. Acad. Sci. USA* **2015**, *112*, 3558–3563. [CrossRef]

47. Morikawa, T.; Uraguchi, Y.; Sanda, S.; Nakagawa, S.; Sawayama, S. Overexpression of DnaJ-Like Chaperone Enhances Carotenoid Synthesis in *Chlamydomonas reinhardtii*. *Appl. Biochem. Biotech.* **2018**, *184*, 80–91. [CrossRef]

48. Tamura, K.; Peterson, D.; Peterson, N.; Stecher, G.; Nei, M.; Kumar, S. MEGA5: Molecular Evolutionary Genetics Analysis Using Maximum Likelihood, Evolutionary Distance, and Maximum Parsimony Methods. *Mol. Biol. Evol.* **2011**, *28*, 2731–2739. [CrossRef]

49. Chen, C.; Chen, H.; Zhang, Y.; Thomas, H.R.; Frank, M.H.; He, Y.; Xia, R. TBtools: An Integrative Toolkit Developed for Interactive Analyses of Big Biological Data. *Mol. Plant* **2020**, *13*, 1194–1202. [CrossRef] [PubMed]

50. Bailey, T.L.; Gribskov, M. Combining evidence using p-values: Application to sequence homology searches. *Bioinformatics* **1998**, *14*, 48–54. [CrossRef]

51. Clough, S.J.; Bent, A.F. Floral dip: A simplified method for Agrobacterium-mediated transformation of Arabidopsis thaliana. *Plant J.* **1998**, *16*, 735–743. [CrossRef] [PubMed]

52. Ding, F.; Wang, R. Amelioration of postharvest chilling stress by trehalose in pepper. *Sci. Hortic.* **2018**, *232*, 52–56. [CrossRef]

53. Gu, C.S.; Liu, L.Q.; Zhao, Y.H.; Deng, Y.M.; Zhu, X.D.; Huang, S.Z. Overexpression of *Iris. lactea* var. *chinensis metallothionein llmt2a enhances cadmium tolerance in Arabidopsis thaliana*. *Ecotoxicol. Environ. Saf.* **2014**, *105*, 22–28.

54. Teng, K.; Chang, Z.H.; Xiao, G.Z.; Guo, W.E.; Xu, L.X.; Chao, Y.H.; Han, L.B. Molecular cloning and characterization of a chlorophyll degradation regulatory gene (*ZjSGR*) from *Zoysia japonica*. *Genet. Mol. Res. Gmr* **2016**, *15*. [CrossRef] [PubMed]

55. Kokkirala, V.R.; Yonggang, P.; Abbagani, S.; Zhu, Z.; Umate, P. Subcellular localization of proteins of Oryza sativa L. in the model tobacco and tomato plants. *Plant Signal. Behav.* **2010**, *5*, 1336–1341. [CrossRef] [PubMed]

56. Fan, Z.; Chen, J.; Kuang, J.; Lu, W.; Shan, W. The Banana Fruit SINA Ubiquitin Ligase MaSINA1 Regulates the Stability of MaICE1 to be Negatively Involved in Cold Stress Response. *Front. Plant Sci.* **2017**, *8*, 995. [CrossRef] [PubMed]

Article

Transcriptome Analysis Revealed a Positive Role of Ethephon on Chlorophyll Metabolism of *Zoysia japonica* under Cold Stress

Jiahang Zhang [1], Zhiwei Zhang [1,2], Wen Liu [1], Lijing Li [1], Liebao Han [1], Lixin Xu [1,*] and Yuhong Zhao [3,*]

[1] College of Grassland Science, Beijing Forestry University, Beijing 100083, China; 15104490515@163.com (J.Z.); zhangzw559@163.com (Z.Z.); bxliuwen@163.com (W.L.); llj15122428388@163.com (L.L.); hanliebao@163.com (L.H.)

[2] CCTEG Ecological Environment Technology Co., Ltd., Beijing 100013, China

[3] Animal Science College, Tibet Agriculture & Animal Husbandry University, Nyingchi 860000, China

* Correspondence: lixinxu@bjfu.edu.cn (L.X.); yuhong19801011@126.com (Y.Z.)

Abstract: *Zoysia japonica* is a warm-season turfgrass with a good tolerance and minimal maintenance requirements. However, its use in Northern China is limited due to massive chlorophyll loss in early fall, which is the main factor affecting its distribution and utilization. Although ethephon treatment at specific concentrations has reportedly improved stress tolerance and extended the green period in turfgrass, the potential mechanisms underlying this effect are not clear. In this study, we evaluated and analyzed chlorophyll changes in the physiology and transcriptome of *Z. japonica* plants in response to cold stress (4 °C) with and without ethephon pretreatment. Based on the transcriptome and chlorophyll content analysis, ethephon pretreatment increased the leaf chlorophyll content under cold stress by affecting two processes: the stimulation of chlorophyll synthesis by upregulating *ZjMgCH2* and *ZjMgCH3* expression; and the suppression of chlorophyll degradation by downregulating *ZjPAO*, *ZjRCCR*, and *ZjSGR* expression. Furthermore, ethephon pretreatment increased the ratio of chlorophyll a to chlorophyll b in the leaves under cold stress, most likely by suppressing the conversion of chlorophyll a to chlorophyll b due to decreased chlorophyll b synthesis via downregulation of *ZjCAO*. Additionally, the inhibition of chlorophyll b synthesis may result in energy redistribution between photosystem II and photosystem I.

Keywords: chlorophyll content; cold stress; ethephon; transcriptome; *Zoysia japonica*

Citation: Zhang, J.; Zhang, Z.; Liu, W.; Li, L.; Han, L.; Xu, L.; Zhao, Y. Transcriptome Analysis Revealed a Positive Role of Ethephon on Chlorophyll Metabolism of *Zoysia japonica* under Cold Stress. *Plants* **2022**, *11*, 442. https://doi.org/10.3390/plants11030442

Academic Editors: Hye Ryun Woo and Georgia Ouzounidou

Received: 31 December 2021
Accepted: 3 February 2022
Published: 5 February 2022

Publisher's Note: MDPI stays neutral with regard to jurisdictional claims in published maps and institutional affiliations.

1. Introduction

Zoysia japonica (Z. japonica) is a warm-season turfgrass species with good traffic stress tolerance and minimal maintenance requirements. It is widely used in China for sports turf, landscaping, and soil and water conservation [1,2]. However, in Northern China, the use of *Z. japonica* is relatively limited due to its short green period [3]. Temperature is the main factor affecting its natural distribution, turf quality, and popularization [4].

Low-temperature stress affects photosynthesis [5], the cell membrane [6,7], antioxidant systems [8], and other physiological and biochemical processes in plants [9]. Low-temperature conditions especially alter photosynthesis functions [10], among which light energy absorption, gas exchange, and carbon assimilation are the most low-temperature susceptible processes [11]. Chlorophyll (Chl) fluorescence is reduced in *Zoysia japonica* (zoysiagrass) under natural chilling stress, suggesting that photosystem (PS) II is impaired by photoinhibition [12,13]. Short-term cold stress (2 and 72 h cold treatment at 4 °C) can also induce oxidative stress and inhibit photosynthesis in *Z. japonica* [14]. However, the effect that a pretreatment with ethephon, a commercially available plant growth regulator, has on *Z. japonica* under long-term cold stress is still unclear.

Cold stress affects both the synthesis and degradation of Chl [15,16]. The decreased Chl content under low-temperature stress may be due to the inhibition of Chl biosynthesis and/or the acceleration of Chl degradation [17]. In rice, the Chl synthesis decrease under low-temperature conditions is triggered by the inhibition of δ-aminolevulinic acid (ALA) synthesis and the suppression of the protochlorophyllide (Pchlide) conversion into Chl due to the downregulation of the *protochlorophyllide oxidoreductase (POR)* gene [18]. A transcriptome study in Z. *japonica* revealed that 72 h of cold stress treatment was associated with the downregulation of several differentially expressed genes (DEGs) involved in Chl synthesis and the upregulation of DEGs involved in Chl degradation. Specifically, two genes were upregulated: the gene encoding ferrochelatase (Hem H), which catalyzes the insertion of ferrous iron into protoporphyrin IX to form protoheme that departs from the Chl biosynthetic pathway, and the gene encoding pheophorbide *a* oxygenase (PAO), which catalyzes the porphyrin macrocycle cleavage of pheophorbide a to generate a primary fluorescent Chl catabolite [14,19].

Ethylene acts as a plant hormone actively involved in plant stress response [20–22]. However, as a gaseous agent, the direct application of ethylene in production practice is difficult [23]. The disadvantages associated with its application can be overcome with ethephon, an ethylene-releasing reagent with great potential in outdoor use [23,24]. Interestingly, using a certain concentration of ethephon can improve the cold resistance of grape (*Cabernet Sauvignon*) [25], larch (*Larix gmelinii*) [26], and banana seedlings (*Musa × paradisiaca*) [27]. At a specific concentration, ethephon can help to maintain a high Chl content in a submergence-tolerant rice (*Oryza sativa* L.) cultivar during submergence [28]. An ethephon pretreatment at a concentration of 150 mg L^{-1} can alleviate stress-related injuries in Z. *japonica* and reduce the loss of Chl under low-temperature conditions [29,30]. However, the mechanism underlying the effect of ethephon pretreatment has not been investigated and analyzed.

RNA sequencing (RNA-Seq) is a next-generation sequencing application with some clear advantages over existing sequencing methods [31,32]. Transcriptome analysis has also been applied to turfgrass species, including the bermudagrass [33], Kentucky bluegrass [34], creeping bentgrass [35], and Z. *japonica* [14]. A transcriptome analysis performed by Wang et al. [36] revealed that the gene families encoding auxin signal transduction proteins, ABA signal transduction proteins, and WRKY and bHLH transcription factors might represent the most critical components for salt-stress regulation in Z. *japonica*. Gene expression changes on the whole transcriptome level associated with ethephon pretreatment under cold stress have been rarely studied in Z. *japonica*.

The objective of this study was to explore the mechanism underlying the effect of ethephon on cold tolerance in Z. *japonica* and to prolong the green period of *Zoysia japonica* under cold stress. We assessed this effect by identifying the genes responding to ethephon treatment and analyzing the ethephon-induced key regulatory genes affecting the Chl metabolism in Z. *japonica* under cold stress.

2. Results

2.1. Chlorophyll Content in Leaves

The Chl analysis showed that CE plants had a significantly higher Chl content than CS plants ($p < 0.05$) (Figure 1A), demonstrating that ethephon pretreatment increased the Chl content under cold stress. Figure 1B shows that the ratio of Chla to Chlb did not change significantly, except for a slight increase in CE plants than in CS plants ($p = 0.147$) (Figure 1B). Moreover, the Chla/Chlb ratio change percentage relative to day 0 of CE plants was significantly lower than that of CS plants (sig = 0.027, 0.001 < 0.05) (Figure 1C), which indicated that ethephon pretreatment suppressed the decline of the Chla/Chlb ratio under cold-stressed conditions. Combined with the data in Figure 1B, it is affirmed that ethephon pretreatment could increase the Chla/Chlb ratio under cold stress.

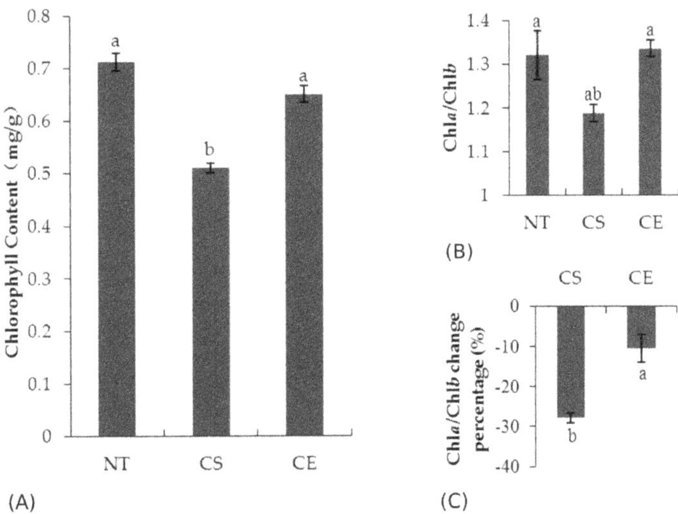

Figure 1. (**A**) Chlorophyll content in *Z. japonica* leaves; (**B**) Ratio of Chla to Chlb in *Z. japonica* leaves; (**C**) Chla/Chlb change percentage on day 28 relative to day 0. Error bars represent standard deviations of three independent samples. Different letters indicate significant differences at $p < 0.05$.

2.2. Sequence Assembly

A total of 75 Gbp of clean sequencing data were obtained by a transcriptome analysis of all combined samples; the smallest clean data set had 6.7 Gbp of sequencing data and the largest had 9.86 Gbp. After removing low-quality reads, an average of 55, 554, and 958 high-quality clean reads per sample was obtained, accounting for 97.57% of the sample's average raw reads. The percentage of Q30 bases reached over 90.51%, and the GC content of the samples varied between 49.73% and 53.48%. The clean reads of each sample were aligned with the reference genome of *Z. japonica*, reaching 93.06% of aligned reads for the sample with the highest efficiency and 88.08% of aligned reads for the sample with the lowest efficiency (Table 1). Thus, the sequencing data cleaning process generated a data set of high-quality reads that were mapped to the reference genome with a high efficiency and met the analysis requirements of subsequent tests.

Table 1. An overview of the RNA-Seq data.

Sample	Raw Reads	Total Reads	Clean Reads	Clean Bases (Gbp)	Q30 (%)	GC Content (%)	Total Map (%)
CE1 [1]	51696468	50530932	50530932	7.58	90.95	51.3	91.45
CE2	67066970	65714084	65714084	9.86	92.08	51.23	92.35
CE3	65392538	64083968	64083968	9.61	92.1	52.98	93.06
CS1 [2]	61806546	60439330	60439330	9.07	90.91	53.48	92.11
CS2	59147316	58002528	58002528	8.7	90.51	53.46	92.19
CS3	61224860	60038194	60038194	9.01	91.67	52.75	92.79
NT1 [3]	50204448	48478352	48478352	7.27	92.51	49.73	88.08
NT2	46169382	44677484	44677484	6.7	92.1	50.06	88.44
NT3	49753502	48029758	48029758	7.2	92.21	49.75	87.66

[1] CE, cold stress (4 °C) with ethephon spray pretreatment. [2] CS, cold stress (4 °C) with water spray pretreatment. [3] NT, normal day/night temperature regimen (28/18 °C) with water spray pretreatment as control.

2.3. Global Gene Expression Analysis

The comparison of CS plants with NT plants (CS vs. NT) identified a total of 16,359 DEGs in the leaves, among which 9199 were upregulated, and 7160 were downregulated (Figure 2a). Gene ontology (GO) and an enrichment analysis identified 516 biological processes, 337 molecular functions, and 128 cellular components (Table S1). Among 16,359 cold-stress-related DEGs identified by the CS vs. NT comparison, there were 1618 DEGs with annotations in 113 KEGG pathways; 924 of those DEGs were upregulated, and 694 were downregulated (Table S2).

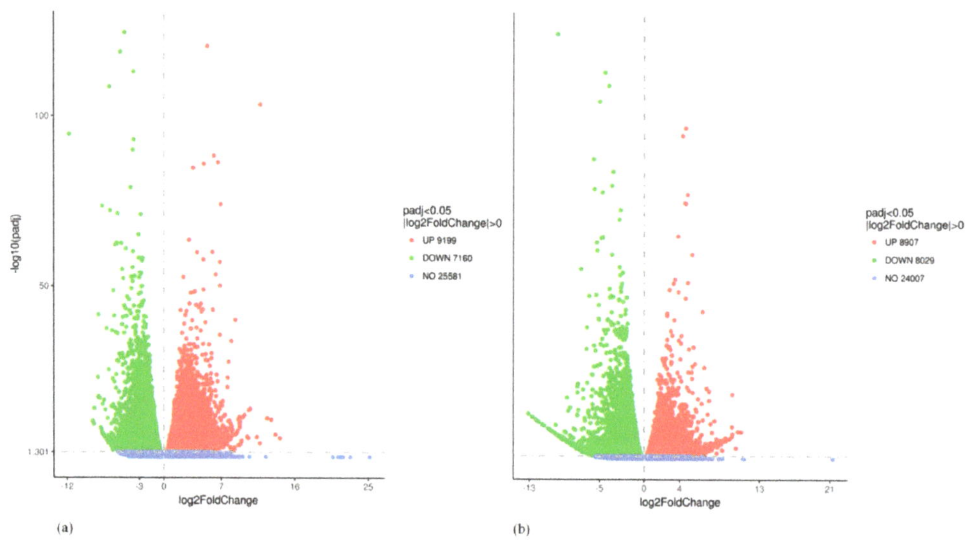

Figure 2. (**a**) Distribution of differential genes in volcanic map (CS vs. NT); (**b**) Distribution of differential genes in volcanic map (CEvs. NT).

The most DEGs in the leaves were identified by comparing CE plants with NT plants (CE vs. NT), with a total of 16,936 DEGs, among which 8907 were upregulated, and 8029 were downregulated (Figure 2b). GO and enrichment analysis identified 519 biological processes, 337 molecular functions, and 128 cellular components (Table S3). Among 16,359 DEGs, 1777 were annotated in 112 KEGG pathways, which included 931 upregulated and 846 downregulated DEGs (Table S4).

The heat map showed the overall effect of ethephon treatment under cold stress on transcription, and visualized how ethephon regulated the effect of cold stress on transcriptome (Figure S1). These results indicated that ethephon induced changes in the gene expression of *Z. japonica* under cold stress. Therefore, only the genes most relevant to ethephon application under cold stress are focused on in the study.

2.4. qRT-PCR Confirmation

These genes displayed a single dissociation peak and linearity between target cDNA and Ct values (Figure 3), showing that the genes used for qRT-PCR were consistent with the RNA-Seq results (Pearson's r = 0.94, $p < 0.001$).

2.5. Differential Gene Expression Related to Photosynthesis-Antenna Proteins and Porphyrin and Chlorophyll Metabolism

In the CE vs. NT comparison, the most-enriched KEGG pathways related to chlorophyll included "porphyrin and Chl metabolism" (Figure 4). The results indicated that ethephon can regulate chlorophyll metabolism to improve cold tolerance. Meanwhile, in land plants, the only depots of Chlb are antenna complexes, which are composed of light harvesting complex (Lhc) proteins, [37] and Chlb content can regulate light-harvesting complexes levels in plants (Ayumi, 2019). Therefore, we analyzed the DEGs, which were derived by matching the DEGs from CE vs. NT with those from CS vs. NT ((CE vs. NT) vs.

(CS vs. NT)), annotated in KEGG pathways of porphyrin and Chl metabolism (osa00860) (Table S5) and photosynthesis-antenna proteins (osa00196) (Table S6).

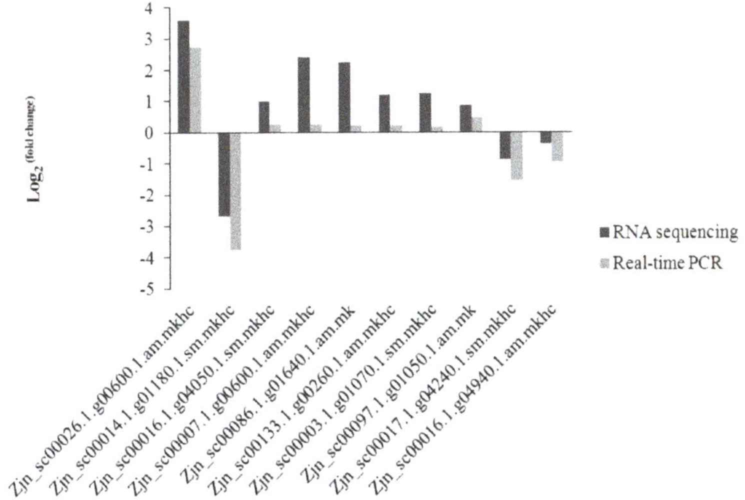

Figure 3. Log$_2$ $^{(fold\ change)}$ of genes based on RNA-Seq and qRT-PCR methods.

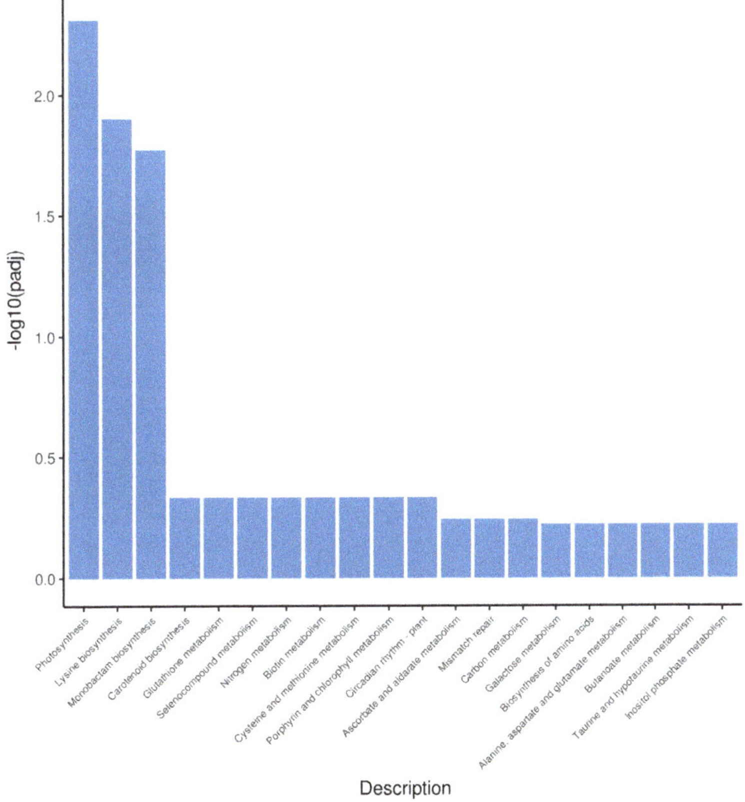

Figure 4. Enriched KEGG terms of CE vs. NT.

The comparison of CS vs. NT identified seventeen DEGs in the porphyrin and Chl metabolism (Figure 5c). Under cold stress, *ZjUroD1*, *ZjUroD2*, *ZjChlP1*, *ZjChlP2*, *ZjNYC1*, *ZjNADPH1*, *ZjPAO*, *ZjGSA*, *ZjPBGD*, *ZjGUS1*, *ZjHemA*, *ZjCLH*, *ZjSGR*, *ZjNADHB*, and putative *ZjCOX10* were upregulated, whereas *ZjHY2* and *ZjCPX* were downregulated (Figure 5a). Twenty-four DEGs were identified by the comparison of CE vs. NT (Figure 5c). Ethephon pretreatment under cold stress was associated with the upregulation of *ZjUroD1*, *ZjUroD2*, *ZjChlP1*, *ZjChlP2*, *ZjMgCH2*, *ZjMgCH3*, *ZjNYC1*, *ZjNADPH1*, *ZjGSA*, *ZjPBGD*, *ZjGUS1*, *ZjHemA*, *ZjNADHB*, and putative *ZjCOX10*, and the downregulation of *ZjNADPG2*, *ZjPAO*, *ZjCAO*, *ZjHY2*, *ZjCPX*, *ZjHMOX1*, *ZjMPE cyclase*, *ZjRCCR*, *ZjTPM*, and *ZjSGR* (Figure 5a). There were sixteen DEGs found both in CS vs. NT and CE vs. NT (Figure 5c).

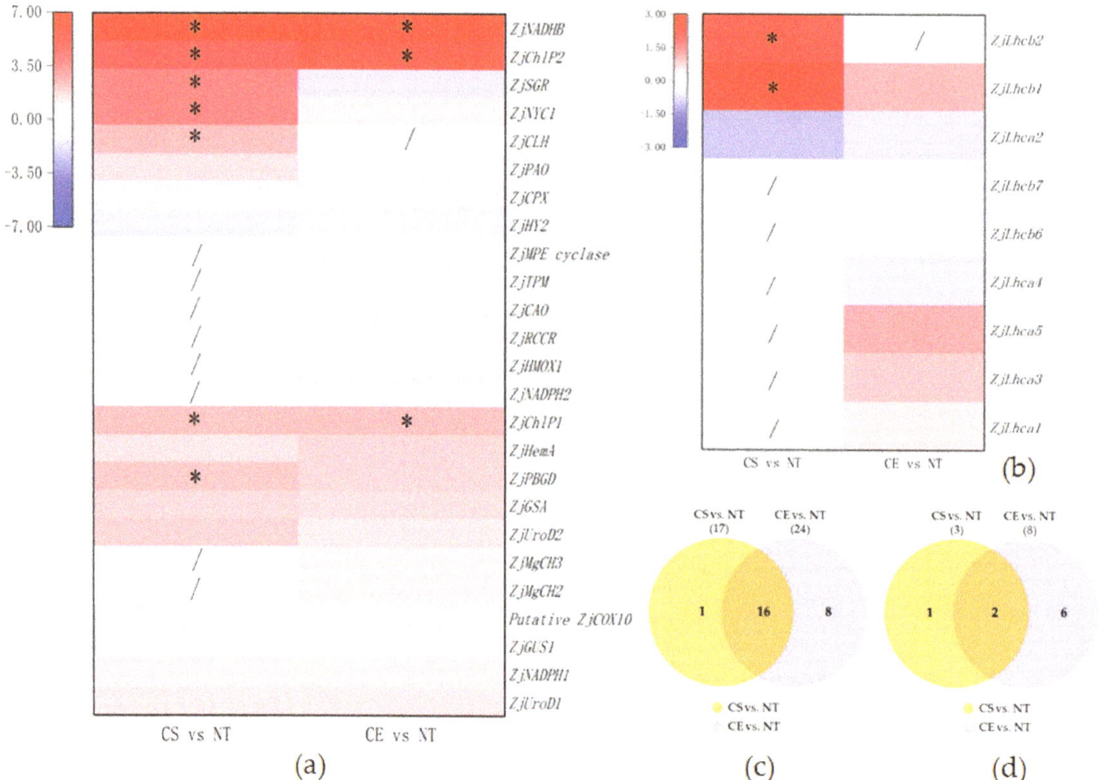

Figure 5. (**a**) Heat map of differentially expressed genes in porphyrin and chlorophyll metabolism pathway; (**b**) Heat map of differentially expressed genes in photosynthesis-antenna proteins pathway; (**c**) Venn diagram for all DEGs in porphyrin and chlorophyll metabolism pathway; (**d**) Venn diagram for all DEGs in photosynthesis-antenna proteins pathway; "/" indicates the DEG was not enriched in this KEGG pathway comparison; and " * " means the $\log_2{}^{(\text{fold change})}$ of this gene is greater than 2. Gene expression in $\log_2{}^{(\text{fold change})}$ scale was elevated with red and decreased with blue.

Based on the CS vs. NT comparison, we identified three DEGs in the photosynthesis-antenna protein pathway (Figure 5d). Under cold stress, *ZjLhcb1* and *ZjLhcb2* were upregulated, and *ZjLhca2* was downregulated (Figure 5b). Eight DEGs were identified by the comparison of CE vs. NT (Figure 5d): *ZjLhca1*, *ZjLhca3*, *ZjLhca5*, *ZjLhcb1* were upregulated, and *ZjLhca2*, *ZjLhca4*, *ZjLhcb6*, *ZjLhcb7* were downregulated by ethephon pretreatment under cold stress (Figure 5b). There were two DEGs found in both CS vs. NT and CE vs. NT (Figure 5d).

3. Discussion

3.1. Gene Expression Analysis of Porphyrin and Chl Metabolism

The Chl content is an important physiological index of plant tolerance to cold stress [38,39]. However, the effects of cold stress on Chl metabolism are modified by ethephon treatment. We found that ethephon pretreatment elevated the Chl content by regulating genes involved in Chl synthesis, the PAO-dependent Chl degradation pathway, and the Chl cycle (Figure 6a,b).

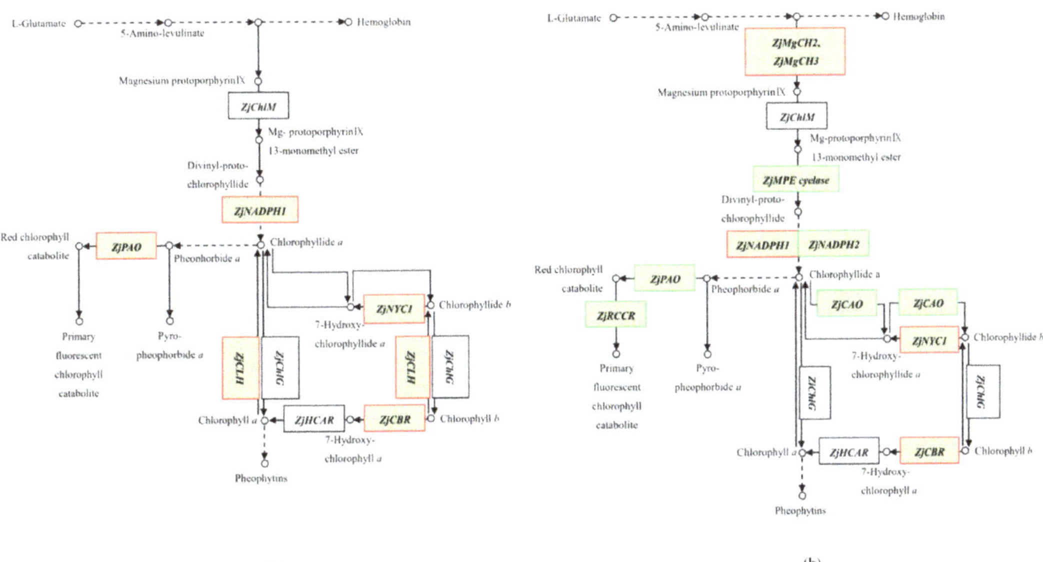

Figure 6. (a) KEGG metabolic pathway based on the comparison of CS vs. NT that matches the chlorophyll metabolic pathway; (b) KEGG metabolic pathway based on the comparison of CE vs. NT that matches the chlorophyll metabolic pathway. KEGG nodes indicating upregulated genes are marked in red, KEGG nodes indicating downregulated genes are marked in green. Symbols: → synthesis, --→ indirect synthesis, ○ synthesized product.

In our DEG analysis, we identified two genes, *ZjMgCH2* and *ZjMgCH3*, encoding the magnesium (Mg)-chelatase enzyme functioning in Mg insertion into protoporphyrin IX (PIX). The Mg-chelatase acts at the branch point between the heme and Chl biosynthetic pathways [40]. Therefore, the enzyme is a crucial site for the regulation of the flow of tetrapyrrole intermediates through the Chl branch of this pathway from the common PIX intermediate [41]. Thus, the observed upregulation of *ZjMgCH2* and *ZjMgCH3* by ethephon pretreatment might have increased Chl synthesis, which eventually led to an elevated Chl content under cold stress.

ZjMPE cyclase encodes the Mg-protoporphyrin IX monomethyl ester (MPE) cyclase, which catalyzes the transformation of MPE into divinyl protochlorophyllide (DV-Pchlide). We found that *ZjMPE* cyclase was downregulated by ethephon pretreatment, which inhibited DV-Pchlide synthesis. The suppression of DV-Pchlide synthesis might indicate the slowdown of plant yellowing because previous research showed that DV-Pchlide is ubiquitous in the etiolated tissues of higher plants [42]. In addition, DV-Pchlide, along with other intermediates of the biosynthetic porphyrin pathway, tends to cause photo-oxidative damage in the chloroplasts after light absorption [43].

CAO encodes the Chla oxygenase, which catalyzes the Chlb formation by the oxygenation of Chla [44]. Ethephon pretreatment might have suppressed the conversion from Chla to Chlb under cold stress by downregulating *ZjCAO*. Therefore, the ethephon mediated an increase in the value of Chla/Chlb ratio, which is suggested to be an important determinant of maximum fluorescence intensity [45]. *CLH* encodes the chlorophyllase involved in Chl catabolism, catalyzing stress-induced Chl breakdown [46]. *ZjCLH* was upregulated under cold stress, whereas its expression level was not changed under cold stress with ethephon pretreatment. Therefore, ethephon pretreatment might have inhibited Chl degradation.

PAO and red Chl catabolite reductase (RCCR) are key enzymes involved in the PAO-dependent Chl degradation pathway [47]. The removal of phytol and Mg from Chla generates pheophorbide a. *ZjPAO* and *ZjRCCR* catalyze the conversion of pheophorbide a to red Chl catabolite, which is used to produce primary fluorescent Chl catabolites [48]. In our study, cold stress decreased Chl content by increasing *ZjPAO* gene expression in *Z. japonica* plants. In contrast, ethephon pretreatment inhibited Chl catabolism under cold stress by downregulating the gene expression of *ZjPAO* and *ZjRCCR*.

SGR (Stay Green) is a regulator gene in Chl degradation [49]. RNA interference in the silencing of a homolog of the rice candidate gene in *Arabidopsis* demonstrated a stay-green phenotype [50]. Under cold stress, we found that *ZjSGR* was upregulated in *Z. japonica* plants, while it was downregulated by ethephon pretreatment. Thus, the downregulation of *ZjSGR* by ethephon pretreatment might have led to an elevated Chl content in *Z. japonica* under cold stress, but the underlying mechanism remains unknown. However, ethephon treatment might have suppressed *ZjSGR* transcription because ethylene is the key regulator for *SGR* transcription [51].

3.2. Effect of Ethephon Application on Chl Content under Cold Stress

CE plants maintained a higher Chl content and a higher Chla/Chlb ratio than control CS plants. The elevated Chl content could be caused by increased Chl synthesis and reduced Chl degradation. Ethephon treatment increased Chl synthesis by upregulating the *ZjMgCH2* and *ZjMgCH3* genes encoding Mg-catalase. Ethephon pretreatment diminished Chl degradation under cold stress by downregulating *ZjPAO*, *ZjRCCR*, and *ZjSGR*.

An elevated Chla/Chlb ratio could be due to a reduced conversion from Chla to Chlb. The Chla/Chlb ratio, which is related to antenna size, is an indicator of functional pigment equipment and light adaptation [45]. The relatively high value of Chla/Chlb ratio is an index of stress adaptation due to a smaller light-harvesting complex, making PSII less susceptible during stress [52]. Thus, in our study, the downregulation of *ZjCAO* by ethephon elevated Chla/Chlb ratio, which might have promoted the *Z. japonica* plants to adapt to stress.

3.3. Gene Expression Analysis of Photosynthesis-Antenna Proteins

Photosynthetic organisms use an array of light-harvesting antenna protein complexes to efficiently harvest solar energy [53]. The Lhca proteins are associated with the light-harvesting complexes of PSI, and the Lhcb proteins are associated with those of PSII [54,55]. The genes encoding the antennas of PSI and PSII are members of the *light-harvesting complex a* (*Lhca*) and the *light-harvesting complex b* (*Lhcb*) multigene family, respectively [54]. The upregulation of *ZjLhca1*, *ZjLhca3*, and *ZjLhca5* expression and downregulation of *ZjLhcb6* and *ZjLhcb7* expression indicate a coordinated PSI-stimulation and PSII-suppression strategy adapted in ethephon-pretreated plants under cold stress.

PSII catalyzes light-induced water oxidation in oxygenic photosynthesis to convert light into chemical energy [56,57]. However, PSII is the primary target of photoinhibition [58,59]. In higher plants, PSII is composed of two moieties: the core complex, which contains all the cofactors of the electron transport chain, and the outer antenna, which increases the light-harvesting capacity of the core [60]. In higher plants, the most abundant PSII-associated light-harvesting complex (LhcII) consists of homo- and heterotrimers of Lhcb1, Lhcb2, and Lhcb3 proteins [61]. The other Chla/b-binding proteins, Lhcb4, Lhcb5, and Lhcb6, also known as CP29, CP26, and CP24, exist as monomers [62–64]. *ZjLhcb2* gene expression was upregulated in plants under cold stress, but its expression was not changed under cold stress with ethephon pretreatment. Furthermore, the expression of *ZjLhcb6* and *ZjLhcb7* was downregulated by ethephon pretreatment under cold stress. Lhcb2 and Lhcb6 encode LhcII components [65], but Lhcb7 encodes a minor antenna protein associated with PSII [66]. Ethephon pretreatment downregulated genes encoding antenna proteins associated with PSII to minimize its light-harvesting capacity and diminish photoinhibition under cold stress.

PSI is the most efficient light energy converter in nature [67]. The efficiency of the complex is based on its capacity to deliver the energy quickly to the reaction center, which minimizes energy loss [68]. Lhca1 and Lhca2 very rapidly transfer energy to the PSI core, but these two complexes also transfer energy to Lhca3 and Lhca4 with a similar transfer rate. Although Lhca3 and Lhca4 also have a transfer energy directed to the core, their transfer rate is slow [68]. Here, we observed the upregulation of *ZjLhca1* and *ZjLhca3* by ethephon, which improved the capacity for energy delivery. Emilie et al. analyzed a PSI-LhcI complex that was identical to the WT complex, except for the substitution of Lhca4 with Lhca5; the comparison with the WT complex indicated that the energy transfer from Lhca5 to the core was faster than that from Lhca4 [69]. Thus, in our study, the upregulation of *ZjLhca5* and the downregulation of *ZjLhca4* might have increased the efficiency of energy delivery to the PSI core, but it only required a minor change in the photosynthetic apparatus.

The coordinated PSI-stimulation and PSII-suppression strategy mediated by ethephon can be further explained by the state transition theory. The state transition is a self-regulating mechanism in which photosynthetic apparatuses balance the excitation energy distribution between PSII and PSI and improve the efficiency of light energy utilization by reversibly changing the association of LhcII with PSII and PSI [70]. During State 1, LhcII is preferentially dephosphorylated and in association with PSII. During State 2, LhcII is partially phosphorylated and preferentially leaves PSII, resulting in an association with PSI [71,72]. When PSII absorbs excessive light, LhcII moves from PSII to PSI; then, the state transition occurs to equalize the electron transport between PSII and PSI, which increases the light absorption, fluorescence intensity, and activity of PSI [73,74]. In general, plants redistribute the light energy between PSI and PSII to improve the utilization for photosynthesis and prevent damage due to excessive light exposure [75]. In our study, ethephon pretreatment most likely promoted energy redistribution between PSI and PSII in *Z. japonica* under cold stress by regulating genes encoding antenna proteins.

Evidence showed that the absence of Chlb could change the orientation of pigment molecules, which could impair energy transfer [37]. In this study, a transcriptome analysis showed that spraying ethephon inhibited the synthesis of Chlb and redistributed the energy of PSI and PSII, which also confirmed the findings of previous studies.

4. Materials and Methods

4.1. Plant Material and Growth Conditions

Zoysia japonica Steud. cv. *Chinese Common* plants were obtained from the Turf Research Field of the Beijing Forestry University. After transplanting, the *Z. japonica* plants were randomly placed in the greenhouse of the Beijing Forestry University. The greenhouse conditions were as follows: light intensity of 400 μmol m^{-2} s^{-1}, daily light period of 14 h, day/night temperature regimen of 28/18 °C, and water 2–3 times per week after transplanting. Fifteen pots of *Z. japonica* plants were transferred and acclimated in a growth chamber (day/night temperature of 28/18 °C, light intensity of 3000 lx, light period of 14 h) for approximately one month. Plants were divided into three treatment groups (five pots per group): NT, normal day/night temperature regimen (28/18 °C) with water spray pretreatment as control; CS, cold stress (4 °C) with water spray pretreatment; and CE, cold stress (4 °C) with ethephon spray pretreatment. Ethephon pretreatment was carried out on day 1. Leaves were thoroughly sprayed with ethephon solution until liquid drops started to fall off. The concentration of ethephon was more than 150 mg/L, which was determined by our team in a pervious study [29]. Starting on day 10, the CS and CE plants were subjected to cold stress for 18 days (4 °C) (Figure 7). The cold treatment temperature was set according to Wei et al. [14]. At the end of treatment, fresh leaves of each group were collected.

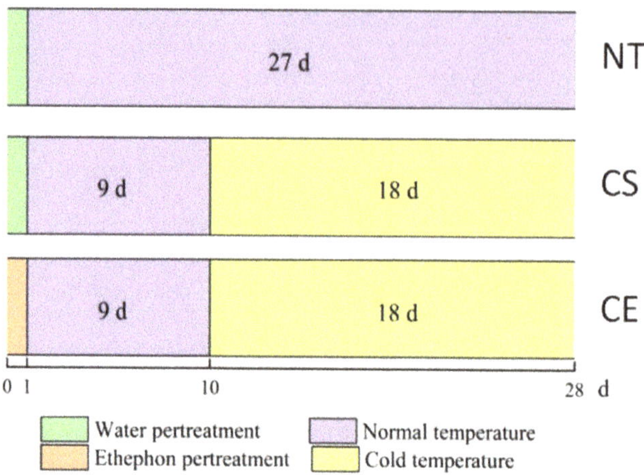

Figure 7. Treatment schedule for *Z. japonica* plants to test the effect of cold stress and ethephon treatment: NT (control), normal day/night temperatures (28/18 °C) with water spray pretreatment as control; CS, cold stress (4 °C) with water spray pretreatment; CE, cold stress (4 °C) with ethephon spray pretreatment.

4.2. Determination of Chl Content in Leaves

Chl content was measured according to the method described previously [76]. Fresh leaves (approximately 0.05 g per sample) were cut into small sections of approximately 5 mm and placed into centrifuge tubes filled with 8 mL of 95% ethanol. Tubes were stored in the dark for 48 h, and the absorption was measured at 665 nm, 649 nm, and 470 nm. The Chl content was calculated according to the following formula:

$$Ca = 13.95A665 - 6.88A649, \quad Cb = 24.96A649 - 7.32A665, \quad CChl = Ca + Cb$$

$$\text{Chla content} = Ca \times \frac{V}{W}, \quad \text{Chlb content} = Cb \times \frac{V}{W}, \quad \text{Chl content} = CChl \times \frac{V}{W}$$

In the formula, C_a refers to the concentration of Chla (mg L^{-1}), C_b to the concentration of Chlb (mg/L), C_{Chl} to the total Chl concentration (mg L^{-1}), V to the volume after extraction (L), and W to the sample weight (g).

We used a Chla/Chlb change percentage to calculate changes in Chla/Chlb ratio. Chla/Chlb change percentage were obtained by calculating the difference between the before-treatment (day 0) and after-treatment (day 28) values of each sample and dividing such a difference by before-treatment (day 0) and after-treatment (day 28). Chla/Chlb change percentage had positive and negative values, which indicated an increase or a decrease in Chla/Chlb after treatment, respectively:

$$\frac{Chla}{Chlb} \text{ratio change percentage\%} = \frac{day28 - day0}{day0} \times 100\%$$

4.3. Total RNA Extraction, RNA-Seq Library Construction, and RNA-Seq

Total RNA was extracted from *Z. japonica* leaves according to the manufacturer's instructions of the TaKaRa MiniBEST Plant RNA Extraction Kit (Takara, Japan). A NanoPhotometer instrument was used to evaluate RNA purity (OD$_{260\,nm}$/OD$_{280\,nm}$ and OD$_{260\,nm}$/OD$_{230\,nm}$ ratios), and an Agilent 2100 Bioanalyzer system was used to determined RNA integrity. The total RNA extracts were processed using magnetic beads with oligo(dT) primers to enrich the mRNA. The first cDNA strands were synthesized from short mRNA fragments, followed by the synthesis of the second strands. The new cDNA fragments were purified and subjected to end repair. The A bases were added to the 3′-ends, and the sequencing adapters were ligated to both ends of the cDNA fragments. The target size fragment fractions were recovered by agarose gel electrophoresis, and the PCR amplification was performed to complete the library preparation. Then, the effective library concentration was accurately quantified for library quality assurance. Illumina HiSeq^{TM}2500 was used for second-generation sequencing after passing the library quality evaluation. We constructed a total of 9 cDNA libraries of *Z. japonica* from the four treatment groups with three biological replicates each: NT (NT1, NT2, NT3), CS (CS1, CS2, CS3), and CE (CE1, CE2, CE3). The samples were submitted for sequencing by Illumina HiSeq^{TM}2500 (Novogene Co., Ltd., Beijing, China) (http://www.novogene.com, accessed on 23 March 2019).

4.4. RNA-Seq Data Processing and Assembly

The original image data files obtained by Illumina HiSeq^{TM} were converted into raw sequencing reads for analysis by CASAVA Base Calling [77]. Clean reads were obtained by filtering the original data and checking the sequencing error rate and GC content distribution. The clean reads were assembled based on sequence overlaps using the Trinity software [78]; the short sequence fragments were extended into longer fragments to obtain the fragment sets. The transcripts and unigene sequences were identified using the De Bruijin graph [79]. Clean reads were aligned with the reference genome of *Z. japonica* (http://zoysia.kazusa.or.jp, accessed on 2 February 2022), and their locations were mapped with the TopHat2 algorithm [80]. The HTSeq software (Baltimore, MD, USA) and union model were used to analyze the gene expression levels of the samples. The Sequence Read Archive (SRA) data were submitted to NCBI (PRJNA741873).

4.5. DEG Analysis

Sequencing quality assessment and gene expression volume analysis were performed based on the sequences located on the genome. We estimated the gene expression levels with the most commonly used method based on the fragment counts per kilobase of gene transcript per one million reads (FPKM) [81]. DEGs were detected by comparing the data from different samples. DEGs were subsequently subjected to cluster analysis and evaluated using DESeq2. The DEGs were verified by correcting the threshold P-value of the DEGs using the Benjamini–Hochberg multiple testing procedure for calibration to obtain the false discovery rate (FDR/Padj). Fold change refers to the ratio of the relative expression levels between two samples, which is processed by the shrinkage model of difference analysis software, and finally takes the logarithm based on 2 samples. Padj < 0.05 | log2FoldChange | >0 are chosen as the screening criteria for DEGs with biological duplication [82]. DEG with log2FoldChange value above 2 was considered to be significant.

4.6. Annotation Analysis

Gene Ontology (GO) is a comprehensive database describing gene function and can be divided into three parts: biological process, cellular component and molecular function. GO functional enrichment takes Padj < 0.05 as the threshold of significant enrichment. The Kyoto Encyclopedia of Genes and Genomes (KEGG) is a comprehensive database that integrates information on the genome, chemistry, and system functions with data of the respective enzymes and genes [83]. KEGG pathway enrichment of DEGs was analyzed by clusterProfiler R software. A pathway satisfying the threshold Padj < 0.05 was defined as a significantly enriched KEGG pathway for a DEG.

4.7. qRT-PCR Confirmation

Confirmation of RNA-Seq results by qRT-PCR was based on ten genes selected randomly from the comparison of CS vs. NT (Zjn_sc00026.1.g00600.1.am.mkhc, Zjn_sc00014.1.g01180.1.sm.mkhc, Zjn_sc00016.1.g04050.1.sm.mkhc, Zjn_sc00007.1.g00600.1.am.mkhc, Zjn_sc00086.1.g01640.1.am.mk) and CE vs. NT (Zjn_sc00133.1.g00260.1.am.mkhc, Zjn_sc00003.1.g01070.1.sm.mkhc, Zjn_sc00097.1.g01050.1.am.mk, Zjn_sc00017.1.g04240.1.sm.mkhc, Zjn_sc00016.1.g04940.1.am.mkhc). The cDNA preparations synthesized from the RNA samples were also used for qRT-PCR analysis based on ten randomly selected genes. Three technical repeats were processed for each of the three biological repeats originally performed per treatment. The primers were designed and synthesized by RuiBiotech Co., Ltd. (http://www.ruibiotech.com/, accessed on 2 February 2022) (Table 2). The instructions for the TB Green Premix Ex Taq kit (Tli RNaseH Plus) were obtained from Takara Bio Inc. (https://www.takarabiomed.com.cn/, accessed on 2 February 2022), and the qRT-PCR was performed using a Bio Rad CFX Connect Real-Time PCR Detection System (Bio-Rad Laboratories Co., Ltd., California, USA) (https://www.bio-rad.com/, accessed on 2 February 2022) with the following parameter settings: 95 °C for 30 s, followed by 40 cycles of 95 °C for 5 s and annealing/extension at 60 °C for 30 s. Single-product amplification was confirmed with a melt curve.

Table 2. The target genes and primer sequences.

Gene ID	Primer Sequence
Actin	F:5′-GGTGTTATGGTTGGGATGG-3′ R:5′-CAGTGAGCAGGACAGGGTG-3′
Zjn_sc00026.1.g00600.1.am.mkhc	F:5′-GCAGCAAGAACGAATGAT-3′ R:5′-CTGAAGAGTGGAAGGAGAA-3′
Zjn_sc00014.1.g01180.1.sm.mkhc	F: GATGACAGAGATGCCAAT R: CGATGAATACACCAGACA
Zjn_sc00016.1.g04050.1.sm.mkhc	F: GGCAAGTGGTATTAGTGAA R: CAGTATGTGTTCCGTTGT
Zjn_sc00007.1.g00600.1.am.mkhc	F:GGACCTTGGACAGCATCTT R:CGGCGACGAAGTAGAGAAT
Zjn_sc00086.1.g01640.1.am.mk	F:CACGGACCAAGGACTCAAG R:CCAGCGTCAGTCACAAGA
Zjn_sc00133.1.g00260.1.am.mkhc	F:5′-GAAGGACACAGGAGTTGATG-3′ R:5′-CCATTACCAAGGCGTCTC-3′
Zjn_sc00003.1.g01070.1.sm.mkhc	F:5′-ATCCTTACACCACTTCCT-3′ R:5′-CTCATCTCGCAACACATT-3′
Zjn_sc00097.1.g01050.1.am.mk	F:5′-CTACCACGCTCAATCCTAT-3′ R:5′-GTCATCCTCCTCTTCATCTT-3′
Zjn_sc00017.1.g04240.1.sm.mkhc	F: GGTGGTCATTGTGGATAA R: GGAGTCAGGTTCAGATAAG
Zjn_sc00016.1.g04940.1.am.mkhc	F: GCAAGAATGGAACCTGTG R: TCAGCAGCAATCTCATCA

5. Conclusions

Based on the transcriptome and chlorophyll content analysis, ethephon pretreatment increased the leaf chlorophyll content under cold stress by affecting two processes: the stimulation of chlorophyll synthesis by upregulating *ZjMgCH2* and *ZjMgCH3* expression and the suppression of chlorophyll degradation by downregulating ZjPAO, *ZjRCCR*, and *ZjSGR* expression. Furthermore, ethephon pretreatment increased the ratio of chlorophyll a to chlorophyll b in the leaves under cold stress, most likely by suppressing the conversion of chlorophyll a to chlorophyll b due to decreased chlorophyll b synthesis via the downregulation of *ZjCAO*. Additionally, the inhibition of chlorophyll b synthesis may result in an energy redistribution between photosystem II and photosystem I.

Supplementary Materials: The following supporting information can be downloaded at: https://www.mdpi.com/article/10.3390/plants11030442/s1, Figure S1: Heat map of all differentially expressed genes in *Z. japonica*; Table S1: GO enrichment of CS vs. NT; Table S2: KEGG enrichment of CS vs. NT; Table S3: GO enrichment of CE vs. NT; Table S4: KEGG enrichment of CE vs. NT. Table S5: DEGs enriched in the KEGG pathway of porphyrin and chlorophyll metabolism; Table S6: DEGs enriched in the KEGG pathway of photosynthesis-antenna proteins.

Author Contributions: Methodology, W.L. and L.X.; formal analysis, J.Z. and W.L.; investigation, Z.Z., W.L. and L.L.; data curation, J.Z., Z.Z., W.L. and L.L.; writing—original draft preparation, J.Z. and W.L.; writing—review and editing, J.Z., L.H., L.X. and Y.Z.; supervision., L.H., L.X. and Y.Z.;

formal analysis, J.Z. and W.L.; investigation, Z.Z., W.L. and L.L.; data curation, J.Z., Z.Z., W.L. and L.L.; writing—original draft preparation, J.Z. and W.L.; writing—review and editing, J.Z., L.H., L.X. and Y.Z.; supervision, L.H., L.X. and Y.Z. All authors have read and agreed to the published version of the manuscript.

Funding: This research was funded by "the Fundamental Research Funds for the Central Universities, grant number 2021ZY83" and "the National Natural Science Foundation of China, grant number 31860151".

Institutional Review Board Statement: Not applicable.

Informed Consent Statement: Not applicable.

Data Availability Statement: The data presented in this study are openly available in NCBI: reference number (PRJNA741873).

Conflicts of Interest: The authors declare no conflict of interest.

References

1. Chen, Y.; Zong, J.; Tan, Z.; Li, L.; Hu, B.; Chen, C.; Chen, J.; Liu, J. Systematic mining of salt-tolerant genes in halophyte-Zoysia matrella through cDNA expression library screening. *Plant. Physiol. Biochem.* **2015**, *89*, 44–52. [CrossRef]
2. Teng, K.; Tan, P.; Xiao, G.; Han, L.; Chang, Z.; Chao, Y. Heterologous expression of a novel *Zoysia japonica* salt-induced glycine-rich RNA-binding protein gene, ZjGRP, caused salt sensitivity in Arabidopsis. *Plant. Cell Rep.* **2017**, *36*, 179–191. [CrossRef]
3. Du, Y.; Yu, L.; Sun, J.; Lu, W.; Han, L. Study on Cold-resistance and Its Mechanism of Different Cultivars of *Zoysia japonica* Steud. *Acta Agrestia Sin.* **2008**, *4*, 347–352. [CrossRef]
4. Teng, K.; Chang, Z.; Li, X.; Sun, X.; Liang, X.; Xu, L.; Chao, Y.; Han, L. Functional and RNA-Sequencing Analysis Revealed Expression of a Novel Stay-Green Gene from Zoysia japonica (ZjSGR) Caused Chlorophyll Degradation and Accelerated Senescence in Arabidopsis. *Front. Plant. Sci.* **2016**, *7*, 1894–1907. [CrossRef]
5. Hu, W.; Wu, Y.; Zeng, J.; He, L.; Zeng, Q. Chill-induced inhibition of photosynthesis was alleviated by 24-epibrassinolide pretreatment in cucumber during chilling and subsequent recovery. *Photosynthetica* **2010**, *48*, 537–544. [CrossRef]
6. Palta, J.P.; Whitaker, B.D.; Weiss, L.S. Plasma membrane lipids associated with genetic variability in freezing tolerance and cold acclimation of solanum species. *Plant. Physiol.* **1993**, *103*, 793–803. [CrossRef]
7. Orvar, B.L.; Sangwan, V.; Omann, F.; Dhindsa, R.S. Early steps in cold sensing by plant cells: The role of actin cytoskeleton and membrane fluidity. *Plant. J.* **2000**, *23*, 785–794. [CrossRef]
8. Nobuhiro, S.; Ron, M. Reactive oxygen species and temperature stresses: A delicate balance between signaling and destruction. *Physiol. Plant.* **2006**, *126*, 45–51. [CrossRef]
9. Hughes, M.A.; Dunn, M.A. The Effect of Temperature on Plant Growth and Development. *Biotechnol. Genet. Eng. Rev.* **2013**, *8*, 161–188. [CrossRef]
10. Mohabbati, F.; Paknejad, F.; Vazan, S.; Habibi, D.; Tookallo, M.R.; Moradi, F. Protective effect of exogenous PGPRs on chlorophyll florescence and membrane integrity of rice seedlings under chilling stress. *Res. J. Appl. Sci. Eng. Technol.* **2013**, *5*, 146–153. [CrossRef]
11. Zhang, X.; Wang, K.; Ervin, E.H.; Waltz, C.; Murphy, T. Metabolic changes during cold acclimation and deacclimation in five bermudagrass varieties. I. Proline, total amino acid, protein, and dehydrin expression. *Crop. Sci.* **2011**, *51*, 847–853. [CrossRef]
12. Ryoji, O.; Seiji, K. Reduction of chlorophyll fluorescence in zoysiagrasses at chilling and high temperatures with moderate light. *Jpn. J. Grassl. Sci.* **1995**, *41*, 31–36. [CrossRef]
13. Ryoji, O.; Seiji, K. Changes in photosynthetic oxygen evolution and chlorophyll fluorescence in some cool season grasses and zoysiagrasses (*Zoysia spp.*) from autumn to winter. *Soil Sci. Plant. Nutr.* **1995**, *41*, 801–806. [CrossRef]
14. Wei, S.; Du, Z.; Gao, F.; Ke, X.; Li, J.; Liu, J.; Zhou, Y. Global Transcriptome Profiles of 'Meyer' Zoysiagrass in Response to Cold Stress. *PLoS ONE* **2015**, *10*, e0131153. [CrossRef]
15. He, J.; Liu, H.; Wang, Y.; Guo, J. Low Temperature and Photosynthesis of Plants. *Plant. Physiol. Commun.* **1986**, *2*, 1–6. [CrossRef]
16. Ma, D.; Pang, J.; Huo, Z. Research Progress on low temperature tolerance of Cucumber. *Tianjin Agric. Sci.* **1997**, *4*, 3–10.
17. Nagata, N.; Tanaka, R.; Tanaka, A. The major route for chlorophyll synthesis includes [3,8-divinyl]-chlorophyllide a reduction in *Arabidopsis thaliana*. *Plant. Cell Physiol.* **2007**, *48*, 1803–1808. [CrossRef]
18. Zhao, Y.; Han, Q.; Ding, C.; Huang, Y.; Liao, J.; Chen, T.; Feng, S.; Zhou, L.; Zhang, Z.; Chen, Y.; et al. Effect of Low Temperature on Chlorophyll Biosynthesis and Chloroplast Biogenesis of Rice Seedlings during Greening. *Int. J. Mol. Sci.* **2020**, *21*, 1390. [CrossRef]
19. Pruzinská, A.; Tanner, G.; Anders, I.; Roca, M.; Hörtensteiner, S. Chlorophyll breakdown: Pheophorbide a oxygenase is a Rieske-type iron-sulfur protein, encoded by the accelerated cell death 1 gene. *Proc. Natl. Acad. Sci. USA* **2003**, *100*, 15259–15264. [CrossRef]
20. Tanimoto, M.; Roberts, K.; Dolan, L. Ethylene is a positive regulator of root hair development in emphArabidopsis thaliana. *Plant. J.* **1995**, *8*, 943–948. [CrossRef]
21. Lynch, J.; Brown, K.M. Ethylene and plant responses to nutritional stress. *Physiol. Plant* **1997**, *100*, 613–619. [CrossRef]

22. Sharp, R.E.; LeNoble, M.E. ABA, ethylene and the control of shoot and root growth under water stress. *J. Exp. Bot.* **2002**, *53*, 33–37. [CrossRef]
23. Maruthasalam, S.; Ling, Y.S.; Murugan, L.; Wei, C.L.; Yi, L.L.; Ching, M.S.; Chih, W.Y.; Shu, H.H.; Yeong, K.; Chin, H.L. Forced flowering of pineapple (*Ananas comosus* cv. Tainon 17) in response to cold stress, ethephon and calcium carbide with or without activated charcoal. *Plant Growth Regul.* **2009**, *60*, 83–90. [CrossRef]
24. Han, L.; Han, L.; Xu, L. Effects of Ethephon Treatment on Plant Drought Tolerance. *Acta Agrestia Sin.* **2013**, *21*, 631–636. [CrossRef]
25. Wang, W.; Wang, Z.; Ping, J.; Zhang, Y. Effect of Ethephon on Several Cold Hardiness Indexes of Carbernet Sauvignon. *Sino-Overseas Grapevine Wine* **2005**, *5*, 14–15. [CrossRef]
26. Niu, H. Effects of DPCA, Paclobutrazol and ethephon on resistance of Larch Seedlings. *Yunnan For. Sci. Technol.* **1998**, *3*, 3–5. [CrossRef]
27. Wei, D.; Li, Y.; Di, N.; Bi, Z.; Bu, Z. Effect of Ethephon on Cold Resistance of Banana (Musa AAA Cavendish subgroup) Seedlings. *Chin. J. Trop. Crops* **2009**, *30*, 1789–1792.
28. Sone, C.; Sakagami, J.I. Physiological mechanism of chlorophyll breakdown for leaves under complete submergence in rice. *Crop. Sci.* **2017**, *57*, 2729–2738. [CrossRef]
29. Liu, W. *Application of Ethephon Impact Cold Tolerance of Zoysiagrass*; Beijing Forestry University: Beijing, China, 2019.
30. Zhang, Z. *Effects of Ethephon on Chlorophyll Metabolic Pathway of Zoysia Japonica under Low Temperature Stress*; Beijing Forestry University: Beijing, China, 2020.
31. Wang, Z.; Gerstein, M.; Snyder, M. RNA-Seq: A revolutionary tool for transcriptomics. *Nat. Rev. Genet.* **2009**, *10*, 57–63. [CrossRef]
32. Wilhelm, B.T.; Landry, J.R. RNA-Seq—Quantitative measurement of expression through massively parallel RNA-sequencing. *Methods* **2009**, *48*, 249–257. [CrossRef]
33. Hu, X.; Makita, S.; Schelbert, S.; Sano, S.; Ochiai, M.; Tsuchiya, T.; Hasegawa, S.F.; Hörtensteiner, S.; Tanaka, A.; Tanaka, R. Reexamination of chlorophyllase function implies its involvement in defense against chewing herbivores. *Plant. Physiol.* **2015**, *167*, 660–670. [CrossRef] [PubMed]
34. Bushman, B.S.; Amundsen, K.L.; Warnke, S.E.; Robins, J.G.; Johnson, P.G. Transcriptome profiling of Kentucky bluegrass (*Poa pratensis* L.) accessions in response to salt stress. *BMC Genom.* **2016**, *17*, 48–59. [CrossRef] [PubMed]
35. Ma, Y.; Shukla, V.; Merewitz, E.B. Transcriptome analysis of creeping bentgrass exposed to drought stress and polyamine treatment. *PLoS ONE* **2017**, *12*, e175848. [CrossRef]
36. Wang, J.; An, C.; Guo, H.; Yang, X.; Chen, J.; Zong, J.; Li, J.; Liu, J. Physiological and transcriptomic analyses reveal the mechanisms underlying the salt tolerance of Zoysia japonica Steud. *BMC Plant. Bio.* **2020**, *20*, 114. [CrossRef]
37. Voitsekhovskaja, O.V.; Tyutereva, E.V. Chlorophyll b in angiosperms: Functions in photosynthesis, signaling and ontogenetic regulation. *J. Plant. Physiol.* **2015**, *189*, 51–64. [CrossRef]
38. Wang, G.; Chen, Z.; Zhang, Y.; Gong, G.; Wen, P. Effects of three plant growth substances on Chilling Resistance of potato. *J. Huizhou Univ.* **2018**, *38*, 21–28. [CrossRef]
39. Zhang, J.; Zhu, W. Effects of chilling stress on contents of chlorophyll and malondialdehyde in tomato seedlings. *Acta Agric. Shanghai* **2012**, *28*, 74–77.
40. Stenbaek, A.; Jensen, P.E. Redox Regulation of Chlorophyll Biosynthesis. *Phytochemistry* **2010**, *71*, 853–859. [CrossRef]
41. Pontier, D.; Albrieux, C.; Joyard, J.; Lagrange, T.; Block, M.A. Knock-out of the Magnesium Protoporphyrin IX Methyltransferase Gene in Arabidopsis: Effects on chloroplast development and on chloroplast-to-nucleus signaling. *J. Biol. Chem.* **2007**, *282*, 2297–2304. [CrossRef]
42. Shioi, Y.; Takamiya, K. Monovinyl and divinyl protochlorophyllide pools in etiolated tissues of higher plants. *Plant Physiol.* **1992**, *100*, 1291–1295. [CrossRef]
43. Reinbothe, S.; Reinbothe, C.; Apel, K.; Lebedev, N. Evolution of Chlorophyll Biosynthesis -The Challenge to Survive Photooxidation. *Cell* **1996**, *86*, 703–705. [CrossRef]
44. Tanaka, A.; Ito, H.; Tanaka, R.; Tanaka, N.K.; Yoshida, K.; Okada, K. Chlorophyll a oxygenase (CAO) is involved in chlorophyll b formation from chlorophyll a. *Proc. Natl. Acad. Sci. USA* **1998**, *95*, 12719–12723. [CrossRef]
45. Dinc, E.; Ceppi, M.G.; Toth, S.Z.; Bottka, S.; Schansker, G. The chla fluorescence intensity is remarkably insensitive to changes in the chlorophyll content of the leaf as long as the chla/b ratio remains unaffected. *Biochim. Biophys. Acta* **2012**, *1817*, 770–779. [CrossRef] [PubMed]
46. Hu, L.; Li, H.; Chen, L.; Lou, Y.; Amombo, E.; Fu, J. RNA-seq for gene identification and transcript profiling in relation to root growth of bermudagrass (*Cynodon dactylon*) under salinity stress. *BMC Genom.* **2015**, *16*, 575–586. [CrossRef]
47. Tang, Y.; Li, M.; Chen, Y.; Wu, P.; Wu, G.; Jiang, H. Knockdown of OsPAO and OsRCCR1 cause different plant death phenotypes in rice. *J. Plant. Physiol.* **2011**, *168*, 1952–1959. [CrossRef] [PubMed]
48. Liu, F.; Guo, F.Q. Nitric oxide deficiency accelerates chlorophyll breakdown and stability loss of thylakoid membranes during dark-induced leaf senescence in Arabidopsis. *PLoS ONE* **2013**, *8*, e56345. [CrossRef] [PubMed]
49. Park, S.Y.; Yu, J.W.; Park, J.S.; Li, J.; Yoo, S.C.; Lee, N.Y.; Lee, S.K.; Jeong, S.W.; Seo, H.S.; Koh, H.J.; et al. The senescence-induced staygreen protein regulates chlorophyll degradation. *Plant Cell* **2007**, *19*, 1649–1664. [CrossRef]
50. Armstead, I.; Donnison, I.; Aubry, S.; Harper, J.; Hörtensteiner, S.; James, C.; Mani, J.; Moffet, M.; Ougham, H.; Roberts, L.; et al. Cross-species identification of Mendel's I locus. *Science* **2007**, *315*, 73. [CrossRef]

51. María, E.G.; Sofía, A.M.; Pedro, M.C.; Gustavo, A.M. Expression of Stay-Green encoding gene (BoSGR) during postharvest senescence of broccoli. *Postharvest Biol. Technol.* **2014**, *95*, 88–94. [CrossRef]

52. Caudle, K.L.; Johnson, L.C.; Baer, S.G.; Maricle, B.R. A comparison of seasonal foliar chlorophyll change among ecotypes and cultivars of Andropogon gerardii (Poaceae) by using nondestructive and destructive methods. *Photosynthetica* **2014**, *52*, 511–518. [CrossRef]

53. Damkjær, J.T.; Kereïche, S.; Johnson, M.P.; Kovacs, L.; Kiss, A.Z.; Boekema, E.J.; Ruban, A.V.; Horton, P.; Jansson, S. The photosystem II Light-Harvesting protein lhcb3 affects the macrostructure of photosystem II and the rate of state transitions in arabidopsis. *Plant Cell* **2009**, *21*, 3245–3256. [CrossRef] [PubMed]

54. Jansson, S. A guide to the Lhc genes and their relatives in Arabidopsis. *Trends Plant. Sci.* **1999**, *4*, 236–240. [CrossRef]

55. Klimmek, F.; Sjödin, A.; Noutsos, C.; Leister, D.; Jansson, S. Abundantly and rarely expressed Lhc protein genes exhibit distinct regulation patterns in plants. *Plant. Physiol.* **2006**, *140*, 793–804. [CrossRef]

56. Goussias, C.; Boussac, A.; Rutherford, A.W. Photosystem II and photosynthetic oxidation of water: An overview. *Philos. Trans. R. Soc. B: Biol. Sci.* **2002**, *357*, 1369–1381. [CrossRef] [PubMed]

57. Kawakami, K.; Shen, J.R. Purification of fully active and crystallizable photosystem II from thermophilic cyanobacteria. *Methods Enzym.* **2018**, *613*, 1–16. [CrossRef]

58. Barber, J.; Andersson, B. Too much of a good thing: Light can be bad for photosynthesis. *Trends Biochem Sci.* **1992**, *17*, 61–66. [CrossRef]

59. Chan, T.; Shimizu, Y.; Pospíšil, P.; Nijo, N.; Fujiwara, A.; Taninaka, Y.; Ishikawa, T.; Hori, H.; Nanba, D.; Imai, A.; et al. Quality control of photosystem II: Lipid peroxidation accelerates photoinhibition under excessive illumination. *PLoS ONE* **2012**, *7*, e52100. [CrossRef]

60. Dekker, J.P.; Boekema, E.J. Supramolecular organization of thylakoid membrane proteins in green plants. *Biochim. Biophys. Acta* **2005**, *1706*, 12–39. [CrossRef]

61. Drop, B.; Webber-Birungi, M.; Yadav, S.K.; Filipowicz-Szymanska, A.; Fusetti, F.; Boekema, E.J.; Croce, R. Light-harvesting complex II (LHCII) and its supramolecular organization in *Chlamydomonas reinhardtii*. *Biochim. Biophys. Acta* **2014**, *1837*, 63–72. [CrossRef]

62. Croce, R.; Canino, G.; Ros, F.; Bassi, R. Chromophore organization in the higher-plant photosystem II antenna protein CP26. *Biochem* **2002**, *41*, 7334–7343. [CrossRef]

63. Passarini, F.; Wientjes, E.; Hienerwadel, R.; Croce, R. Molecular basis of light harvesting and photoprotection in CP24: Unique features of the most recent antenna complex. *J. Biol. Chem.* **2009**, *284*, 29536–29546. [CrossRef] [PubMed]

64. Pan, X.; Li, M.; Wan, T.; Wang, L.; Jia, C.; Hou, Z.; Zhao, X.; Zhang, J.; Chang, W. Structural insights into energy regulation of light-harvesting complex CP29 from spinach. *Nat. Struct. Mol. Biol.* **2011**, *18*, 309–315. [CrossRef] [PubMed]

65. Peng, X.; Deng, X.; Tang, X.; Tan, T.; Zhang, D.; Liu, B.; Lin, H. Involvement of Lhcb6 and Lhcb5 in Photosynthesis Regulation in Physcomitrella patens Response to Abiotic Stress. *Int. J. Mol. Sci.* **2019**, *20*, 3665. [CrossRef] [PubMed]

66. Peers, G.; Niyogi, K.K. Pond scum genomics: The genomes of chlamydomonas and ostreococcus. *Plant Cell* **2008**, *20*, 502–507. [CrossRef]

67. Nelson, N. Plant photosystem I-the most efficient Nano-Photochemical machine. *J. Nanosci. Nanotechnol.* **2009**, *9*, 1709–1713. [CrossRef]

68. Croce, R.; van Amerongen, H. Light-harvesting in photosystem I. *Photosynth Res.* **2013**, *116*, 153–166. [CrossRef]

69. Wientjes, E.; van Stokkum, I.H.; van Amerongen, H.; Croce, R. The role of the individual Lhcas in photosystem I excitation energy trapping. *Biophys. J.* **2011**, *101*, 745–754. [CrossRef]

70. Bellafiore, S.; Barneche, F.; Peltier, G.; Rochaix, J. State transitions and light adaptation require chloroplast thylakoid protein kinase STN7. *Nature* **2005**, *433*, 892–895. [CrossRef]

71. Allen, J.F. BOTANY: State transitions–a question of balance. *Science* **2003**, *299*, 1530–1532. [CrossRef]

72. Kanervo, E.; Suorsa, M.; Aro, E.M. Functional flexibility and acclimation of the thylakoid membrane. *Photochem. Photobiol. Sci.* **2005**, *4*, 1072–1080. [CrossRef]

73. Haldrup, A.; Jensen, P.E.; Lunde, C.; Scheller, H.V. Balance of power: A view of the mechanism of photosynthetic state transitions. *Trends Plant. Sci.* **2001**, *6*, 301–305. [CrossRef]

74. Nawrocki, W.J.; Santabarbara, S.; Mosebach, L.; Wollman, F.A.; Rappaport, F. State transitions redistribute rather than dissipate energy between the two photosystems in Chlamydomonas. *Nature Plants.* **2016**, *2*, 16031. [CrossRef] [PubMed]

75. Allen, J.F. Protein phosphorylation in regulation of photosynthesis. *Biochim. Biophys. Acta* **1992**, *1098*, 275–335. [CrossRef]

76. Sartory, D.P.; Grobbelaar, J.U. Extraction of chlorophyll a from freshwater phytoplankton for spectrophotometric analysis. *Hydrobiologia* **1984**, *114*, 177–187. [CrossRef]

77. Chi, Y.; Ren, T.; Shi, X.; Jin, X.; Jin, P. Mechanism of nutrient removal enhancement in low carbon/nitrogen wastewater by a novel high-frequency micro-aeration/anoxic (HMOA) mode. *Chemosphere* **2021**, *263*, 128003. [CrossRef] [PubMed]

78. Li, B.; Dewey, C.N. RSEM: Accurate transcript quantification from RNA-Seq data with or without a reference genome. *BMC Bioinform.* **2011**, *12*, 323–338. [CrossRef]

79. Grabherr, M.G.; Haas, B.J.; Yassour, M.; Levin, J.Z.; Thompson, D.A.; Amit, I.; Adiconis, X.; Fan, L.; Raychowdhury, R.; Zeng, Q.; et al. Full-length transcriptome assembly from RNA-Seq data without a reference genome. *Nat. Biotechnol.* **2011**, *29*, 644–652. [CrossRef]

80. Kim, D.; Pertea, G.; Trapnell, C.; Pimentel, H.; Kelley, R.; Salzberg, S.L. TopHat2: Accurate alignment of transcriptomes in the presence of insertions, deletions and gene fusions. *Genome Biol.* **2013**, *14*, R36. [CrossRef]
81. Mortazavi, A.; Williams, B.A.; McCue, K.; Schaeffer, L.; Wold, B. Mapping and quantifying mammalian transcriptomes by RNA-Seq. *Nat. Methods* **2008**, *5*, 621–628. [CrossRef]
82. Love, M.I.; Huber, W.; Anders, S. Moderated estimation of fold change and dispersion for RNA-seq data with DESeq2. *Genome Biol.* **2014**, *15*, 550–570. [CrossRef]
83. Hadadi, N.; MohammadiPeyhani, H.; Miskovic, L.; Seijo, M.; Hatzimanikatis, V. Enzyme annotation for orphan and novel reactions using knowledge of substrate reactive sites. *Proc. Natl. Acad. Sci. USA* **2019**, *116*, 7298–7307. [CrossRef]

Article

Turf Quality and Physiological Responses to Summer Stress in Four Creeping Bentgrass Cultivars in a Subtropical Zone

Zhou Li [1,2,†], Weihang Zeng [1,3,†], Bizhen Cheng [1], Jie Xu [3], Liebao Han [2,*] and Yan Peng [1,*]

1   Department of Turf Science and Engineering, College of Grassland Science and Technology, Sichuan Agricultural University, Chengdu 611130, China; lizhou1986814@163.com (Z.L.); zengwh0123@163.com (W.Z.); chengbizhengrass@163.com (B.C.)
2   Institute of Turfgrass Science, Beijing Forestry University, Beijing 100083, China
3   Chengdu Times Green Spaces Horticulture Co., Ltd., Chengdu 610300, China; xujie8231881@163.com
*   Correspondence: hanliebao@163.com (L.H.); pengyanlee@163.com (Y.P.)
†   These authors contributed equally to this work.

Abstract: Cool-season creeping bentgrass (*Agrostis stolonifera*) has the ability to form fine sports turf, but high temperatures result in summer bentgrass decline (SBD), especially in transitional and subtropical zones. Physiological responses in combination with the alteration in turf quality (TQ) will contribute to a better understanding of SBD in a subtropical zone. Field experiments were conducted from 2017 to 2019 to test the adaptability to summer stress among four cultivars (13M, Penncross, Seaside II, and PA-1). A constant ambient high temperature above 30 °C significantly decreased the TQ of the four cultivars during the summer months in 2017, 2018, and 2019. Significant declines in the chlorophyll content, photochemical efficiency of photosystem II (Fv/Fm and PIABS), leaf relative water content (RWC), and osmotic potential (OP) were induced by summer stress, whereas gradual increases in water-soluble carbohydrates, proline, hydrogen peroxide (H_2O_2), malondialdehyde (MDA), and electrolyte leakage (EL) were observed in the four cultivars during the summer months. The 13M and Penncross cultivars exhibited better performance than Seaside II and PA-1 in response to summer stress from 2017 to 2019, which is associated with better maintenance of photosynthesis, water status, WSC and proline accumulation, and cell membrane stability. The 13M and Penncross cultivars could be used as potential candidates for turf establishment in a subtropical zone. Physiological responses together with alterations in TQ also provided critical information for the breeding and development of germplasm with heat tolerance in creeping bentgrass species.

Keywords: high temperature; oxidative damage; photosynthesis; osmotic adjustment; photochemical efficiency

Citation: Li, Z.; Zeng, W.; Cheng, B.; Xu, J.; Han, L.; Peng, Y. Turf Quality and Physiological Responses to Summer Stress in Four Creeping Bentgrass Cultivars in a Subtropical Zone. *Plants* **2022**, *11*, 665. https://doi.org/10.3390/plants11050665

Academic Editor: Josefa M. Alamillo

Received: 16 January 2022
Accepted: 23 February 2022
Published: 28 February 2022

Publisher's Note: MDPI stays neutral with regard to jurisdictional claims in published maps and institutional affiliations.

1. Introduction

Cool-season turfgrasses such as creeping bentgrass (*Agrostis stolonifera*), perennial ryegrass (*Lolium perenne*), and Kentucky bluegrass (*Poa pratensis*) have an optimal range of growth temperature from 15 to 24 °C and are highly susceptible to sustained high temperatures in the summer [1]. Creeping bentgrass, characterized by rapid thatch accumulation, fine texture, and low mowing height, is widely used in sports turfs such as golf green and tennis lawns. In transitional and subtropical zones, the high temperature during the summer months is a critical stress factor resulting in summer bentgrass decline (SBD) [2,3]. Maintenance practices including water and fertilizer management, the alteration of mowing height, and the application of plant growth regulators have been utilized to alleviate SBD [3–6]. In addition to managerial actions, the identification and selection of creeping bentgrass cultivars that adapt to one particular climate type are of primary importance to reduce maintenance cost because environmental adaptability to heat stress varies with creeping bentgrass cultivars or genotypes [7,8]. Previous studies have identified some heat-tolerant creeping bentgrass cultivars such as 'L-93' in controlled growth chambers or under field conditions [2,9]. The field performance of different creeping bentgrass cultivars, especially for those newly developed or promoted cultivars, should be further assessed during the summer months in different climatic regions.

Various metabolic processes in plants are interrupted by heat stress. For example, high temperatures limit carbohydrate production and supply for growth maintenance in cool-season turfgrass, mainly due to accelerated respiration and reduced photosynthesis [10]. Carbohydrates and other osmolytes such as proline (Pro) exhibit a variety of roles in osmotic adjustment, antioxidant, and metabolic homeostasis [11]. It has been suggested that carbohydrate production is beneficial in alleviating SBD [2]. Proline accumulation and metabolism are also associated with better adaptation to enhanced heat tolerance in creeping bentgrass [12–14]. In addition, the heat-induced decline in the turf quality (TQ) of creeping bentgrass is related to accelerated leaf senescence, as demonstrated by the chlorosis of turf [15]. One of the core factors responsible for heat-induced leaf senescence in bentgrass species is reactive oxygen species (ROS) overaccumulation leading to membrane peroxidation [16–19]. However, most of the controlled studies in growth chambers limit a deep understanding of SBD in relation to the multiple physiological responses of different creeping bentgrass genotypes in a subtropical zone.

As one of the most commonly used cool-season turfgrasses in golf greens, the heat-induced decline in the TQ of creeping bentgrass is the most intractable problem to turf managers during hot summers. The objectives of this study were to evaluate the summer performance of four different creeping bentgrass cultivars ('Penncross', '13M', 'Seaside II', and 'PA-1') and to further examine common or different physiological responses, including leaf water relation, photosynthetic performance, osmolytes accumulation, and oxidative damage associated with heat tolerance among these cultivars during the summer months under field conditions. Physiological responses in combination with the alteration in TQ will contribute to a better understanding of SBD in a subtropical zone.

2. Materials and Methods

2.1. Plant Materials and Treatments

The experiment was conducted in the Research Farm of Sichuan Agricultural University which is located in southwest China (Chongzhou, Sichuan, east longitude 103°07′–103°49′ and north latitude 30°30′–30°53′). The area has a typical subtropical-monsoon climate with an annual mean temperature between 16 °C and 17 °C. Reclaimed soils (loams: sands, 2:1) were used as the plant layer with the supply of 2.0 g m^{-2} fertilizers (nitrogen: phosphorus: potassium, 3:1:1). The seeds of four different creeping bentgrass cultivars (Penncross, 13M, Seaside II, or PA-1) were sown evenly in each 2 m × 2 m plot in October 2016. The seeding rate was 10 g m^{-2} for each cultivar. After nearly 8 months of establishment from October 2016 to May 2017, the turfgrass coverage of all cultivars reached nearly 100%, and mowing height was maintained at 2 cm. The summer tolerance of the four cultivars was evaluated from May to August 2017, June to September 2018, and June to September 2019. During summer stress, all turfs were irrigated daily with city water to avoid drought stress. The irrigation time changed within the year of the experiment in relation to seasonal temperatures and soil conditions. The maximum, minimum, and average daily air temperatures are demonstrated in Figure 1A–C. The total number of days where the maximum air temperatures were above 30 °C were 64, 59, and 45 days in the summers of 2017, 2018, and 2019, respectively (Figure 1A–C), which indicated that all materials suffered from a long period of high temperature stress in the summers of 2017, 2018, and 2019. Dead spots in each turf plot were replaced by new sods in the spring to make all turfs uniform before the summer stress of the next year.

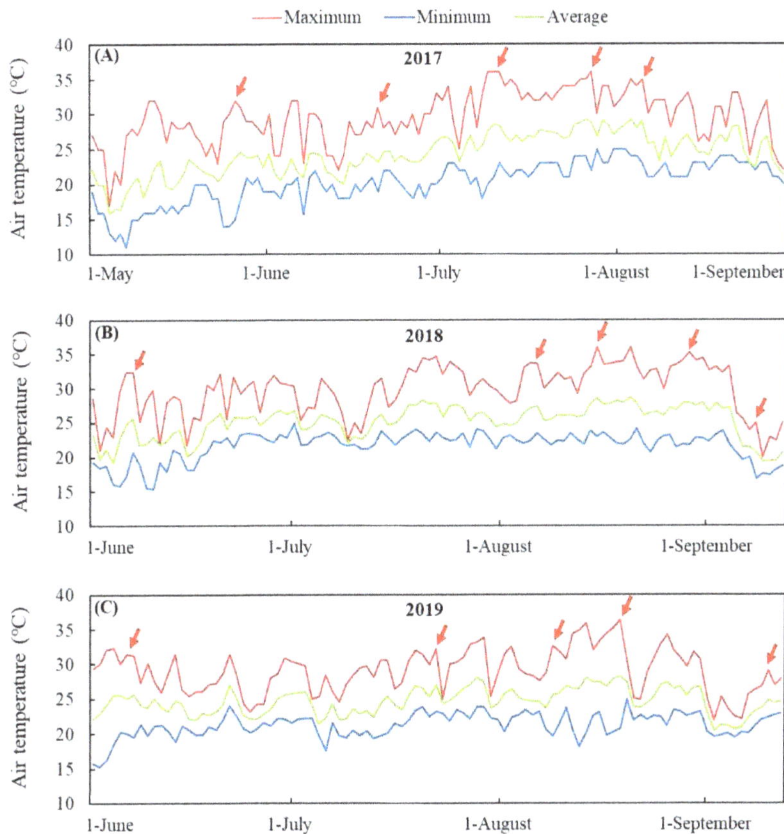

Figure 1. Daily maximum, minimum, and average air temperature in (**A**) 2017, (**B**) 2018, and (**C**) 2019 in the Research Farm of Sichuan Agricultural University (Chongzhou, Sichuan, China, east longitude 103°07′–103°49′ and north latitude 30°30′–30°53′). Red arrows indicate sampling dates.

2.2. Measurements of Turf Quality and Photosynthetic Parameters

TQ was evaluated based on presentation quality, which included uniformity, color, and density, and was rated a 1, 6, or 9 if the presentation quality of the turf was the worst (brown and desiccated), minimally acceptable (pale green and a 10% decline in denseness), or excellent (green and dense), respectively [20]. For the Chl content, fresh leaves (0.2 g) were collected from each turf and soaked in a 20 mL of 80% acetone and 95% ethanol (1:1, *v:v*) solution for 48 h in the dark. The absorbance of the extraction liquid was detected at 645 and 663 nm with a spectrophotometer (Spectronic Instruments, Rochester, NY, USA) [21]. A portable Chl fluorescence system (Pocket PEA, Hansatech, Norfolk, UK) was used to determine the photochemical efficiency (Fv/Fm) and the performance index on absorption basis (PIABS). In brief, a single layer of leaves was covered by leaf clips for 30 min for dark adaptation, and then, the Fv/Fm and PIABS data were recorded. The net photosynthetic rate (Pn), transpiration rate, and instantaneous water use efficiency (WUE) were detected by using a portable photosynthetic system (CIRAS-3, PP Systems, Amesbury, MA, USA) that provided stable and continuous light and carbon dioxide conditions (400 μL/L CO_2 and 800 μmol photon m^{-2} red and blue lights).

2.3. Measurements of Water Status and Osmolyte Contents

The water status in leaves was evaluated with the leaf relative water content (RWC) and the osmotic potential (OP). Fresh leaves were collected from the turf and were weighed immediately to obtain the fresh weight (FW), and then, these leaves were soaked in distilled water for 12 h to weigh turgid weight (TW). Dry weight (DW) was obtained after leaves were dried in an oven at 105 °C for 2 h and then at 80 °C for 72 h. The leaf RWC (%) was calculated based on the formula (RWC (%) = [(FW − DW)/(TW − DW)] × 100) [22]. Osmotic potential (OP) was determined by using an osmometer 4420 (Wescor, Inc., Logan, UT, USA), and the assay method has been recorded in detail in the studies of [23,24]. For the analysis of the proline content, fresh leaves (0.2 g) were extracted in boiling water for 20 min in a 10 mL mixture containing 3% sulfosalicylic acid, and then, 2 mL of glacial acetic acid and 3 mL of 2.5% ninhydrine were added. The reaction mixture was boiled for 30 min. The 5 mL of toluene was added and shaken up. The absorbance of upper toluene was recorded at 520 nm using a Genesys 2PC spectrophotometer (Spectronic Instruments, Rochester, NY, USA) [25]. Water-soluble carbohydrate (WSC) levels were determined according to the method of [26] with some modification, which has been clearly demonstrated in our previous study [27].

2.4. Measurements of Oxidative Damage and Membrane Stability

The hydrogen peroxide (H_2O_2) content was detected by using 0.2 g of fresh leaves which were homogenized in 10 mL of 0.1% TCA. After being centrifuged at $10,000 \times g$ for 15 min, 0.5 mL of the supernatant was mixed with 0.5 mL of 10 mM potassium phosphate and 1 mL of 1 M potassium iodide and then placed in the dark for 5 min. The absorbance of the reaction mixture was read at 390 nm [28]. For the analysis of malondialdehyde (MDA) content, 3 mL of 50 mM cold phosphate buffer (pH 7.8) was used to extract 0.2 g of fresh leaves. The supernatant was obtained after the homogenate was centrifuged at $10,000 \times g$ for 30 min at 4 °C, and then, 0.5 mL of the supernatant was mixed with 1 mL of the reaction solution (20% w/v trichloroacetic acid and 0.5% w/v thiobarbituric acid). After being heated in a boiling water bath for 15 min, the reaction mixture was centrifuged at $8000 \times g$ for 10 min. The absorbance of 1.5 mL of supernatant was measured at 532 and 600 nm [29]. Leaf electrolyte leakage (EL) was calculated based on the formula (%) = $C_{initial}/C_{max} \times 100$, where $C_{initial}$ indicated the initial conductivity and C_{max} presented the max conductivity. Fresh leaves (0.2 g) were soaked instantly in 50 mL of distilled water for 24 h to detect the $C_{initial}$. These samples were autoclaved at 120 °C for 20 min and cooled down to room temperature to detect the C_{max} using a conductivity meter (YSI Model 32, Yellow Spring, OH, USA) [30].

2.5. Experimental Design and Statistical Analysis

The experimental design was a randomized blocks design, and each treatment (each cultivar) was replicated four times (four test plots) in the field. All measurements were sampled in each plot including five subsamples, and the average value of the five subsamples was regarded as the effective value in each plot. Variations among the four cultivars in response to summer stress were analyzed by the general linear model procedure of Statistical Product and Service Solutions 24 (SPSS Institute, IBM, Armonk, NY, USA, 2018). Differences among treatments (cultivars) were determined by using the least significant difference (LSD) at $p \leq 0.05$.

3. Results

3.1. Turf Quality and Photosynthetic Parameters Affected by Summer Stress

The TQ of the four cultivars declined gradually during the summer months in 2017, 2018, and 2019, but Penncross and 13M showed higher TQ than Seaside II and PA-1 in response to summer stress (Figure 2A–C). On 7 August 2017, the highest TQ was observed in the Penncross cultivar (Figure 2A); however, the 13M cultivar exhibited the highest TQ out of the four cultivars on 8 September 2018 and 2019 (Figure 2B,C). The Chl content in all cultivars decreased with the development of summer stress from June to September in 2017, 2018, and 2019 (Figure 3A–C). The Penncross, 13M, and PA-1 cultivars had a significantly higher TQ than Seaside II on July 26th, whereas no significant difference in the Chl content among the four cultivars was detected on other sampling dates during the summer in 2017 (Figure 3A). On 8 September 2018 and 2019, significantly lower Chl content was observed in PA-1 and Seaside II as compared to that in Penncross and 13M (Figure 3B,C).

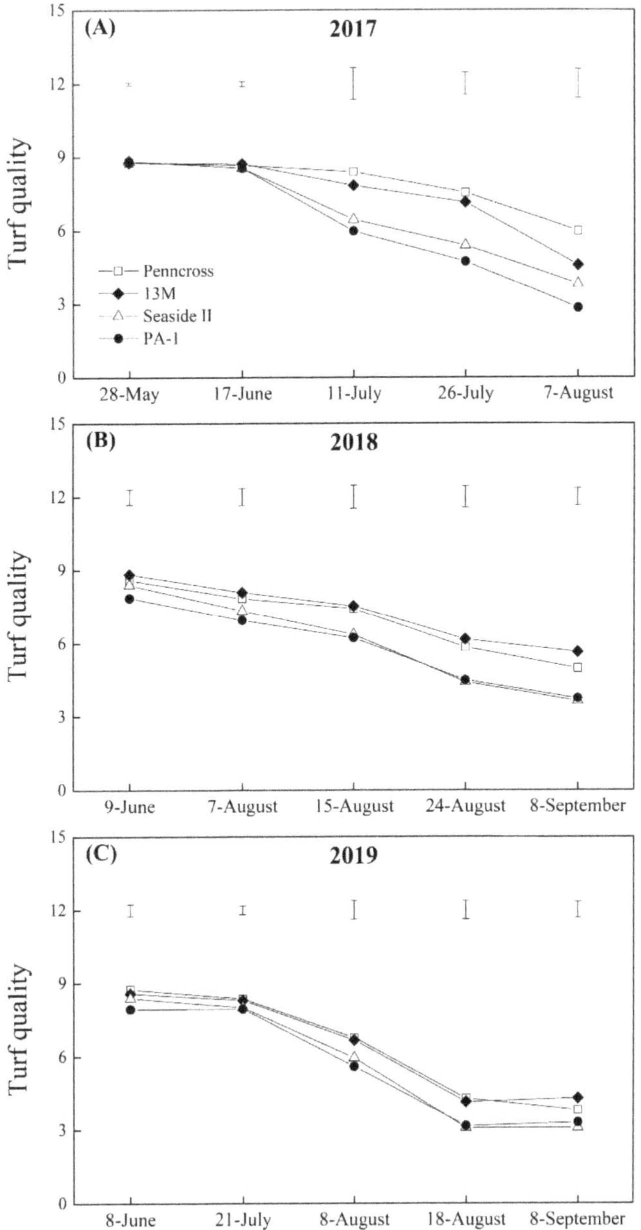

Figure 2. Change in turf quality (TQ) of four different creeping bentgrass cultivars (Penncross, 13M, Seaside II, and PA-1) during summer in (**A**) 2017, (**B**) 2018, and (**C**) 2019. Vertical bars above curves indicate least significant difference (LSD) values ($p < 0.05$) at a given day.

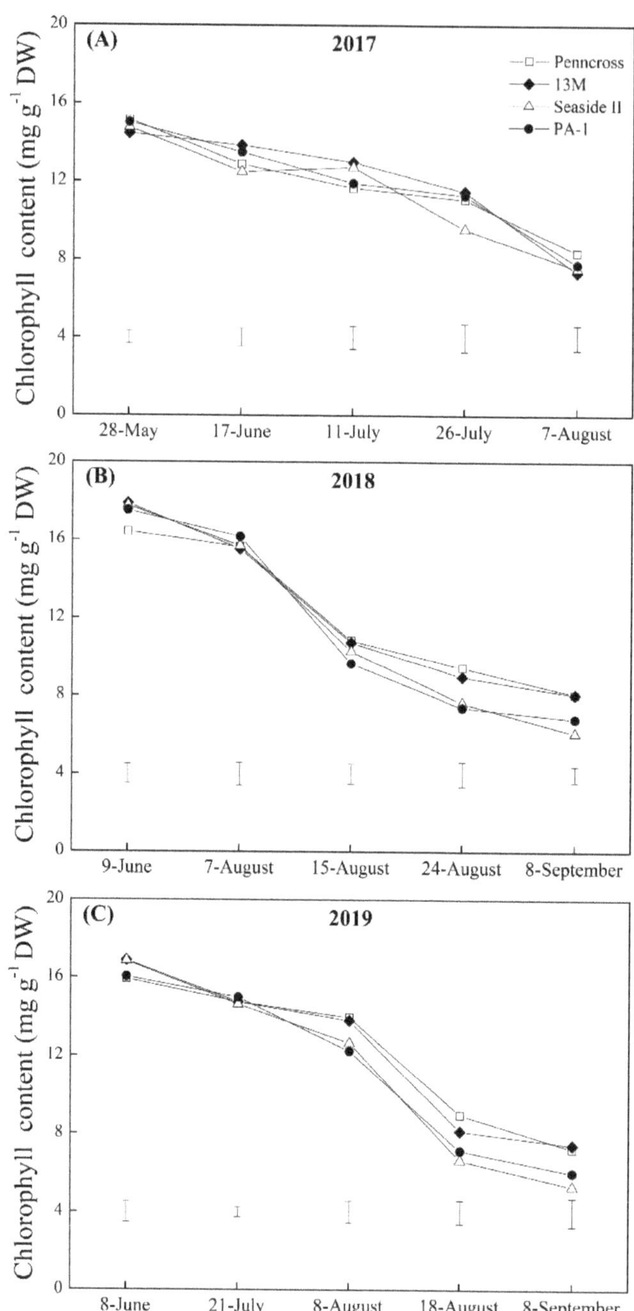

Figure 3. Change in chlorophyll (Chl) content of four different creeping bentgrass cultivars (Penncross, 13M, Seaside II, and PA-1) during summer in (**A**) 2017, (**B**) 2018, and (**C**) 2019. Vertical bars below curves indicate least significant difference (LSD) values ($p < 0.05$) at a given day.

Summer stress induced a gradual decline in the Fv/Fm of the four cultivars in 2017 (Figure 4A). The decline in the Fv/Fm was at its maximum in PA-1 on 11 July 2017 (Figure 4A). Penncross and 13M had significantly higher Fv/Fm than Seaside II and

PA-1 on 6 August 2017 (Figure 4A). No cultivar differences in the Fv/Fm were detected on the first two sampling dates in 2018 (9 June and 15 August) and 2019 (8 June and 8 August) (Figure 4B,C). The effects of summer stress on the PIABS were more pronounced than the Fv/Fm among the four cultivars in 2017, 2018, and 2019 (Figure 4D–F). Penncross and 13M maintained a significantly higher PIABS than Seaside II and PA-1 during the summer; however, there was no significant difference in the PIABS between Penncross and 13M or between Seaside II and PA-1 (Figure 4D–F). The Tr of the four cultivars did not show significant differences in 2017 and 2018, but Penncross and 13M maintained significantly higher Tr as compared to Seaside II and PA-1 in 2019 (Figure 5A). In 2017, Penncross and 13M exhibited a 28% increase in Pn over Seaside II and PA-1 in the summer (Figure 5B). In the summer of 2018, 13M showed the highest Pn of the four cultivars (Figure 5B). The 13M cultivar also had a 17%, 69%, and 31% significantly higher Pn than Penncross, Seaside II, or PA-1 in the summer of 2019, respectively (Figure 5B). A significantly higher WUE was detected in Penncross and 13M than in Seaside II and PA-1 in the summer of 2017 and 2019 (Figure 5C). In 2018, 13M maintained 12%, 21%, and 20% higher WUE than Penncross, Seaside II, and PA-1 in the summer, respectively (Figure 5C).

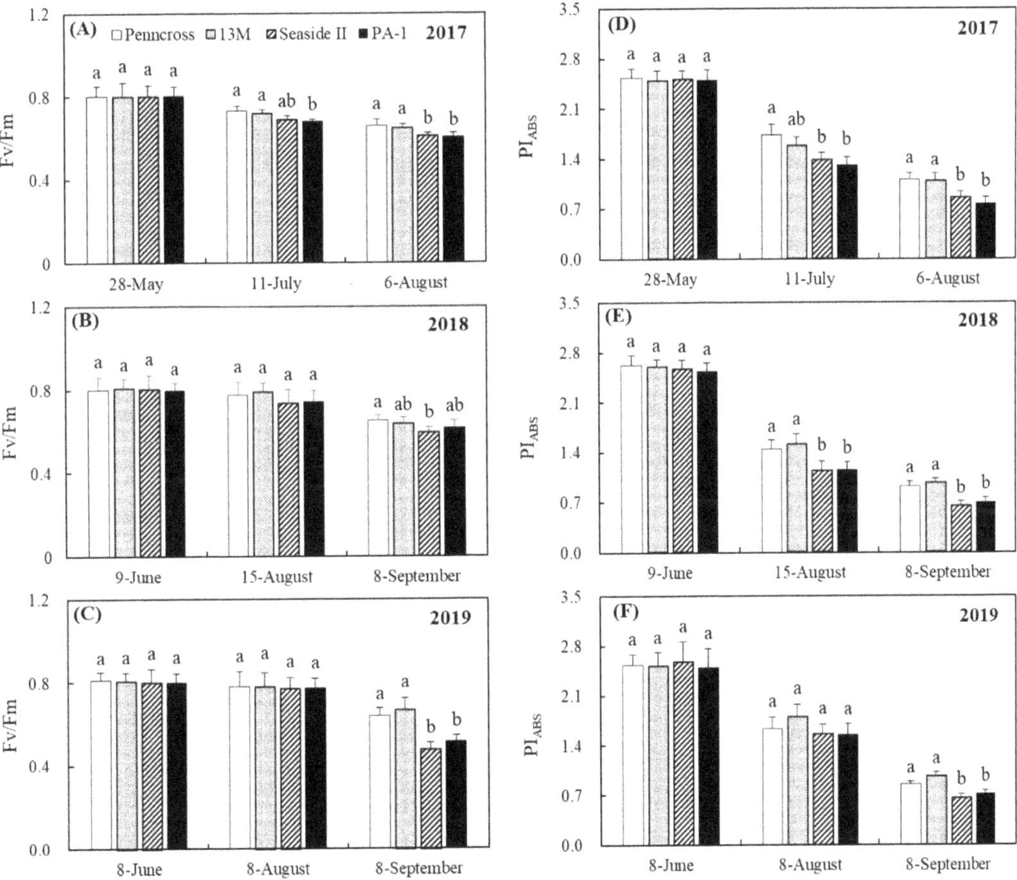

Figure 4. Change in (**A–C**) photochemical efficiency (Fv/Fm) and (**D–F**) performance index on absorption basis (PIABS) of four different creeping bentgrass cultivars (Penncross, 13M, Seaside II, and PA-1) during summer in (**A**) 2017, (**B**) 2018, and (**C**) 2019. Vertical bars indicate ±SE of mean (n = 4). Different letters indicate significant differences (*p* < 0.05) at a given day.

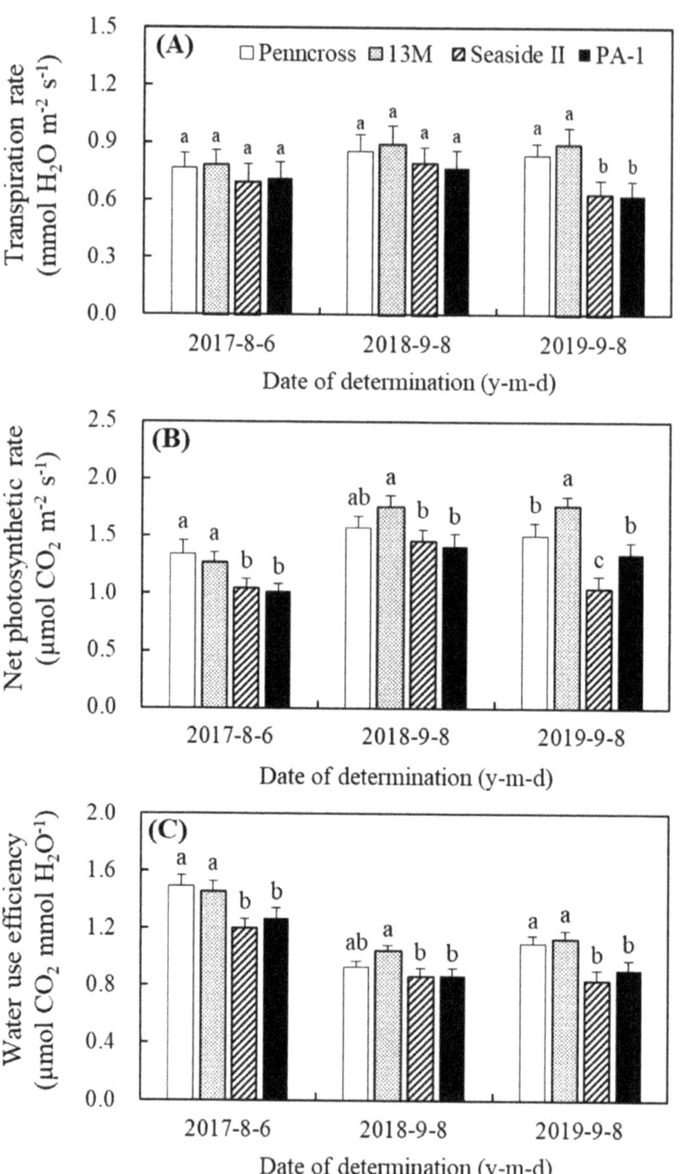

Figure 5. Change in (**A**) transpiration rate (Tr), (**B**) net photosynthetic rate (Pn), and (**C**) water use efficiency (WUE) of four different creeping bentgrass cultivars (Penncross, 13M, Seaside II, and PA-1) during summer in (**A**) 2017, (**B**) 2018, and (**C**) 2019. Vertical bars indicate ±SE of mean (n = 4). Different letters indicate significant differences ($p < 0.05$) at a given day.

3.2. Water Status and Osmolytes Affected by Summer Stress

The Pro and WSC contents in the four cultivars increased gradually from May to August in 2017, 2018, and 2019 (Figure 6A–F). There were no significant differences in Pro and WSC contents among the four cultivars in May 2017, 2018, or 2019 (Figure 6A–F). The 13M and the PA-1 cultivars had the maximum and second-highest Pro content when compared to Penncross and Seaside II in July 2017 (Figure 6A). The Pro content was significantly higher in 13M in August 2017 than in the other cultivars (Figure 6A). Significantly higher Pro content was detected in the leaves of Penncross and 13M as compared to that in the leaves of Seaside II and PA-1 in July and August of 2018 (Figure 6B). The lowest Pro content was observed in PA-1 and Seaside II in July and August 2019, respectively (Figure 6C). The 13M cultivar showed the highest WSC accumulation out of the four cultivars in July 2017, 2018, and 2019 (Figure 6D,E). In August 2017, 2018, and 2019, Penncross and 13M accumulated more WSC than Seaside II and PA-1 (Figure 6D–F). Changes in the RWC and OP showed similar trends in the four cultivars, as demonstrated by the gradual declines during summer stress in 2017, 2018, and 2019 (Figure 7A–F). The 13M cultivar maintained a higher RWC than the other three cultivars during summer stress in 2017 and 2019 (Figure 7A,C). On 8 September 2018, Penncross and 13M had a 13% increase in RWC compared to Seaside II and PA-1 (Figure 7B). OP in 13M was maintained at the lowest levels when compared to the other three cultivars in response to high temperatures during the summer in 2017, 2018, and 2019 (Figure 7D–F).

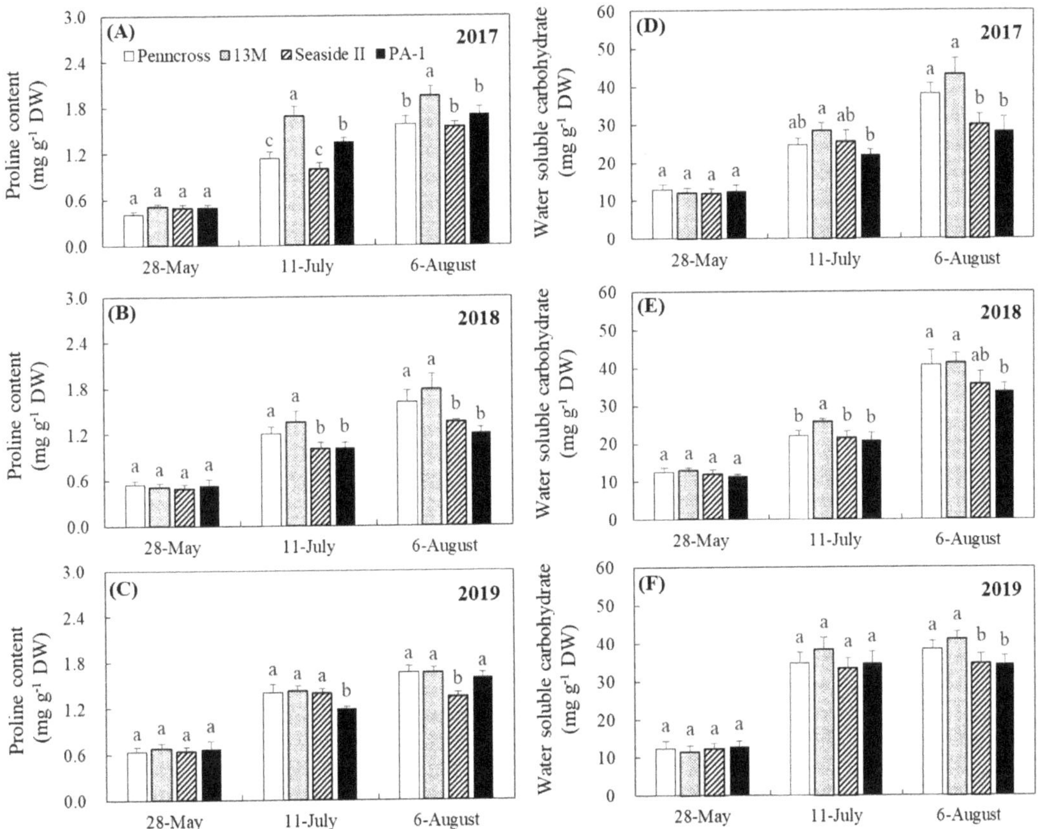

Figure 6. Change in (**A–C**) proline (Pro) content and (**D–F**) water-soluble carbohydrate (WUE) of four different creeping bentgrass cultivars (Penncross, 13M, Seaside II, and PA-1) during summer in (**A**) 2017, (**B**) 2018, and (**C**) 2019. Vertical bars indicate ±SE of mean (n = 4). Different letters indicate significant differences (*p* < 0.05) at a given day.

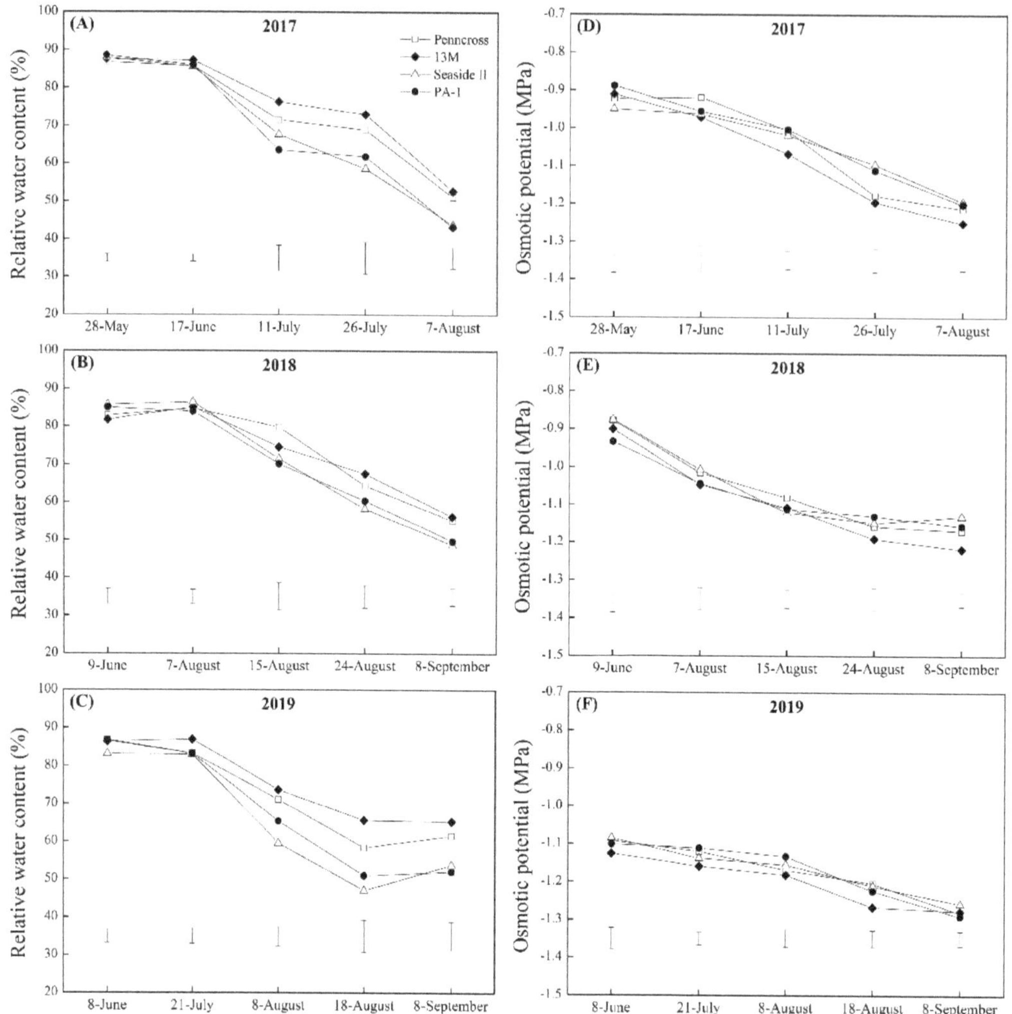

Figure 7. Change in (**A–C**) relative water content (RWC) and (**D–F**) osmotic potential (OP) of four different creeping bentgrass cultivars (Penncross, 13M, Seaside II, and PA-1) during summer in (**A**) 2017, (**B**) 2018, and (**C**) 2019. Vertical bars below curves indicate least significant difference (LSD) values ($p < 0.05$) at a given day.

3.3. Oxidative Damage and Membrane Stability Affected by Summer Stress

The H_2O_2 and the MDA contents in the four cultivars increased significantly from May to September in 2017, 2018, and 2019 (Figure 8A–F). Seaside II and PA-1 accumulated significantly higher H_2O_2 content than Penncross and 13M from July to August in 2017 and 2018 (Figure 8A,B). The 13M cultivar maintained the lowest H_2O_2 content out of the cultivars during summer stress in 2018 (Figure 8B). An 18% increase in H_2O_2 content was detected in Seaside II and PA-1 over Penncross and 13M on August 18 and September 8, 2019 (Figure 8C). In the year 2019, PA-1 exhibited the highest MDA content of the cultivars from May to August (Figure 8D), and Seaside II and 13M had the highest and lowest MDA content of the cultivars during summer stress in 2018 and 2019, respectively (Figure 8E,F). The EL in the leaves of the four cultivars increased significantly from June to September in 2017, 2018, and 2019 (Figure 9A–C). The EL was the lowest in Penncross when compared to the other three cultivars on 7 August 2017 (Figure 9A). Seaside II and PA-1 exhibited an 18% and 16% increase in EL over Penncross and 13M on 8 September 2018 and 2019, respectively (Figure 9B,C). The 13M cultivar maintained the lowest EL level in its leaves during summer stress in 2018 and 2019 (Figure 9B,C).

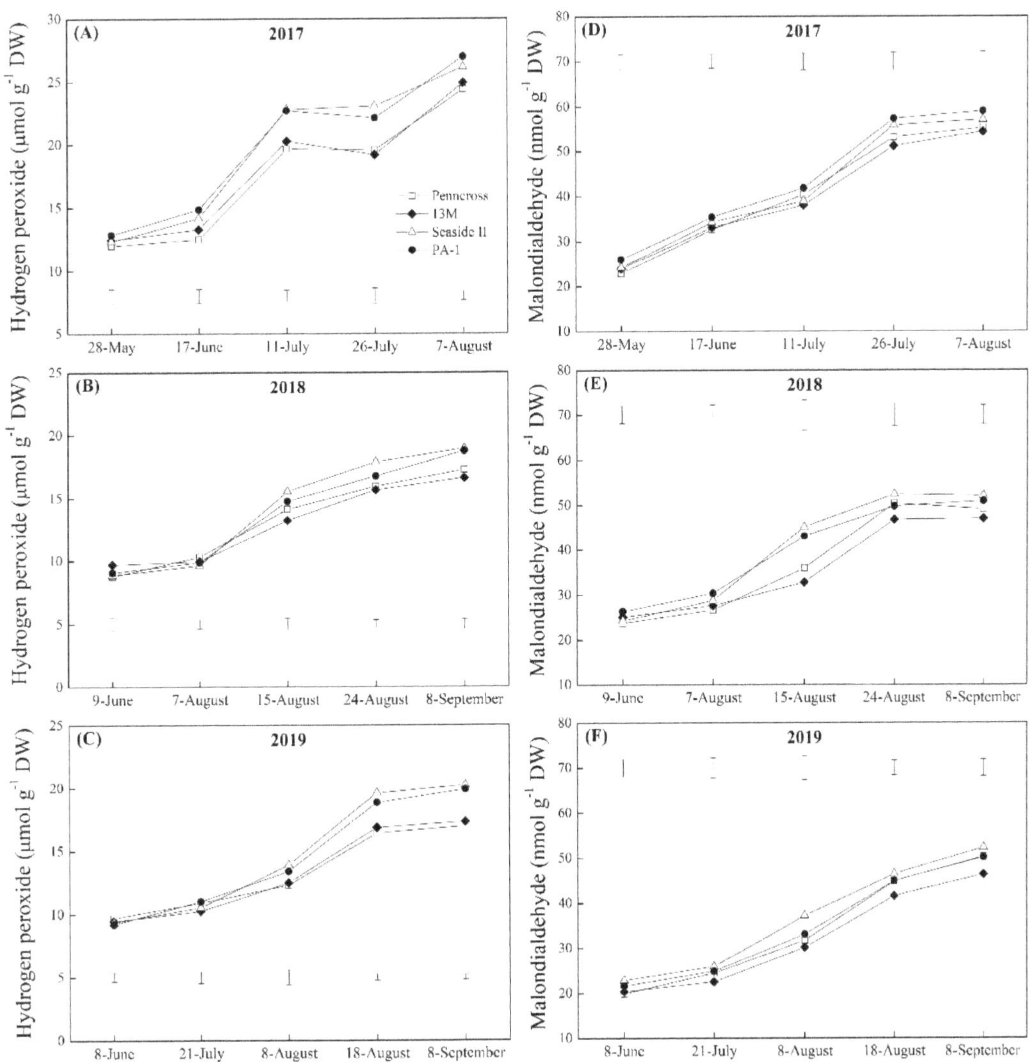

Figure 8. Change in (**A–C**) hydrogen peroxide (H_2O_2) content and (**D–F**) malondialdehyde (MDA) content of four different creeping bentgrass cultivars (Penncross, 13M, Seaside II, and PA-1) during summer in (**A**) 2017, (**B**) 2018, and (**C**) 2019. Vertical bars below or above curves indicate least significant difference (LSD) values ($p < 0.05$) at a given day.

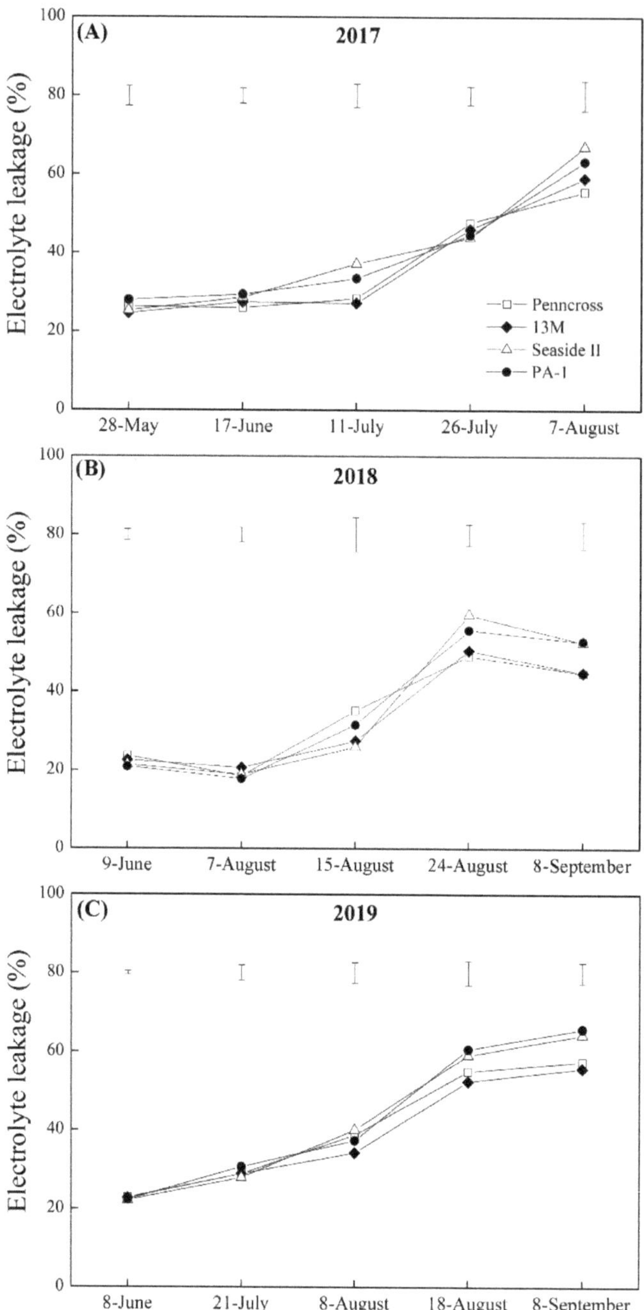

Figure 9. Change in electrolyte leakage (EL) of four different creeping bentgrass cultivars (Penncross, 13M, Seaside II, and PA-1) during summer in (**A**) 2017, (**B**) 2018, and (**C**) 2019. Vertical bars above curves indicate least significant difference (LSD) values ($p < 0.05$) at a given day.

4. Discussion

Creeping bentgrass is used as sports turfs and often needs high cultural inputs for adequate performance and functionality. The maintenance of higher TQ on golf greens during hot summer months provided better playability to golfers [6]. However, the TQ of creeping bentgrass significantly declines during the summer months when the ambient temperature exceeds its optimum growth temperature [1]. Our current study demonstrated that heat waves in the summer significantly decreased the TQ of four creeping bentgrass cultivars (Penncross, 13M, Seaside II, and PA-1) in 2017, 2018, and 2019, but 13M and Penncross could maintain a higher TQ than Seaside II and PA-1 during the summer months. A decline in TQ is characterized by a reduced grass density and yellowing leaf tissue due to accelerated leaf senescence under high-temperature stress [31,32]. Gradual declines in Chl content were observed in the four cultivars, which was consistent with the significant decreases in Fv/Fm, PIABS, and Pn. Previous studies have found that higher Chl content was a key indicator for selecting heat-tolerant genotypes in creeping bentgrass species [7,8]. Delayed Chl degradation and the higher photochemical efficiency of PSII and Pn were beneficial for better adaptation to high temperatures in creeping bentgrass during the summer months, because metabolites produced from photosynthesis provided an available energy supply for growth maintenance [13,15]. The 13M and Penncross cultivars exhibited significantly higher Chl content and photosynthesis than Seaside II and PA-1 in the summer, which could indicate that those two cultivars had better adaptability to high temperatures in the subtropical zone.

Reduced available carbohydrates limited energy supply for plant growth and development associated with heat-induced photoinhibition in plants [33]. It has been found that carbohydrate accumulation benefits creeping bentgrass against drought or heat stress [12,16,34]. An earlier study on putting green also showed that the heat-tolerant creeping bentgrass cultivar L-93 exhibited significantly higher carbohydrate contents than Penncross in response to summer heat stress in Manhattan [2]. In addition, proline accumulation and metabolism are important survival strategies against heat stress, owing to the protective functions of osmotic adjustment and ROS scavenging ability in plants [11]. High temperatures cause physiological drought mainly due to reduced water absorption in roots and accelerated transpiration in leaves. The positive effects of the accumulation of proline and carbohydrates on water homeostasis have been proved in creeping bentgrass and other plant species under high-temperature stress [35–38]. Leaf RWC and OP gradually declined with the development of summer stress, whereas proline and carbohydrates significantly accumulated in the four cultivars during the summer months. Interestingly, 13M and Penncross accumulated more proline and carbohydrates as well as a better leaf water status and osmotic adjustment ability than Seaside II and PA-1 in the summers of 2017, 2018, and 2019. These findings indicated that the cultivars' variations in their adaptations to high temperatures could be associated with the modification of water homeostasis during the summer.

The two major indicators of membrane peroxidation were H_2O_2 and MDA accumulation, and their accumulations with high levels of toxicity to cells accelerated senescence under heat stress [39]. In addition, the EL level was inversely correlated to heat tolerance in creeping bentgrass species [7]. Persistent high-temperature stress induced gradual increases in H_2O_2 content, MDA accumulation, and EL levels in the four creeping bentgrass cultivars from June to August in 2017, 2018, and 2019. These findings indicated that the four cultivars suffered from serious oxidative damage to their cell membranes during the summer months. However, 13M and Penncross could maintain lower EL, H_2O_2, and MDA than Seaside II and PA-1 in the summer. Similar results were found in a previous study, which demonstrated that a lower lipid membrane peroxidation level was beneficial to the alleviation of SBD [18]. The better maintenance of cell membrane stability was also propitious to photosynthesis and metabolic activity in plants under high-temperature environmental conditions [40]. Proline accumulation and metabolism have been known to confer heat tolerance in plants associated with ROS detoxifying and delayed senescence [38,41].

An exogenous application of proline could also mitigate the detrimental effects of high temperatures on creeping bentgrass in relation to delayed leaf senescence [15].

5. Conclusions

Summer stress significantly decreased the TQ of four creeping bentgrass cultivars (13M, Penncross, Seaside II, and PA-1) in 2017, 2018, and 2019. A variety of physiological processes were affected by heat stress in the summer, including significant declines in Chl content, photosynthesis, leaf RWC, and OP as well as obvious increases in carbohydrates, proline, H_2O_2, MDA, and EL in the four cultivars. The 13M and Penncross cultivars exhibited better performance than Seaside II and PA-1 in response to summer stress from 2017 to 2019, which was associated with the maintenance of better photosynthesis, water status, osmolytes accumulation, and cell membrane stability. The 13M and Penncross cultivars could be used as potential candidates for turf establishment in a subtropical zone. An in-depth understanding of physiological responses to summer stress also provided critical information for the breeding and development of germplasm with heat tolerance in creeping bentgrass species.

Author Contributions: Conceptualization, Z.L., J.X., L.H. and Y.P.; methodology, Z.L., L.H. and Y.P.; formal analysis, Z.L. and W.Z.; investigation, W.Z. and B.C.; writing—original draft preparation, Z.L.; writing—review and editing, L.H., J.X. and Y.P.; funding acquisition, Z.L. and Y.P. All authors have read and agreed to the published version of the manuscript.

Funding: This research was funded by the China Postdoctoral Science Foundation (2021T140058) and the Sichuan Forage Innovation Team Project of the Industrial System Construction of Modern Agriculture of China (sccxtd-2020-16).

Institutional Review Board Statement: Not applicable.

Informed Consent Statement: Not applicable.

Data Availability Statement: All data presented in this study are available in the article.

Conflicts of Interest: All authors declare no conflict of interest.

References

1. Fry, J.; Huang, B. *Applied Turfgrass Science and Physiology*; John Wiley Sons: Hoboken, NJ, USA, 2004.
2. Xu, Q.; Huang, B. Seasonal changes in carbohydrate accumulation for two creeping bentgrass cultivars. *Crop Sci.* **2003**, *43*, 266–271. [CrossRef]
3. Miller, G.L.; Brotherton, M.A. Creeping bentgrass summer decline as influenced by climatic conditions and cultural practices. *Agron. J.* **2020**, *112*, 3500–3512. [CrossRef]
4. Krishnan, S.; Ma, Y.; Merewitz, E. Leaf trimming and high temperature regulation of phytohormones and polyamines in creeping bentgrass leaves. *J. Am. Soc. Hortic. Sci.* **2016**, *141*, 66–75. [CrossRef]
5. Fu, J.; Dernoeden, P.H. Creeping bentgrass putting green turf responses to two irrigation practices: Quality, chlorophyll, canopy temperature, and thatch–mat. *Crop Sci.* **2009**, *49*, 1071–1078. [CrossRef]
6. McCullough, P.E.; Liu, H.; McCarty, L.B.; Toler, J.E. Trinexapac-ethyl application regimens influence growth, quality, and performance of bermuda grass and creeping bentgrass putting greens. *Crop Sci.* **2007**, *47*, 2138–2144. [CrossRef]
7. Li, Z.; Tang, M.; Hassan, M.J.; Zhang, Y.; Han, L.; Peng, Y. Adaptability to high temperature and stay-green genotypes associated with variations in antioxidant, chlorophyll metabolism, and γ-aminobutyric acid accumulation in creeping bentgrass species. *Front. Plant Sci.* **2021**, *12*, 2341. [CrossRef]
8. Jespersen, D.; Ma, X.; Bonos, S.A.; Belanger, F.C.; Raymer, P.; Huang, B. Association of SSR and candidate gene markers with genetic variations in summer heat and drought performance for creeping bentgrass. *Crop Sci.* **2018**, *58*, 2644–2656. [CrossRef]
9. Xu, Q.; Huang, B. Morphological and physiological characteristics associated with heat tolerance in creeping bentgrass. *Crop Sci.* **2001**, *41*, 127–133. [CrossRef]
10. Huang, B.; Dacosta, M.; Jiang, Y. Research advances in mechanisms of turfgrass tolerance to abiotic stresses: From physiology to molecular biology. *Crit. Rev. Plant Sci.* **2014**, *33*, 141–189. [CrossRef]
11. Yancey, P.H. Organic osmolytes as compatible, metabolic and counteracting cytoprotectants in high osmolarity and other stresses. *J. Exp. Biol.* **2005**, *208*, 2819–2830. [CrossRef]
12. Liu, Z.; Liu, T.; Liang, L.; Li, Z.; Hassan, M.J.; Peng, Y.; Wang, D. Enhanced photosynthesis, carbohydrates, and energy metabolism associated with chitosan-induced drought tolerance in creeping bentgrass. *Crop Sci.* **2020**, *60*, 1064–1076. [CrossRef]

13. Rossi, S.; Chapman, C.; Yuan, B.; Huang, B. Improved heat tolerance in creeping bentgrass by γ-aminobutyric acid, proline, and inorganic nitrogen associated with differential regulation of amino acid metabolism. *Plant Growth Regul.* **2021**, *93*, 231–242. [CrossRef]
14. Xu, Y.; Du, H.; Huang, B. Identification of metabolites associated with superior heat tolerance in thermal bentgrass through metabolic profiling. *Crop Sci.* **2013**, *53*, 1626–1635. [CrossRef]
15. Rossi, S.; Chapman, C.; Huang, B. Suppression of heat-induced leaf senescence by γ-aminobutyric acid, proline, and ammonium nitrate through regulation of chlorophyll degradation in creeping bentgrass. *Environ. Exp. Bot.* **2020**, *177*, 104116. [CrossRef]
16. Liang, L.L.; Cao, Y.Q.; Wang, D.; Peng, Y.; Zhang, Y.; Li, Z. Spermine alleviates heat-induced senescence in creeping bentgrass by regulating water and oxidative balance, photosynthesis, and heat shock proteins. *Biol. Plant.* **2021**, *65*, 184–192. [CrossRef]
17. Li, Z.; Yu, J.; Peng, Y.; Huang, B. Metabolic pathways regulated by γ-aminobutyric acid (GABA) contributing to heat tolerance in creeping bentgrass (*Agrostis stolonifera*). *Sci. Rep.* **2016**, *6*, 30338. [CrossRef]
18. Xu, Q.; Huang, B. Antioxidant metabolism associated with summer leaf senescence and turf quality decline for creeping bentgrass. *Crop Sci.* **2004**, *44*, 553–560. [CrossRef]
19. Zhang, X.; Ervin, E.H. Impact of seaweed extract-based cytokinins and zeatin riboside on creeping bentgrass heat tolerance. *Crop Sci.* **2008**, *48*, 364–370. [CrossRef]
20. Beard, J.B. *Turf Management for Golf Courses*, 2nd ed.; Ann Arbor Press: Chelsea, MI, USA, 2002.
21. Amnon, D.I. Copper enzymes in isolated chloroplasts. Polyphenoloxidase in *Beta vulgaris*. *Plant Physiol.* **1949**, *24*, 1–15.
22. Barrs, H.; Weatherley, P. A re-examination of the relative turgidity technique for estimating water deficits in leaves. *Aust. J. Biol. Sci.* **1962**, *15*, 413–428. [CrossRef]
23. Blum, A. Osmotic adjustment and growth of barley genotypes under drought stress. *Crop Sci.* **1989**, *29*, 230–233. [CrossRef]
24. Burgess, P.; Huang, B. Effects of sequential application of plant growth regulators and osmoregulants on drought tolerance of creeping bentgrass (*Agrostis stolonifera*). *Crop Sci.* **2013**, *54*, 837–844. [CrossRef]
25. Bates, L.S.; Waldren, R.P.; Teare, I.D. Rapid determination of free proline for water-stress studies. *Plant Soil* **1973**, *39*, 205–207. [CrossRef]
26. Jan, B.; Roel, M. An improved colorimetric method to quantify sugar content of plant tissue. *J. Exp. Bot.* **1993**, *44*, 1627–1629.
27. Li, Z.; Peng, Y.; Huang, B. Physiological effects of γ-aminobutyric acid application on improving heat and drought tolerance in creeping bentgrass. *J. Am. Soc. Hortic. Sci.* **2016**, *141*, 76–84. [CrossRef]
28. Velikova, V.; Yordanov, I.; Edreva, A. Oxidative stress and some antioxidant systems in acid rain-treated bean plants: Protective role of exogenous polyamines. *Plant Sci.* **2000**, *151*, 59–66. [CrossRef]
29. Dhindsa, R.; Plumb-Dhindsa, P.; Thorpe, T. Leaf senescence: Correlated with increased levels of membrane permeability and lipid peroxidation, and decreased levels of superoxide dismutase and catalase. *J. Exp. Bot.* **1981**, *32*, 93–101. [CrossRef]
30. Blum, A.; Ebercon, A. Cell membrane stability as a measure of drought and heat tolerance in wheat. *Crop Sci.* **1981**, *21*, 43–47. [CrossRef]
31. Xin, J.; Yan, X.; Jiang, T.; Gianfagna, T.; Huang, B. Suppression of shade- or heat-induced leaf senescence in creeping bentgrass through transformation with the IPT gene for cytokinin synthesis. *J. Am. Soc. Hortic. Sci.* **2009**, *134*, 602–609.
32. Jespersen, D.; Zhang, J.; Huang, B. Chlorophyll loss associated with heat-induced senescence in bentgrass. *Plant Sci.* **2016**, *249*, 1–12. [CrossRef]
33. Zhang, Z.; Hu, M.; Xu, W.; Wang, Y.; Huang, K.; Zhang, C.; Wen, J. Understanding the molecular mechanism of anther development under abiotic stresses. *Plant Mol. Biol.* **2021**, *105*, 1–10. [CrossRef] [PubMed]
34. Liu, X.; Huang, B. Carbohydrate accumulation in relation to heat stress tolerance in two creeping bentgrass cultivars. *J. Am. Soc. Hortic. Sci.* **2000**, *125*, 442–447. [CrossRef]
35. Wahid, A.; Close, T.J. Expression of dehydrins under heat stress and their relationship with water relations of sugarcane leaves. *Biol. Plant.* **2007**, *51*, 104–109. [CrossRef]
36. Zeng, W.; Hassan, M.J.; Kang, D.; Peng, Y.; Li, Z. Photosynthetic maintenance and heat shock protein accumulation relating to γ-aminobutyric acid (GABA)-regulated heat tolerance in creeping bentgrass (*Agrostis stolonifera*). *S. Afr. J. Bot.* **2021**, *141*, 405–413. [CrossRef]
37. Dobra, J.; Motyka, V.; Dobrev, P.; Malbeck, J.; Prasil, I.T.; Haisel, D.; Gaudinova, A.; Havlova, M.; Gubis, J.; Vankova, R. Comparison of hormonal responses to heat, drought and combined stress in tobacco plants with elevated proline content. *J. Plant Physiol.* **2010**, *167*, 1360–1370. [CrossRef]
38. Batcho, A.A.; Sarwar, M.B.; Rashid, B.; Hassan, S.; Husnain, T. Heat shock protein gene identified from Agave sisalana (AsHSP70) confers heat stress tolerance in transgenic cotton (*Gossypium hirsutum*). *Theor. Exp. Plant Physiol.* **2021**, *33*, 141–156. [CrossRef]
39. Belhadj Slimen, I.; Najar, T.; Ghram, A.; Dabbebi, H.; Ben Mrad, M.; Abdrabbah, M. Reactive oxygen species, heat stress and oxidative-induced mitochondrial damage. A review. *Int. J. Hyperth.* **2014**, *30*, 513–523. [CrossRef]
40. Allakhverdiev, S.I.; Kreslavski, V.D.; Klimov, V.V.; Los, D.A.; Carpentier, R.; Mohanty, P. Heat stress: An overview of molecular responses in photosynthesis. *Photosynth. Res.* **2008**, *98*, 541. [CrossRef]
41. Hanif, S.; Saleem, M.F.; Sarwar, M.; Irshad, M.; Shakoor, A.; Wahid, M.A.; Khan, H.Z. Biochemically triggered heat and drought stress tolerance in rice by proline application. *J. Plant Growth Regul.* **2021**, *40*, 305–312. [CrossRef]

Article

Reallocation of Soluble Sugars and IAA Regulation in Association with Enhanced Stolon Growth by Elevated CO_2 in Creeping Bentgrass

Jingjin Yu [1], Meng Li [1], Qiuguo Li [1], Ruying Wang [2], Ruonan Li [1] and Zhimin Yang [1,*]

[1] College of Agro-Grassland Science, Nanjing Agricultural University, Nanjing 210095, China; nauyjj@njau.edu.cn (J.Y.); 2018120014@stu.njau.edu.cn (M.L.); 2020820040@stu.njau.edu.cn (Q.L.); 2021120015@stu.njau.edu.cn (R.L.)

[2] Department of Horticulture, Oregon State University, 4017 Agriculture and Life Sciences Building, Corvallis, OR 97331, USA; ruying.wang@oregonstate.edu

[*] Correspondence: nauyzm@njau.edu.cn

Abstract: Extensive stolon development and growth are superior traits for rapid establishment as well as post-stress regeneration in stoloniferous grass species. Despite the importance of those stoloniferous traits, the regulation mechanisms of stolon growth and development are largely unknown. The objectives of this research were to elucidate the effects of the reallocation of soluble sugars for energy reserves and endogenous hormone levels for cell differentiation and regeneration in regulating stolon growth of a perennial turfgrass species, creeping bentgrass (*Agrostis stolonifera* L.). Plants were grown in growth chambers with two CO_2 concentrations: ambient CO_2 concentration (400 ± 10 μmol mol^{-1}) and elevated CO_2 concentration (800 ± 10 μmol mol^{-1}). Elevated CO_2 enhanced stolon growth through increasing stolon internode number and internode length in creeping bentgrass, as manifested by the longer total stolon length and greater shoot biomass. The content of glucose, sucrose, and fructose as well as endogenous IAA were accumulated in stolon nodes and internodes but not in leaves or roots under elevated CO_2 concentration. These results illustrated that the production and reallocation of soluble sugars to stolons as well as the increased level of IAA in stolon nodes and internodes could contribute to the enhancement of stolon growth under elevated CO_2 in creeping bentgrass.

Keywords: elevated CO_2; stolon growth; soluble sugars; hormone; creeping bentgrass

Citation: Yu, J.; Li, M.; Li, Q.; Wang, R.; Li, R.; Yang, Z. Reallocation of Soluble Sugars and IAA Regulation in Association with Enhanced Stolon Growth by Elevated CO_2 in Creeping Bentgrass. *Plants* **2022**, *11*, 1500. https://doi.org/10.3390/plants11111500

Academic Editor: Bernhard Huchzermeyer

Received: 28 April 2022
Accepted: 1 June 2022
Published: 2 June 2022

Publisher's Note: MDPI stays neutral with regard to jurisdictional claims in published maps and institutional affiliations.

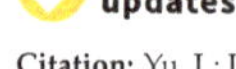

1. Introduction

Stolon is an elongated axillary shoot composed of nodes, internodes, and leaves [1,2]. Stolon nodes contain meristematic tissues which are capable of producing adventitious roots and offspring ramets from nodes [3,4]. The benefits of clonality include rapid local spread through stolon growth as well as a high stand establishment rate due to the physiological connections between ramets to share resources of carbohydrates, nutrients and water [3,5,6]. Stoloniferous plants can be clonally propagated and have advantageous traits such as rapid establishment and recovery from stresses. In turfgrass, vegetative propagation (stolon cuttings) or sprigging is a commonly and extensively used method for rapid turf production and establishment utilizing stolon cuttings in warm-season grass species [7–9]. Therefore, rapid stolon growth is one of the highly desirable characteristics of stoloniferous turfgrass species.

The atmospheric CO_2 concentration has risen from the pre-industrialized era 280 μmol mol^{-1} to the present 400 μmol mol^{-1} and will reach over 1000 μmol mol^{-1} by the end of this century according to IPCC [10,11]. Therefore, research interests in plant responses to elevated CO_2 are increasing [12]. A large number of studies reported that elevated CO_2 could promote plant growth and development, including perennial grass species, such as tall fescue (*Festuca arundinacea* Schreb.) [13–16], Kentucky bluegrass (*Poa pratensis* L.) [17,18], and bermudagrass (*Cynodon dactylon* (L.) Pers.) [19,20]. However, few previous studies about elevated CO_2-induced effects were found focusing on stolon growth.

Many studies documented that elevated CO_2-induced promotion in plant growth was associated with changes in carbohydrates content via stimulating photosynthetic capacity for synthesizing carbohydrates in plants [18,20,21]. For example, elevated CO_2 led to a significant increase in total non-structural carbohydrates in the leaves of Kentucky bluegrass [17]. Kinmonth-Schultz and Kim [22] found that elevated CO_2 improved fructan accumulation in the underground rhizomes in order to overwinter and spread in reed canary grass (*Phalaris arundinacea* L.). In stolon tips of creeping bentgrass (*Agrostis stolonifera*), there was a significant decrease in fructose and sucrose as well as an increase in maltose under elevated CO_2 conditions through GC-MS analysis [2]. Burgess and Huang [21] found that elevated CO_2 caused an increase in the total stolon length as well as net photosynthetic rate in creeping bentgrass. Such an increase in total stolon length was attributed to the increased stolon internode number [2].

Apart from carbohydrates, plant hormones are also of great importance in affecting growth and development via biosynthesis, degradation, transport, and signaling to regulate multiple biological processes in plants [23]. Among several common endogenous hormones, auxins (including indole-3-acetic acid; IAA), cytokinins (including isopentenyl adenosine; iPA), and gibberellic acids (GAs) are the most well-known ones in controlling cell division and elongation during plant growth and development due to their regulatory roles in each biological process from embryogenesis to maturity in various plant species [24–26]. In potato (*Solanum tuberosum* L.), IAA and GA_3 were found to be essential for stolon elongation [25]. In creeping bentgrass, Burgess et al. [27] reported that elevated CO_2 did not alter endogenous iPA or IAA in the leaves under well-watered conditions but increased the content of iPA and decreased IAA under drought stress. The GA regulation under elevated CO_2 is still unknown because the contents of GAs were not measured in that study [27]. As illustrated in our previous study, elevated CO_2 caused increases in total stolon length by some metabolites involved in carbohydrate reserves, respiratory metabolism, and membrane maintenance in the stolon tips of creeping bentgrass [2]. Nevertheless, very limited knowledge is currently available about the effects of elevated CO_2 on stolon growth with respect to soluble sugars and endogenous hormones allocation in different perennial plant organs such as root, leaf, node, and internode in stoloniferous grass species.

We hypothesized that elevated CO_2 improvement on stolon growth might be associated with the reallocation of soluble sugars for energy reserves and endogenous hormone levels for cell differentiation and regeneration in regulating stolon growth of a perennial turfgrass species, creeping bentgrass. Understanding the specific soluble sugars and hormones in different organs in response to elevated CO_2 concentration will provide some new insights into mechanisms about how elevated CO_2 enhances stolon growth of stoloniferous plants under the scenario of climate change in the future.

2. Results

2.1. Effects of Elevated CO_2 on Morphological Parameters in Creeping Bentgrass

The phenotypic responses of creeping bentgrass to CO_2 levels were dramatically different as shown in Figure 1A,B, indicating that elevated CO_2 significantly enhanced stolon growth compared to ambient CO_2. Shoot biomass was positively correlated with total stolon length, internode length, internode number, and root biomass (Table 1) and significantly increased by 1.10-fold due to elevated CO_2 (Figure 1C).

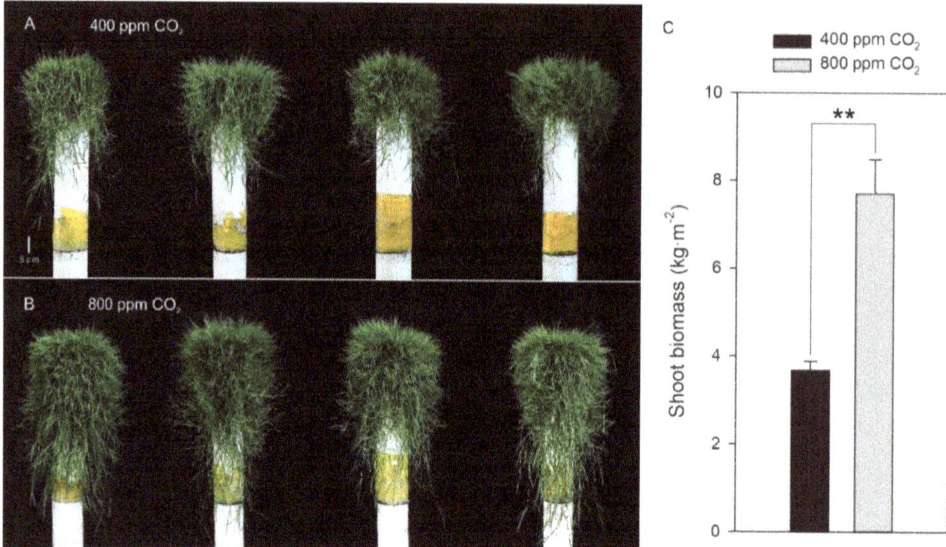

Figure 1. Effects of elevated CO_2 concentration on shoot phenotype (**A**,**B**) and shoot biomass (**C**) of creeping bentgrass at 42 d of experiment. Four hundred μmol mol^{-1} (ppm) CO_2, ambient CO_2 concentration; 800 ppm CO_2, elevated CO_2 concentration. ** indicates a significant difference between ambient and elevated CO_2 concentrations according to Student's *t*-test at $p \leq 0.01$. Error bars represent standard error (SE).

Table 1. Pearson Correlation analysis among growth parameters in creeping bentgrass.

	Total Stolon Length	Internode Length	Internode Number	Shoot Biomass	Root Biomass	Longest Root Length
Total stolon length	1	0.938 **	0.973 **	0.693 *	0.939 **	0.121
Internode length	0.938 **	1	0.883 **	0.686 *	0.831 **	0.121
Internode number	0.973 **	0.883 **	1	0.587	0.920 **	0.138
Shoot biomass	0.693 *	0.686 *	0.587	1	0.811 **	0.383
Root biomass	0.939 **	0.831 **	0.920 **	0.811 **	1	0.186
Longest root	0.121	0.121	0.138	0.383	0.186	1

Note: * and ** indicate significant correlation at 0.05 and 0.01 probability levels, respectively.

Total stolon length was positively correlated with stolon internode length, internode number, and shoot and root biomass (Table 1). Elevated CO_2 significantly enhanced stolon internode number and total stolon length from 7 to 42 d, and stolon internode length from 21 to 42 d of the experimental period (Figure 2). The stolon internode number of creeping bentgrass grown under elevated CO_2 was consistently greater than under ambient CO_2 and the differences in internode were increased from an average of 0.5 at 7 d to 2.9 at 42 d (Figure 2A). At the conclusion of the study, the elevated CO_2-caused increase in stolon internode length reached 12.1 mm at 42 d in comparison with the ambient CO_2 concentration (Figure 2B).

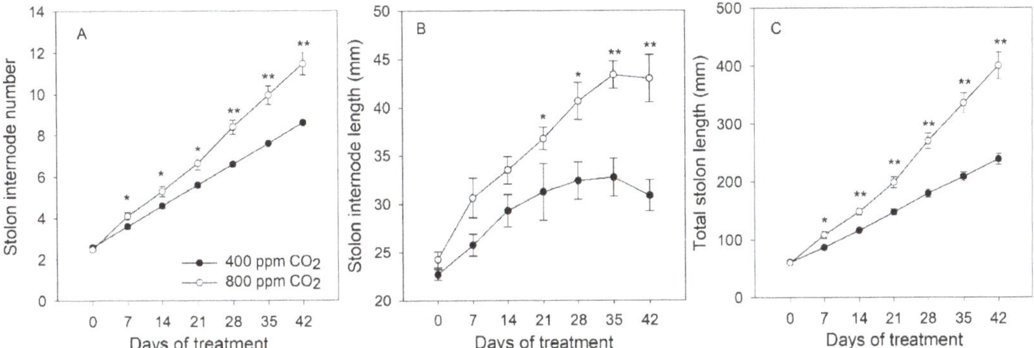

Figure 2. Effects of elevated CO_2 concentration on stolon internode number (**A**), stolon internode length (**B**), and total stolon length (**C**) of creeping bentgrass at 42 d of experiment. Four hundred μmol mol^{-1} (ppm) CO_2, ambient CO_2 concentration; 800 ppm CO_2, elevated CO_2 concentration. * and ** indicate a significant difference between ambient and elevated CO_2 concentrations according to Student's *t*-test at $p \leq 0.05$ and $p \leq 0.01$, respectively. Error bars represent standard error (SE).

In addition to proliferated shoot growth, elevated CO_2 also stimulated substantial root growth of creeping bentgrass (Figure 3A,B). Root biomass was positively correlated with total stolon length, internode length, internode number, and shoot biomass (Table 1). Elevated CO_2 significantly increased root biomass by 1.64-fold at 42 d of experiment in comparison with ambient CO_2 (Figure 3C). However, no difference was found in the longest root length of creeping bentgrass between elevated and ambient CO_2 concentrations (Figure 3D).

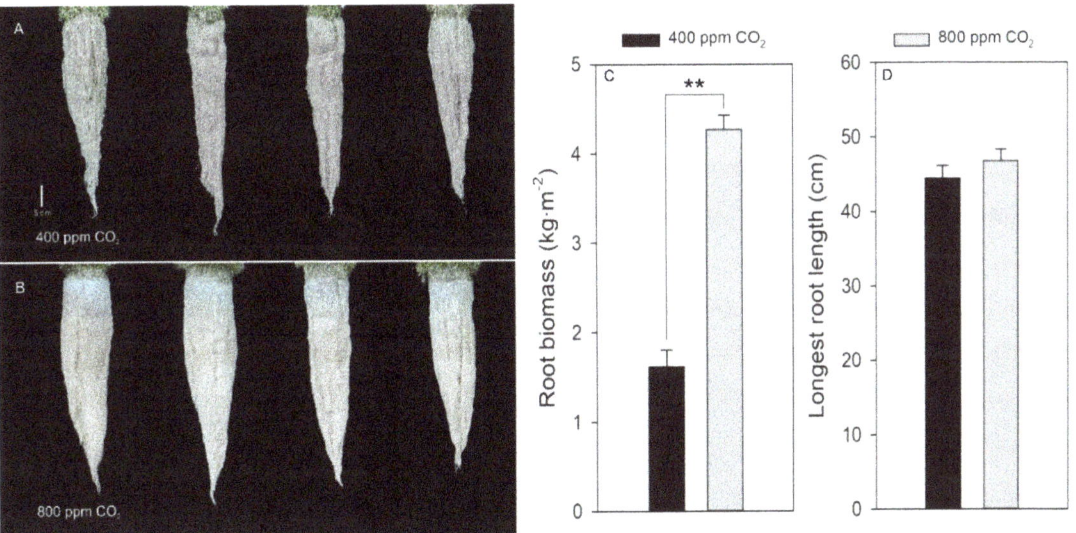

Figure 3. Effects of elevated CO_2 concentration on root phenotype (**A,B**), root biomass (**C**), and longest root length (**D**) of creeping bentgrass at 42 d of experiment. Four hundred μmol mol^{-1} (ppm) CO_2, ambient CO_2 concentration; 800 ppm CO_2, elevated CO_2 concentration. ** indicates a significant difference between ambient and elevated CO_2 concentrations according to Student's *t*-test at $p \leq 0.01$. Error bars represent standard error (SE).

2.2. Effects of Elevated CO_2 on Shoot Soluble Sugars

In creeping bentgrass, soluble sugar levels were generally lowered in the leaf tissue than in the node and internode (Figure 4). In the leaves, elevated CO_2 caused significant decreases in all soluble sugars measured in this study from 7 to 42 d of experiment (Figure 4A–D). More specifically, glucose content in the leaves in response to elevated CO_2 was reduced by 29.0%, 32.6%, 17.8%, and 33.9% at 7, 21, 35, and 42 d, respectively, in comparison with ambient CO_2 (Figure 4A); and reduction in fructose content was 31.5%, 30.6%, 18.3%, and 35.0% (Figure 4B) and in sucrose content was 29.1%, 32.6%, 17.8%, and 33.9% (Figure 4C) at 7, 21, 35, and 42 d, respectively. Therefore, the content of total soluble sugars was significantly decreased by 22.8%, 24.1%, 15.2%, and 25.4% at 7, 21, 35, and 42 d, respectively, under elevated CO_2 concentration (Figure 4D).

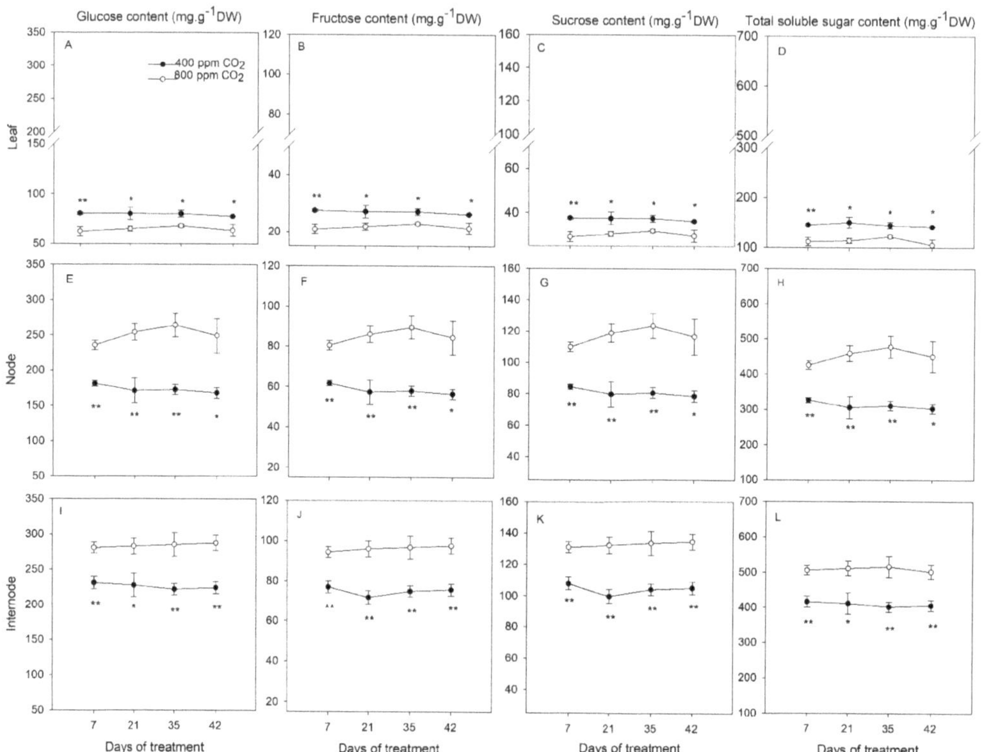

Figure 4. Effects of elevated CO_2 concentration on content of glucose (**A,E,I**), fructose (**B,F,J**), sucrose (**C,G,K**) and total soluble sugar (**D,H,L**) in the leaf (**A–D**), node (**E–H**), and internode (**I–L**) tissues of creeping bentgrass at 42 d of experiment. Four hundred μmol mol^{-1} (ppm) CO_2, ambient CO_2 concentration; 800 ppm CO_2, elevated CO_2 concentration. Sugar contents are presented in the unit of mg g^{-1} dry weight (DW). * and ** indicate a significant difference between ambient and elevated CO_2 concentrations according to Student's *t*-test at $p \leq 0.05$ and $p \leq 0.01$, respectively. Error bars represent standard error (SE).

Three soluble sugars and total soluble sugars exhibited greater accumulation in the stolon nodes in response to elevated CO_2, which was the opposite change observed in the leaves of creeping bentgrass (Figure 4). In the stolon nodes, glucose content under elevated CO_2 was 30.0%, 48.7%, 53.2%, and 48.4% higher (Figure 4E) and fructose content was 30.4%, 50.1%, 54.6%, and 49.8% higher (Figure 4F) than that under ambient CO_2 at 7, 21, 35, and 42 d, respectively. Similarly, elevated CO_2 resulted in a significant increase in sucrose content from 7 to 42 d of treatment time in comparison with ambient CO_2 (Figure 4G). Hence, total soluble sugar in the stolon nodes of creeping bentgrass was significantly enhanced by 30.1%, 49.8%, 53.5%, and 48.7% at 7, 21, 35, and 42 d, respectively, under elevated CO_2 concentration (Figure 4H).

Similar to the responses observed in the stolons, soluble sugars including glucose, fructose, and sucrose as well as total soluble sugar contents also increased under elevated CO_2 compared with ambient CO_2 in the stolon internodes of creeping bentgrass (Figure 4). Elevated CO_2 significantly increased the glucose content by 21.4%, 24.2%, 28.4%, and 28.1% (Figure 4I), and fructose content by 22.5%, 33.5%, 29.0%, and 28.6% at 7, 21, 35, and 42 d of treatments, respectively (Figure 4J). The sucrose content was significantly enhanced by elevated CO_2 in consistence with glucose and fructose at 7, 21, 35, and 42 d of treatments (Figure 4K). Therefore, total soluble sugars in the internodes increased by 21.6%, 24.3%, 28.5%, and 28.2% at 7, 21, 35, and 42 d, respectively, due to elevated CO_2 compared with ambient CO_2 concentration (Figure 4L).

2.3. Effects of Elevated CO_2 on Root Soluble Sugars

Soluble sugars, including glucose, fructose, sucrose, and total soluble sugars, in the roots, decreased when plants were exposed to elevated CO_2 (Figure 5). In response to elevated CO_2, root glucose and fructose contents significantly declined by 49.7% and 50.4%, respectively (Figure 5A,B). Elevated CO_2 also resulted in a reduction in root sucrose content by 48.0% in comparison with ambient CO_2 (Figure 5C). Collectively, the total soluble sugar contents in the roots were significantly decreased by 49.8% under elevated CO_2 compared with ambient CO_2 concentration (Figure 5D).

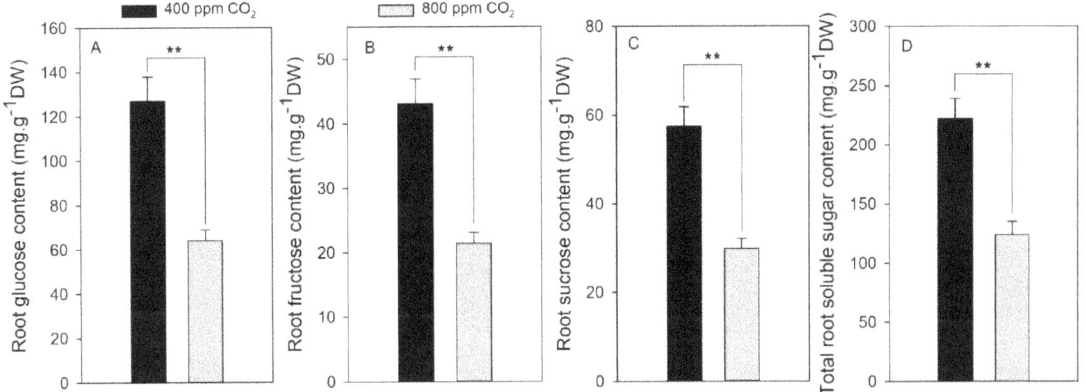

Figure 5. Effects of elevated CO_2 concentration on root content of glucose (**A**), fructose (**B**), sucrose (**C**) and total soluble sugar (**D**) in creeping bentgrass at 42 d of experiment. Four hundred µmol mol^{-1} (ppm) CO_2, ambient CO_2 concentration; 800 ppm CO_2, elevated CO_2 concentration. Sugar contents are presented in the unit of mg g^{-1} dry weight (DW). ** indicates a significant difference between ambient and elevated CO_2 concentrations according to Student's *t*-test at $p \leq 0.01$. Error bars represent standard error (SE).

2.4. Effects of Elevated CO_2 on Endogenous Hormone Content

Endogenous IAA, iPA, GA_1, GA_3, and GA_4 levels in different plant tissues including root, node, internode, and leaf are shown in Figure 6. Compared with ambient CO_2, the content of IAA was significantly increased by elevated CO_2 in both nodes and internodes by 39.0% and 22.1%, respectively, but not in roots or leaves (Figure 6A). No difference was found in the contents of iPA, GA_3, or GA_4 in plants grown under elevated CO_2 (Figure 6B,D,E). Among all the tissues tested, only leaf exhibited a significant decrease in GA_1 content due to elevated CO_2 in comparison with ambient CO_2 (Figure 6C).

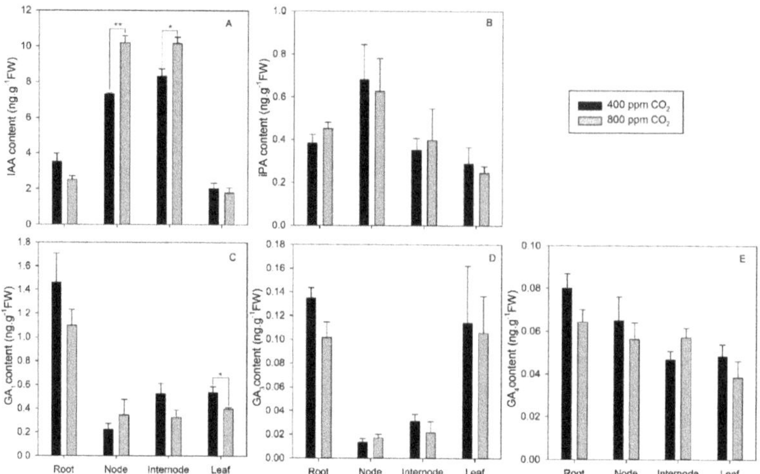

Figure 6. Effects of elevated CO_2 concentration on content of IAA (**A**), iPA (**B**), GA_1 (**C**), GA_3 (**D**), and GA_4 (**E**) in the root, node, internode, and leaf tissues of creeping bentgrass at 42 d of experiment. Hormone contents are presented in the unit of ng g^{-1} fresh weight (FW). Four hundred μmol mol^{-1} (ppm) CO_2, ambient CO_2 concentration; 800 ppm CO_2, elevated CO_2 concentration. * and ** indicate a significant difference between ambient and elevated CO_2 concentrations according to Student's *t*-test at $p \leq 0.05$ and $p \leq 0.01$, respectively. Error bars represent standard error (SE).

3. Discussion

Extensive stolon development and growth are superior traits for the rapid establishment as well as post-stress regeneration for survival in stoloniferous grass species. Although stolon initiation and formation are mainly controlled by genetic factors, the growth and development of stolon are often influenced by diverse factors. Previous reports have demonstrated that changes in stolon internode length and rhizome length were associated with several factors such as temperature, nitrogen application, water availability, stolon internode position, burial depth of stolon internode as well as elevated CO_2 [28–33]. For example, the total rhizome length of Kentucky bluegrass exposed to drought stress was significantly lower under ambient CO_2 but unchanged under elevated CO_2 concentration [33]. In our study, elevated CO_2 stimulated the aboveground stolon growth through increases in stolon internode number and length, these morphological changes lead to more than double the shoot biomass of creeping bentgrass grown under ambient CO_2 levels (Figures 1 and 2). In other stoloniferous plants, increased stolon internode length as well as stolon thickness could enhance the survival rate and regeneration capacity due to the increased amount of reserves such as soluble proteins, starch, and soluble sugars in the stolons [34–36]. In this study, longer stolon internode length suggested the tendency of creeping bentgrass to spread horizontally to sustain the enhanced photosynthetic capacity under elevated CO_2. The greater shoot biomass was a result of the dramatic stolon elongation as well as the increase in leaf number (data not shown) which was due to the increased stolon internode number. The potential mechanisms of CO_2-induced stolon elongation involved in metabolic pathways in stoloniferous creeping bentgrass are discussed below, including soluble sugars (glucose, fructose, and sucrose), endogenous hormones (IAA, iPA, and GAs), and root growth and development.

Elevated CO_2-enhanced plant growth is a common response and has been well documented in various plant species without stolons [13,17,20]. In this study, the proliferative shoot growth under elevated CO_2 was in fact a result of the significant increase in stolon growth (Figures 1 and 2). Interestingly, the contents of soluble sugars glucose, fructose, and sucrose were increased significantly in stolon nodes and internodes but decreased in leaves and roots in creeping bentgrass subjected to elevated CO_2 concentration (Figures 4 and 5), suggesting that soluble carbohydrates were allocated to stolons for storage rather than to leaves and roots in a stoloniferous plant. In other plants without stolons, elevated CO_2 did not lead to a decline in soluble sugars in leaves such as radish (*Raphanus sativus* L.) [37], barley (*Hordeum vulgare* L.) cultivars [38], cork oak (*Quercus suber* L.) [39], and perennial Kentucky bluegrass [17,18]. Therefore, the results of soluble sugars revealed that the positive effects of elevated CO_2 in stolon growth are likely to be attributed to the increased photosynthetic carbon acquisitor as well as the alteration in carbon reallocation [40].

Resources including soluble carbohydrates, starch, mineral nutrients, and soluble protein stored in stolon nodes and internodes may be responsible for plant survival and regeneration in order to cope with severe disturbance when plants are disturbed by various biotic and abiotic factors [4,41–43]. The content of carbohydrates in stolon nodes and internodes was positively correlated with the survival rate of stoloniferous plants [44]. In order to quickly establish in the soil, zoysiagrass (*Zoysia* spp.) genotypes with greater total stolon length were demonstrated to distribute more dry matter to stolons and rhizomes instead of leaves [45]. In this study, the soluble sugars from leaves and roots were reallocated to stolon nodes and internodes to support the enhanced stolon growth under elevated CO_2 concentration. The increased stolon growth and carbohydrates storage may explain the mechanism behind the elevated CO_2 enhanced survival and recovery traits under abiotic stresses such as heat [19], drought [27,46], and salinity [20] in grass species with stolons.

It is interesting to find that root biomass under elevated CO_2 was significantly higher than that under ambient CO_2 conditions, although the content of soluble sugars in root was decreased by elevated CO_2 in this study (Figures 3 and 5). The increase in root biomass was attributed to the greater root density but not the root length as indicated by root phenotype under elevated CO_2 (Figure 3). Our observation of root biomass was in accordance with other studies which were also conducted in creeping bentgrass [21,47]. In other plants, the improvement of elevated CO_2 on root density was mainly due to elevated CO_2-induced formation and development of lateral roots and fine roots as reported in *Sedum alfredii* Hance. [48] and maize (*Zea mays* L.) [49]. The lower root soluble sugars content was likely due to the consumption for producing greater root biomass in combination with the reallocation of soluble sugars from roots to stolon nodes and internodes under elevated CO_2 conditions. Our study is the first report that examined the stimulation of elevated CO_2 on stolon growth from carbon reallocation among root, leaf, node, and internode tissues in stoloniferous plant species.

Hormones are crucial regulators of plant growth and development; hence, plants might alter their hormone levels to regulate plant growth in response to elevated CO_2 conditions. Early research has demonstrated that IAA plays important roles in regulating stolon growth and development by cell division and cell elongation [50]. Exogenous IAA applied at the distal end of decapitated stolons in *Saxifraga sarmentosa* L. enhanced the translocation of ^{14}C assimilates from the leaf into the stolon [50]. Exogenous cytokinin increased auxin content in the stolon tips of potato resulting in tuber initiation [22]. In our study, elevated CO_2 increased the endogenous level of IAA in both stolon nodes and internodes but not in roots or leaves in creeping bentgrass (Figure 6). The result indicated that elevated CO_2 not only directly promoted carbon fixation through photosynthesis but also regulated growth by controlling endogenous auxin levels. This would explain why the higher content of IAA in stolon nodes and internodes but not in leaves or roots was in consistence with the allocation of soluble sugars. The unchanged IAA level in response to elevated CO_2 in leaf was also observed by Burgess et al. [27] in creeping bentgrass under unstressed conditions. The accumulation of IAA in stolon node and internode implied that increased endogenous IAA content could have provided a great contribution to the rapid stolon elongation and growth in creeping bentgrass exposed to elevated CO_2 conditions.

Apart from auxins, cytokinins and GAs are generally believed to serve as positive regulators of plant growth and development [51,52]. In this study, no significant difference was found in the content of iPA and GAs (GA_1, GA_3, and GA_4) in the stolons or roots of creeping bentgrass grown under ambient and elevated CO_2 concentrations (Figure 6B–E). Similarly, exogenous kinetin and GA to the distal zone of stolons in *Saxifraga sarmentosa* generated a small insignificant effect in promoting stolon growth [50]. Auxin was reported to inflict a negative effect on cytokinins by inhibiting IPT expression and enhancing CYTOKININ OXIDASE/DEHYDROGENASE (CKX) expression to reduce the content of cytokinins in different species [53]. Therefore, the significant increase in IAA may have inhibited the production of iPA in stolon nodes and internodes under elevated CO_2 conditions. Furthermore, we also observed a reduction in GA_1 in leaves under elevated CO_2 (Figure 6C). Adjusting the GA_1 concentration in plants has great practical uses. Plant growth regulators, such as trinexapac-ethyl, were utilized to inhibit GA_1 production. In particular, trinexapac-ethyl blocks the conversion of metabolically inactive GA_{20} to active GA_1 [54]. Trinexapac-ethyl is one of the most widely used plant growth regulators in turfgrass management and numerous research reports have demonstrated its benefits to turfgrass with improved tolerance to biotic and abiotic stresses [55]. In creeping bentgrass, trinexapac-ethyl improved drought and heat tolerance [56,57]. The reduction in GA_1 in the leaves due to elevated CO_2 could have contributed to increased tolerance to other stresses (such as heat and drought) in a similar way as regulated by trinexapac-ethyl. However, this speculation will require further investigation. Our study quantified endogenous hormones in different tissue types of creeping bentgrass and therefore provided important evidence suggesting that elevated CO_2-induced stolon elongation resulted from IAA increase but not iPA or GAs in stolon nodes and internodes. In response to elevated CO_2 concentration, the decreased GA_1 level from this research also supported the shorter leaf length observed by Burgess and Huang [21] in the same species, creeping bentgrass. Therefore, the proposed hormone regulation model for creeping bentgrass was that elevated CO_2 promoted lateral growth but not vertical growth by increasing the IAA level in stolons and decreasing the GA_1 level in leaves.

4. Materials and Methods

4.1. Plant Material and Growth Conditions

Creeping bentgrass (cv. 'Penn-A4') stolons with the same number of nodes were planted in polyvinyl chloride (PVC) tubes (10 cm in diameter and 50 cm in depth) filled with sand. Plants were established for about three months from July to September 2020 in a greenhouse with an average temperature of 25/20 °C (day/night), PAR of 450 μmol m$^{-2} \cdot$ s^{-1}, and 14 h photoperiod. Plants were trimmed twice a week to promote density and irrigated with Hoagland solution [58] once a week. After establishment, plants were acclimated in a growth chamber (Xubang, Jinan, China) with the temperature set at 25/20 °C (day/night), 70% relative humidity, PAR of 600 μmol m$^{-2} \cdot$ s^{-1} at the canopy level, and a 14 h photoperiod for one week before treatments initiation.

4.2. Experimental Design and Treatments

The experiment was initiated on 23rd October in 2020 with five replications of two CO_2 treatments: ambient CO_2 concentration (400 $\pm$ 10 μmol mol^{-1}) and elevated CO_2 concentration (800 $\pm$ 10 μmol mol^{-1}). The CO_2 concentration of growth chambers was automatically controlled through an open-chamber control system via computer programs connected to a CO_2 gas tank with 100% CO_2 [2]. During the experiment, PVC pots were randomly relocated every other day within and across chambers to avoid spatial environmental variations in chambers.

4.3. Growth and Physiological Measurements

The impacts of elevated CO_2 on stolon growth were evaluated by measuring stolon internode length and stolon internode number on each individual stolon as well as the total stolon length of plants in each pot according to Xu et al. [2] with minor modifications. Four individual stolons were labeled in each pot at 0 d of treatments. The internode length for each labeled stolon and longest root length were measured by a ruler. The internode numbers of each labeled stolon were counted on every sampling day. The total stolon length was measured from the labeled point to the tip of each stolon.

Biomass of shoot and root was measured by drying the total tissues from each pot to a constant weight at 70 °C for 3 days at 42 d of experiment. The dry biomass weights were divided by the surface area of the PVC pot to report the sample biomass weight per unit area (kg m^{-2}).

4.4. Sugar Extraction and Quantification

Soluble sugars including glucose, fructose, and sucrose were quantified using the phenol-sulfuric acid method described by Liu et al. [59] with modifications. Leaf samples at 7, 21, 35, and 42 days of treatment were collected and dried, then ground to a fine powder with a pestle. To extract soluble sugars, 25 mg of fine powder was mixed with 5.0 mL of 80% (v/v) aqueous ethyl alcohol in a 15 mL microcentrifuge tube and incubated in a water bath at 30 °C for 30 min. Microcentrifuge tubes were then centrifuged at 4500 rpm for 10 min to obtain supernatant. The supernatant was transferred to 50 mL microcentrifuge tubes and 2.5 mL of 80% (v/v) aqueous ethanol was added and extracted two times with the same method to obtain the final extractant. A subsample of 1 mL extractant was mixed with 1 mL 23% (v/v) phenol solution, then 5 mL 98% (v/v) concentrated sulfuric acid was added to the solution and mixed well. The reaction solution was cooled down to room temperature for 15 min and then incubated in a water bath at 30 °C for 30 min. The absorbance of the reaction solution at 490 nm was measured with a spectrophotometer (Ultrospec 2100 pro, Biochrom Ltd., Cambridge, UK). Glucose, fructose, and sucrose contents were quantified by comparing their standard curves. The total soluble sugars reported in this study were calculated as the sum of glucose, fructose, and sucrose.

4.5. Hormone Measurement

The extraction procedure of hormones (IAA, iPA, GA$_1$, GA$_3$, and GA$_4$) was conducted according to the modified method by Pan et al. [60]. One gram of each leaf, node, internode, and root fresh sample was collected at 42 d from plants grown under different CO$_2$ concentrations and ground to a fine powder in liquid nitrogen and then transferred into microcentrifuge tubes. A 10 mL isopropanol/hydrochloric extract buffer was added to tubes and shaken at 4 °C for 30 min, and 20 mL dichloromethane was added for an additional 30 min shaking at 4 °C. The solution was centrifuged at 4 °C, 12,000 rpm for 5 min, and the lower phase was concentrated by nitrogen evaporator into the dried precipitate which was dissolved in 200 µL methanol containing 0.1% formic acid. Then, the extraction was filtered by a 0.22 µm filter membrane for further hormones measurement.

Plant hormone samples were quantified using HPLC-MS/MS by 1290 HPLC (Agilent, Santa Clara, CA, USA) and SCIEX-6500 Qtrap (AB *Sciex*, Foster, CA, USA), following the parameters setup as described by Pan et al. [60]. Standards of plant hormones including IAA, iPA, GA$_1$, GA$_3$, and GA$_4$ were ordered from Sigma-Aldrich and dissolved in methyl alcohol with 0.1% methanoic acid for the external standard curves. The HPLC conditions were: reverse-phase poroshell 120 SB-C^{18} chromatographic column (Agilent, Palo Alto, CA, USA) with a column temperature of 30 °C. Mobile phases A:B = (0.1% formic acid in methanol): (0.1% formic acid in water) was used for separation. The elution gradient was set as follows: 0–1 min A = 20%; 1–3 min A increased from 20% to 50%; 3–9 min A increased from 50% to 80%; 9–10.5 min A = −80%; 10.5–10.6 min A decreased from 80% to 20%; 10.6–13.5 min A = 20%. The injection volume was 2 µL. The MS conditions were set as follows: ionspray voltage 4500 v, source temperature 400 °C, curtain gas 15 psi, nebulizing gas 65 psi, auxiliary gas 70 psi.

4.6. Statistical Analyses

Data were analyzed using SPSS statistics software (SPSS 18.0; SPSS Inc., Chicago, IL, USA). The Pearson correlation analysis was used to analyze the effects of elevated CO$_2$ on all parameters including shoot and root biomass, stolon internode number and length, total stolon length, and longest root length. The means $\pm$ standard error (SE) was summarized in charts for shoot biomass, root biomass, longest root length, internode number, internode length, total stolon length, shoot soluble sugar contents, root soluble sugar contents, and hormone contents. Student's *t*-tests were used to determine significant differences at confidence levels of 0.05 and 0.01.

5. Conclusions

In conclusion, elevated CO_2 enhanced the stolon growth by promoting stolon internode number, internode length, and root biomass in creeping bentgrass, as manifested by the longer total stolon length and greater shoot biomass. The regulatory model of the aforementioned carbohydrates and hormones which may be associated with stolon growth are summarized in Figure 7. The content of soluble sugars including glucose, sucrose, and fructose as well as endogenous IAA was accumulated in stolon nodes and internodes but not in leaves or roots under elevated CO_2 concentration. These results illustrated that the accumulation and reallocation of glucose, sucrose, and fructose to stolons as well as the increased IAA level in stolon nodes and internodes could contribute to the enhancement of stolon growth under elevated CO_2 in creeping bentgrass. Our study is an important step further in understanding the endogenous hormones and soluble sugars reallocation involved in elevated CO_2-enhanced stolon growth. However, the molecular mechanism underlying the enhanced stolon development is still unknown. Research is needed to explore the detailed mechanisms as to how CO_2-responsive soluble carbohydrates and IAA in stolon node and internode regulate stolon growth in creeping bentgrass in order to provide further insights into survival strategies by promoting stolon growth and biomass production of above-ground shoots.

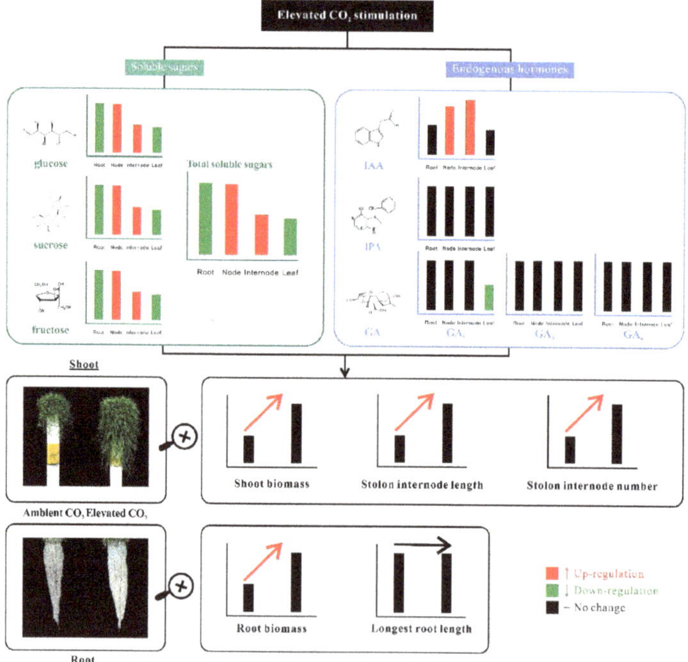

Figure 7. Working model for elevated CO_2-responsive metabolic pathways associated with soluble sugars and endogenous hormones in regulating stolon growth in creeping bentgrass.

Author Contributions: Conceptualization, Z.Y. and J.Y.; methodology, M.L. and Q.L.; formal analysis, J.Y. and M.L.; data curation, J.Y., Q.L., R.L. and R.W.; writing—original draft preparation, J.Y. and R.W.; writing—review and editing, J.Y., M.L., Q.L., R.W., R.L. and Z.Y. All authors revised and edited the manuscript; funding acquisition, Z.Y. and J.Y. All authors have read and agreed to the published version of the manuscript.

Funding: This study was supported by the program of the National Natural Science Foundation of China (31971771), Chongqing Key Laboratory of Germplasm Innovation and Utilization of Native Plants (XTZW2021-KF04) and Chinese Postdoctoral Science Foundation (2019M651866).

Institutional Review Board Statement: Not applicable.

Informed Consent Statement: Not applicable.

Data Availability Statement: Data are available upon request from the first author.

Conflicts of Interest: The authors declare no conflict of interest.

References

1. Savini, G.; Giorgi, V.; Scarano, E.; Neri, D. Strawberry plant relationship through the stolon. *Physiol. Plant.* **2008**, *134*, 421–429. [CrossRef] [PubMed]
2. Xu, Q.; Fan, N.; Zhuang, L.; Yu, J.; Huang, B. Enhanced stolon growth and metabolic adjustment in creeping bentgrass with elevated CO_2 concentration. *Environ. Exp. Bot.* **2018**, *155*, 87–97. [CrossRef]
3. Pitelka, L.E.; Ashmun, J.W. Physiology and integration of ramets in clonal plants. In *Population Biology and Evolution of Clonal Organisms*; Jackson, J.B.C., Ed.; Yale University Press: New Haven, CT, USA, 1985.
4. Stuefer, J.F.; Huber, H. The role of stolon internodes for ramet survival after clone fragmentation in *Potentilla anserina*. *Ecol. Lett.* **1999**, *2*, 135–139. [CrossRef]
5. Fahrig, L.; Coffin, D.P.; Lauenroth, W.K.; Shugart, H.H. The advantage of long-distance clonal spreading in highly disturbed habitats. *Evol. Ecol.* **1994**, *8*, 172–187. [CrossRef]
6. Song, Y.B.; Yu, F.H.; Li, J.M.; Keser, L.H.; Fischer, M.; Dong, M.; Kleunen, M.V. Plant invasiveness is not linked to the capacity of regeneration from small fragments: An experimental test with 39 stoloniferous species. *Biol. Invasions* **2013**, *15*, 1367–1376. [CrossRef]
7. Hanna, W.W. Centipedegrass—Diversity and vulnerability. *Crop Sci.* **1995**, *35*, 332–334. [CrossRef]
8. Pessarakli, M. *Handbook of Turfgrass Management and Physiology*; CRC Press: Boca Raton, FL, USA, 2007.
9. Turgeon, A. *Turfgrass Management*, 8th ed.; Pearson Prentice Hall: Upper Saddle River, NJ, USA, 2008.
10. Zheng, Y.; Li, F.; Hao, L.; Yu, J.; Guo, L.; Zhou, H.; Ma, C.; Zhang, X.; Xu, M. Elevated CO_2 concentration induces photosynthetic down-regulation with changes in leaf structure, non-structural carbohydrated and nitrogen content of soybean. *BMC Plant Biol.* **2019**, *19*, 255. [CrossRef]
11. Pedersen, J.; Santos, F.D.; Vuuren, D.V.; Gupta, J.; Swart, R. An assessment of the performance of scenarios against historical global emissions for IPCC reports. *Global Environ. Chang.* **2021**, *66*, 102199. [CrossRef]
12. Huang, B.; Xu, Y. Cellular and molecular mechanisms for elevated CO_2-regulation of plant growth and stress adaptation. *Crop Sci.* **2015**, *55*, 1405–1424. [CrossRef]
13. Yu, J.J.; Chen, L.H.; Xu, M.; Huang, B.R. Effects of elevated CO_2 on physiological responses of tall fescue to elevated temperature, drought stress, and the combined stresses. *Crop Sci.* **2012**, *52*, 1848–1858. [CrossRef]
14. Yu, J.J.; Yang, Z.M.; Jespersen, D.; Huang, B.R. Photosynthesis and protein metabolism associated with elevated CO_2-mitigation of heat stress damages in tall fescue. *Environ. Exp. Bot.* **2014**, *99*, 75–85. [CrossRef]
15. Yu, J.; Fan, N.; Li, R.; Zhuang, L.; Xu, Q.; Huang, B. Proteomic profiling for metabolic pathways involved in interactive effects of elevated carbon dioxide and nitrogen on leaf growth in a perennial grass species. *J. Proteome Res.* **2019**, *18*, 2446–2457. [CrossRef]
16. Chen, Y.J.; Yu, J.J.; Huang, B.R. Effects of elevated CO_2 concentration on water relations and photosynthetic responses to drought stress and recovery during rewatering in tall fescue. *J. Am. Soc. Hortic. Sci.* **2015**, *140*, 19–26. [CrossRef]
17. Song, Y.L.; Yu, J.J.; Huang, B. Elevated CO_2-mitigation of high temperature stress associated with maintenance of positive carbon balance and carbohydrate accumulation in kentucky bluegrass. *PLoS ONE* **2014**, *9*, e89725. [CrossRef]
18. Zhuang, L.; Yang, Z.; Fan, N.; Yu, J.; Huang, B. Metabolomic changes associated with elevated CO_2-regulation of salt tolerance in Kentucky bluegrass. *Environ. Exp. Bot.* **2019**, *165*, 129–138. [CrossRef]
19. Yu, J.; Li, R.; Fan, N.; Yang, Z.; Huang, B. Metabolic pathways involved in carbon dioxide enhanced heat tolerance in bermudagrass. *Front. Plant Sci.* **2017**, *8*, 1506. [CrossRef]
20. Yu, J.; Sun, L.; Fan, N.; Yang, Z.; Huang, B. Physiological factors involved in positive effects of elevated carbon dioxide concentration on bermudagrass tolerance to salinity stress. *Environ. Exp. Bot.* **2015**, *115*, 20–27. [CrossRef]
21. Burgess, P.; Huang, B. Growth and physiological responses of creeping bentgrass (*Agrostis stolonifera*) to elevated carbon dioxide concentrations. *Hortic. Res.* **2014**, *1*, 14021. [CrossRef]

22. Kinmonth-Schultz, H.; Kim, S.H. Carbon gain, allocation, and storage in rhizomes in response to elevated CO_2 and fertilization in an invasive perennial C_3 grass, *Phalaris arundinacea*. *Func. Plant Biol.* **2011**, *38*, 797–807. [CrossRef]
23. Kondhare, K.R.; Patil, A.B.; Giri, A.P. Auxin: An emerging regulator of tuber and storage root development. *Plant Sci.* **2021**, *306*, 110854. [CrossRef]
24. Mu, X.; Chen, Q.; Wu, X.; Chen, F.; Yuan, L.; Mi, G. Gibberellins synthesis is involved in the reduction of cell flux and elemental growth rate in maize leaf under low nitrogen supply. *Environ. Exp. Bot.* **2018**, *150*, 198–208. [CrossRef]
25. Liu, D.; Xu, M.; Hu, Y.; Wang, R.; Tong, J.; Xiao, L. Dynamic changes of key plant hormones during potato tuber development. *Mol. Plant Breed.* **2019**, *6*, 1998–2003.
26. Wu, W.; Kang, D.; Kang, X.; Wei, H. The diverse roles of cytokinins in regulating leaf development. *Hortic. Res.* **2021**, *8*, 118. [CrossRef]
27. Burgess, P.; Chapman, C.; Zhang, X.; Huang, B. Stimulation of growth and alteration of hormones by elevated carbon dioxide for creeping bentgrass exposed to drought. *Crop Sci.* **2019**, *59*, 1672–1680. [CrossRef]
28. Elgersma, A.; Li, F. Effects of cultivar and cutting frequency on dynamics of stolon growth and leaf appearance in white clover in mixed swards. *Grass Forage Sci.* **1997**, *52*, 370–380. [CrossRef]
29. Tworkoski, T.J.; Benassi, T.E.; Takeda, F. The effect of nitrogen on stolon and ramet growth in four genotypes of *Fragaria chiloensis* L. *Sci. Hortic.* **2001**, *88*, 106. [CrossRef]
30. O'Neal, S.W.; Prince, J.S. Seasonal effects of light, temperature, nutrient concentration and salinity on the physiology and growth of *Caulerpa paspaloides* (Chlorophyceae). *Marine Biol.* **1988**, *97*, 17–24. [CrossRef]
31. Dong, B.C.; Yu, G.L.; Wei, G.; Zhang, M.X.; Dong, M.; Yu, F.H. How internode length, position and presence of leaves affect survival and growth of *Alternanthera philoxeroides* after fragmentation? *Ecol. Evol.* **2010**, *24*, 1447–1461. [CrossRef]
32. Dong, B.; Liu, R.; Zhang, Q.; LI, H.; Zhang, M.; Lei, G.; Yu, F. Burial depth and stolon internode length independently affect survival of small clonal fragments. *PLoS ONE* **2011**, *6*, e23942. [CrossRef]
33. Chapman, C.; Burgess, P.; Huang, B. Effects of elevated carbon dioxide on drought tolerance and post-drought recovery involving rhizome growth in kentucky bluegrass (*Poa pratensis* L.). *Crop Sci.* **2021**, *61*, 3219–3231. [CrossRef]
34. Lawson, A.R.; Kelly, K.B.; Sale, P. Defoliation frequency and cultivar effects on the storage and utilisation of stolon and root reserves in white clover. *Aust. J. Agr. Res.* **2000**, *51*, 1039–1046. [CrossRef]
35. Goulas, E.; Le Dily, F.; Teissedre, L.; Corbel, G.; Robin, C.; Christophe, R.; Ourry, A. Vegetative storage proteins in white clover (*Trifolium repens* L.): Quantitative and qualitative features. *Ann. Bot.* **2001**, *88*, 789–795.
36. Huang, Q.; Shen, Y.; Li, X.; Zhang, G.; Huang, D.; Fan, Z. Regeneration capacity of the small clonal fragments of the invasive *Mikania micrantha* H.B.K.: Effects of the stolon thickness, internode length and presence of leaves. *Weed Biol. Manag.* **2014**, *15*, 70–77. [CrossRef]
37. Urbonaviciute, A.; Samuoliene, G.; Sakalauskaite, J.; Duchovskis, P.; Brazaityte, A.; Siksnianiene, J.B.; Ulinskaite, R.; Sabajeviene, G.; Baranauskis, K. The effect of elevated CO_2 concentrations on leaf carbohydrate, chlorophyll contents and photosynthesis in radish. *Pol. J. Environ. Stud.* **2006**, *15*, 921–925.
38. Pérez-López, U.; Robredo, A.; Lacuesta, M.; Mena-Petite, A.; Muñoz-Rueda, A. Elevated CO_2 reduces stomatal and metabolic limitations on photosynthesis caused by salinity in *Hordeum vulgare*. *Photosynth. Res.* **2012**, *111*, 269–283. [CrossRef]
39. Faria, T.; Wilkins, D.; Besford, R.T.; Vaz, M.; Pereira, J.S.; Chaves, M.M. Growth at elevated CO_2 leads to down-regulation of photosynthesis and altered response to high temperature in *Quercus suber* L. seedlings. *J. Exp. Bot.* **1996**, *47*, 1755–1761. [CrossRef]
40. Jach, M.E.; Ceulemans, R. Effects of elevated atmospheric CO_2 on phenology, growth and crown structure of Scots pine (*Pinus sylvestris*) seedlings after two years of exposure in the field. *Tree Physiol.* **1999**, *19*, 289–300. [CrossRef]
41. Li, X.; Shen, Y.; Huang, Q.; Fan, Z.; Huang, D. Regeneration capacity of small clonal fragments of the invasive *Mikania micrantha* H.B.K.: Effects of burial depth and stolon internode length. *PLoS ONE* **2013**, *8*, e84657. [CrossRef]
42. Zhou, Y.; Lambrides, C.; Fukai, S. Associations between drought resistance, regrowth and quality in a perennial C_4 grass. *Eur. J. Agron.* **2015**, *65*, 1–9. [CrossRef]
43. He, L.; Xiao, X.; Zhang, X.; Jin, Y.; Pu, Z.; Lei, N.; He, X.; Chen, J. Clonal fragments of stoloniferous invasive plants benefit more from stolon storage than their congeneric native species. *Flora* **2021**, *281*, 151877. [CrossRef]
44. Fry, J.D.; Lang, N.S.; Clifton, R.; Maier, F.P. Freezing tolerance and carbohydrate content of low-temperature-acclimated and nonacclimated centipedegrass. *Crop Sci.* **1993**, *33*, 1051–1055. [CrossRef]
45. Patton, A.J.; Volenec, J.J.; Reicher, Z.J. Stolon growth and dry matter partitioning explain differences in zoysiagrass establishment rates. *Crop Sci.* **2007**, *47*, 1237–1245. [CrossRef]
46. Chapman, C.; Burgess, P.; Huang, B. Responses to elevated carbon dioxide for postdrought recovery of turfgrass species differing in growth characteristics. *Crop Sci.* **2021**, *61*, 4436–4446. [CrossRef]
47. Burgess, P.; Huang, B. Root protein metabolism in association with improved root growth and drought tolerance by elevated carbon dioxide in creeping bentgrass. *Field Crop Res.* **2014**, *165*, 80–91. [CrossRef]
48. Li, T.; Di, Z.; Han, X.; Yang, X. Elevated CO_2 improves root growth and cadmium accumulation in the hyperaccumulator *Sedum alfredii*. *Plant Soil* **2012**, *354*, 325–334. [CrossRef]
49. Hiltpold, I.; Moore, B.; Johnson, S. Elevated atmospheric carbon dioxide concentrations alter root morphology and reduce the effectiveness of entomopathogenic nematodes. *Plant Soil* **2020**, *477*, 29–38. [CrossRef]

50. Da Cruz, G.S.; Audus, L.J. Studies of hormone-directed transport in decapitated stolons of *Saxifraga sarmentosa*. *Ann. Bot.* **1978**, *42*, 1009–1027. [CrossRef]
51. Braun, J.; Kender, W. Correlative bud inhibition and growth habit of the strawberry as influenced by application of gibberellic acid, cytokinin, and chilling during short daylength. *J. Am. Soc. Hortic. Sci.* **1985**, *110*, 28–34.
52. Tyagi, S.; Kumar, S. Exogenous supply of IAA, GA and cytokinin to salinity stressed seeds of chickpea improve the seed germination and seedling growth. *Int. J. Plant Sci.* **2016**, *11*, 88–92. [CrossRef]
53. Kotov, A.A.; Kotova, L.M.; Romanov, G.A. Signaling network regulating plant branching: Recent advances and new challenges. *Plant Sci.* **2021**, *307*, 110880. [CrossRef]
54. Adams, R.E.; Kerber, K.; Pfister, E.W.; Weiler, E.W. Studies on the action of the new growth retardant CGA 163′935 (Cimectacarb). In *Progress in Plant Growth Regulation*; Karssen, C.M., van Loon, L.C., Vreugdenhil, D., Eds.; Kluwer Academic Publishers: Dordrecht, The Netherlands, 1992; pp. 818–827.
55. Reicher, Z.J.; Dernoeden, P.H.; Richmond, D.S. Insecticides, fungicides, herbicides, and growth regulators used in turfgrass systems. In *Turfgrass: Biology, Use, and Management*; Stier, J.C., Horgan, B.P., Bonos, S.A., Eds.; American Society of Agronomy Soil Science Society of America Crop Science Society of America, Inc.: Madison, WI, USA, 2013.
56. McCann, S.E.; Huang, B. Effects of trinexapac-ethyl foliar application on creeping bentgrass responses to combined drought and heat stress. *Crop Sci.* **2007**, *47*, 2121–2128. [CrossRef]
57. McCann, S.E.; Huang, B. Drought responses of Kentucky bluegrass and creeping bentgrass as affected by abscisic acid and trinexapac-ethyl. *J. Am. Soc. Hortic. Sci.* **2008**, *133*, 20–26. [CrossRef]
58. Hoagland, D.R.; Arnon, D.I. *The Water-Culture Method for Growing Plans without Soil*; California Agricultural Experiment Station; The College of Agriculture University of California Berkeley: Davis, CA, USA, 1950; Volume 347, pp. 1–32.
59. Liu, N.; Shen, Y.; Huang, B. Osmoregulants involved in osmotic adjustment for differential drought tolerance in different bentgrass genotypes. *J. Am. Soc. Hortic. Sci.* **2015**, *140*, 605–613. [CrossRef]
60. Pan, X.; Welti, R.; Wang, X. Quantitative analysis of major plant hormones in crude plant extracts by high-performance liquid chromatography-mass spectrometry. *Nat. Protoc.* **2010**, *5*, 986–992. [CrossRef]

Article

Seed Priming with Chitosan Improves Germination Characteristics Associated with Alterations in Antioxidant Defense and Dehydration-Responsive Pathway in White Clover under Water Stress

Yao Ling [†], Yue Zhao [†], Bizhen Cheng, Meng Tan, Yan Zhang and Zhou Li *

Department of Turf Science and Engineering, College of Grassland Science and Technology, Sichuan Agricultural University, Chengdu 611130, China; ly9729752@163.com (Y.L.); zhaoyue19971120@126.com (Y.Z.); chengbizhengrass@163.com (B.C.); tanmeng194@163.com (M.T.); zhangyan1111zy@126.com (Y.Z.)
* Correspondence: lizhou1986814@163.com
† These authors contributed equally to this work.

Abstract: Water stress decreases seed-germination characteristics and also hinders subsequent seedling establishment. Seed priming with bioactive compounds has been proven as an effective way to improve seed germination under normal and stressful conditions. However, effect and mechanism of seed priming with chitosan (CTS) on improving seed germination and seedling establishment were not well-understood under water-deficit conditions. White clover (*Trifolium repens*) seeds were pretreated with or without 5 mg/L CTS before being subjected to water stress induced by 18% (w/v) polyethylene glycol 6000 for 7 days of germination in a controlled growth chamber. Results showed that water stress significantly decreased germination percentage, germination vigor, germination index, seed vigor index, and seedling dry weight and also increased mean germination time and accumulation of reactive oxygen species, leading to membrane lipid peroxidation during seed germination. These symptoms could be significantly alleviated by the CTS priming through activating superoxide dismutase, catalase, and peroxidase activities. In addition, seeds pretreated with CTS exhibited significantly higher expression levels of genes encoding dehydration-responsive transcription factors (*DREB2*, *DREB4*, and *DREB5*) and dehydrins (*Y2K*, *Y2SK*, and *SK2*) than those seeds without the CTS priming. Current findings indicated that the CTS-induced tolerance to water stress could be associated with the enhancement in dehydration-responsive pathway during seed germination.

Keywords: antioxidant enzyme; dehydrins; *DREB* transcription factor; oxidative damage; seed vigor; reactive oxygen species

Citation: Ling, Y.; Zhao, Y.; Cheng, B.; Tan, M.; Zhang, Y.; Li, Z. Seed Priming with Chitosan Improves Germination Characteristics Associated with Alterations in Antioxidant Defense and Dehydration-Responsive Pathway in White Clover under Water Stress. *Plants* **2022**, *11*, 2015. https://doi.org/10.3390/plants11152015

Academic Editor: Anis Limami

Received: 17 June 2022
Accepted: 29 July 2022
Published: 2 August 2022

Publisher's Note: MDPI stays neutral with regard to jurisdictional claims in published maps and institutional affiliations.

1. Introduction

With the development of global warming, drought stress has become one of the destructive environmental factors affecting seed germination and plant growth worldwide [1]. Seed germination and seedling establishment are key stages of plant growth and development but are also very vulnerable to drought stress [2]. Drought reduced seed-germination rate and subsequent seedlings establishment, leading to yield loss and quality deterioration [3]. Cell dehydration is one of main adverse effects induced by drought. Alteration of dehydration-responsive pathway is a universal response to water deficit in the plant kingdom [4]. For example, dehydration-responsive element-binding proteins (DREBs) recognize and bind to the dehydration-responsive element (DRE) of many downstream stress-responsive genes such as dehydrins (DHNs), which is an important adaptive strategy when plants survive under drought stress [5]. It has been widely reported that the overexpression of *DREBs* up-regulated transcriptional levels of different types of *DHNs*, thereby enhancing drought tolerance in many plants such as *Arabidopsis thaliana*, rice (*Oryza sativa* L.), tobacco (*Nicotiana tabacum* L.), and wheat (*Triticum aestivum* L.) [6,7]. Wheat seed pretreated with microbe *Bacillus* sp. or *Klebsiella* sp. effectively mitigated drought-induced declines in seedling biomass and root growth associated with significant up-regulation of *DHN* and *DREB* [8].

Oxidative damage induced by overaccumulation of reactive oxygen species (ROS) such as superoxide anion ($O_2{}^-$) and hydrogen peroxide (H_2O_2) is another serious consequence when seeds germinate and seedlings establish under drought condition [9]. Rapid detoxification of $O_2{}^-$ and H_2O_2 by regulating antioxidant defense has been recognized as one of pivotal adaptive mechanisms of drought tolerance in plants. As key components of antioxidant defense, superoxide dismutase (SOD) catalyzes dismutation of $O_2{}^-$ into H_2O_2 and O_2, and catalase (CAT), peroxidase (POD), and ascorbate peroxidase (APX) reduce H_2O_2 to nontoxic H_2O [10]. It has been found that drought-tolerant alfalfa (*Medicago sativa* L.) cultivar Xinmu No.1 accumulated lower H_2O_2 and malondialdehyde (MDA) contents through activating SOD, CAT, POD, and APX activities during seed germination under drought stress [11]. In response to drought, better antioxidant capacity and less ROS accumulation in soybean (*Glycine max* L.) seedlings were positively correlated with higher seed-germination rate [12]. Exogenous silicon could improve tomato (*Lycopersicon esculentum* L.) seed germination in relation to enhanced antioxidant enzymes activities and reduced oxidative stress [13]. These findings indicated the importance of effective antioxidant defense during seed germination and seedling establishment under water deficit condition.

Chitosan (CTS) is a bioactive compound from plants and marine crustaceans, such as crab shells and waste shrimp. In recent years, the CTS has been widely used in agricultural and horticultural fields for the improvement in crop quality and stress adaptation due to its non-toxic and biodegradable property [14]. It has been reported that the CTS could trigger many defensive responses to drought in plants. For example, exogenous application of CTS helped to maintain functional and structural integrity of biological membranes associated with increases in CAT and APX activities and the accumulation of secondary metabolite in periwinkle (*Catharanthus roseus* L.) [15]. Seed soaking with CTS could increase the accumulation of indoleacetic acid and free amino acids in favor of subsequent lupine (*Lupinus termis* L.) growth and yield under drought stress [16]. Foliar application of CTS also could effectively alleviate drought-induced growth inhibition of lettuce (*Lactuca sativa* L.) plants [17]. In addition, the exogenous CTS significantly increased photosynthetic rate, water use efficiency, and CAT, POD, and SOD activities in pot marigold (*Calendula officinalis* L.) plants, thereby mitigating deleterious effect of drought stress on growth [18].

Seed priming with bioactive compounds or elements such as zinc, γ-aminobutyric acid (GABA), putrescine (Put), diethyl aminoethyl hexanoate (DA-6), or spermidine (Spd) has been proven as an effective way to improve seed germination under normal and stressful conditions [19–23]. Previous studies have found that chitosan-black soybean seed-coat extract exhibited strong antioxidant property, and CTS coating could effectively improve seed germination, seedling growth, and resistance to pests under normal condition [24–26]. However, research has still not fully elucidated the effect of seed priming with CTS on alleviating drought-induced damage to seed germination and seedling establishment. White clover (*Trifolium repens* L.) is an important forage for feeding livestock and also used as an ornamental plant in horticulture. Objects of this study were to investigate the effect of CTS priming on seed-germination characteristics and to further elucidate the underlying mechanism involved in antioxidant defense and the dehydration-responsive pathway during white clover seed germination under water stress. Current findings will be beneficial to better understand the CTS-regulated adaptability to water stress during seed germination.

2. Materials and Methods

2.1. Plant Materials and Treatments

Seeds (white clover cv. Haifa) were surface-sterilized in 0.1% $HgCl_2$ solution for 5 min and then rinsed four times in distilled water (ddH_2O). These seeds were divided into two groups: one group was soaked in ddH_2O for 3 h (seeds without the CTS priming), and another group was firstly soaked in ddH$_2$O for 1 h and then transferred into 5 mg/L CTS (Sigma-Aldrich, 900344, St. Louis, MO, USA) solution for 2 h (seeds priming with the CTS). Seeds primed with or without the CTS were then germinated in Petri dishes. Three sheets of filter papers were laid in each Petri dish and moistened with 15 mL of ddH$_2$O (normal germination condition) or 18% (*w/v*) polyethylene glycol 6000 (PEG 6000) solution (germination under water stress). Each treatment included six biological replications, and each Petri dish included 50 seeds. All Petri dishes were placed randomly in a growth chamber (average day/night temperature of 23/19 °C, 75% relative humidity, and 700 μmol·m^{-2}·s^{-1} photosynthetically active radiation (PAR) at 12 h photoperiod) for 7 days. Seedlings were sampled on the 7th day of germination for determination of growth, physiological parameters, and gene expression levels.

2.2. Measurements of Seed-Germination and Growth Parameters

Germination vigor (GV) or germination percentage (GP) was calculated as a percentage of those seeds that had germinated on the 3rd or 7th day after the start of H$_2$O or CTS pretreatment, respectively. The germination index (GI) was calculated based on the formula:

$$\sum(Gt/Dt) \tag{1}$$

Gt indicates the number of germinated seeds, and Dt indicates the corresponding time to Gt in days.

Mean germination time (MGT) was calculated based on the formula:

$$MGT=\sum(D \times n)/\sum n \tag{2}$$

D indicates the number of days, and N indicates the number of germinations in the corresponding days.

For root length (RL), shoot length (SL), fresh weight (FW), and dry weight (DW), 10 seedlings were randomly selected from each treatment after 7 days of germination. Seed vigor index (SVI) was the product of FW and GI [22].

2.3. Measurements of Reactive Oxygen Species and Antioxidant Enzyme Activities

Superoxide anion (O$_2{}^-$) or hydrogen peroxide (H$_2$O$_2$) content was determined according to the method of Elstner and Heupel [27] or Velikova et al. [28], respectively. For malondialdehyde (MDA) content and antioxidant enzyme activity, 0.2 g of fresh seedlings were homogenized with 3 mL of 50 mM cold phosphate buffer (pH 7.8) and then centrifuged at 10,000× g for 15 min at 4 °C. The supernatant was collected for MDA determination and also as enzyme extract. The MDA content was determined by using 0.5 mL of the supernatant and 1 mL of the reaction solution (20% *w/v* trichloroacetic acid and 0.5% *w/v* thiobarbituric acid). After being heated in a boiling water for 15 min, the reaction mixture was cooled down to room temperature, and the absorbance of reaction solution was measured at 532 and 600 nm by using a spectrophotometer (Spectronic Instruments, Rochester, NY, USA) [29].

For SOD activity, 0.05 mL of supernatant was mixed with 1.45 mL of 50 mM phosphate solution (pH 7.8) containing 1.125 μM NBT, 60 μM riboflavin, 195 mM methionine, and 3 μM EDTA. After being placed under 600 μmol m^{-2}·s^{-1} PAR for 10 min, the absorbance of reaction solution were detected at 560 nm [30]. POD and CAT activities were detected based on the method of Chance and Maehly [31]. Briefly, 0.05 mL of supernatant was mixed with 1 mL of 50 mM phosphate buffer (pH 7.0) containing 45 mM H$_2$O$_2$ solution, and then, the absorbance of reaction solution was detected at 240 nm for the CAT activity. The 0.025 mL of supernatant was mixed with 0.05 mL of H$_2$O$_2$ solution, 0.5 mL guaiacol solution, and 0.925 mL phosphate buffer (pH 7.0). The absorbance of reaction solution was detected at 470 nm for the POD activity. For APX activity, 0.05 mL of supernatant was mixed with 100 mM of sodium acetate buffer (pH 5.8), 10 mM ascorbic acid, 5 mM

H$_2$O$_2$, and 0.003 mM ethylenediaminetetraacetic acid, and then, the absorbance of reaction solution was detected at 290 nm [32].

2.4. Measurements of Genes Expression Levels

Fresh seedlings (0.15 g) were sampled for total RNA extraction using a total RNA extraction kit (Tiangen, China), and then, the RNA were reverse-transcribed into cDNA (PrimeScript™ RT reagent Kit with gDNA Eraser, TaKaRa, Japan). Primers of β-Actin (internal reference gene) and genes encoding different types of dehydrins and dehydration-responsive element-binding proteins (Table 1) were used for real-time quantitative fluorescence PCR (qRT-PCR). The PCR procedure for all genes was: 94 °C for 5 min, denaturation at 95 °C for 30 s (40 repeats), annealing at 58 or 60 °C (Table 1) for 30 s, and extension at 72 °C for 30 s. Genes' relative expression levels were calculated by using the formula $2^{-\Delta\Delta Ct}$ [33].

Table 1. Primer sequences and GeneBank accession numbers of genes.

Target Gene	Accession No.	Forward Primer (5′-3′)	Reverse Primer (5′-3′)	Tm (°C)
SK2	GU443960.1	TGGAACAGGAGTAACAACAGGTGGA	TGCCAGTTGAGAAAGTTGAGGTTGT	58
Y2K	JF748410.1	AGCCACGCAACAAGGTTCTAA	TTGAGGATACGGGATGGGTG	60
Y2SK	GU443965.1	GTGCGATGGAGATGCTGTTTG	CCTAATCCAACTTCAGGTTCAGC	60
DREB2	EU846194.1	CAAGAACAAGATGATGATGGTGAAC	AAGAAGAAGAATTGGAGGAGTCATG	58
DREB3	EU846196.1	GCTCAATAGGACTCAACCAACTCAC	TGACGTTGTCTAACTCCACGGTAA	58
DREB4	EU846198.1	CTTGGTTGTGGAGATAATGGAGC	AAGTTGCAATCTGAATTCTGAGGAC	58
DREB5	EU846200.1	GCGATAGGTTCAAAGAAAGGGTG	AGAGCAGCATCTTGAGCAGTAGG	58
β-Actin	JF968419	TTACAATGAATTGCGTGTTG	AGAGGACAGCCTGAATGG	58

2.5. Statistical Analysis

Variations among four treatments were analyzed by the general linear model procedure of Statistical Product and Service Solutions 24 (SPSS Institute, IBM, Armonk, NY, USA, 2018). Differences among treatments were determined by using the least significant difference (LSD) at $p \leq 0.05$.

3. Results

3.1. Seeds Priming with CTS Affected Germination Characteristics under Water Stress

GP, GV, GI, and MGT were not significantly affected by the CTS priming under normal conditions (Figure 1A–D). PEG-induced water stress resulted in significant declines in GP, GV, GI, and MGT of seeds primed with or without CTS. Seeds primed with CTS exhibited a 16%, 54%, or 26% greater increase in GP, GV, or GI than those seeds without the CTS priming under water stress, respectively (Figure 1A–C). Seeds primed with CTS showed significantly lower MGT than seeds without the CTS priming under water stress (Figure 1D). As compared to normal condition, water stress also significantly decreased SVI, FW, DW, RL, and SL (Figures 2A–C and 3A,B). However, seeds primed with CTS had significantly higher SVI, DW, RL, and SL than those seeds without the CTS priming after 7 days of germination under water stress (Figures 2A,C and 3A,B).

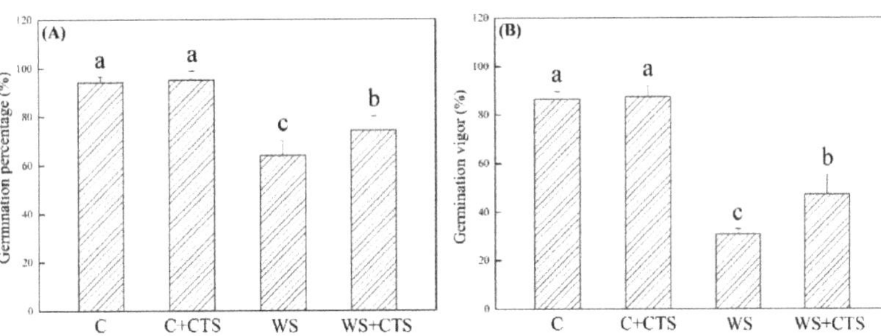

Figure 1. *Cont.*

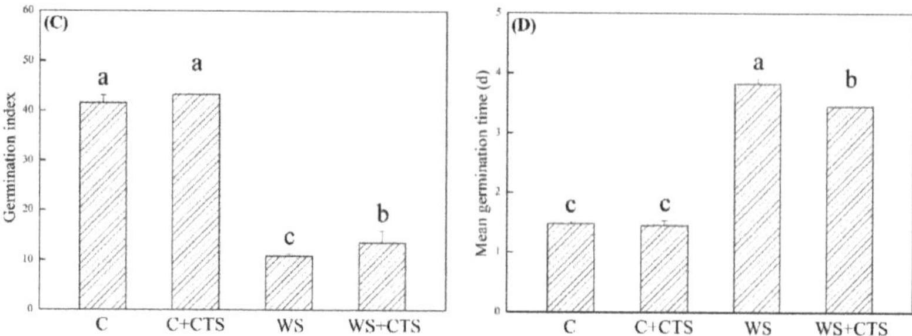

Figure 1. Seeds priming with chitosan affected (**A**) germination percentage, (**B**) germination vigor, (**C**) germination index, and (**D**) mean germination time under water stress. Vertical bars above columns indicate ± SE of mean, and different letters above columns indicate significant difference ($p < 0.05$). C, control (seeds pretreated with H_2O germinated under normal condition); C+CTS, control + CTS (seeds pretreated with CTS germinated under normal condition); WS, water stress (seeds pretreated with H_2O germinated under water stress condition); WS+CTS, water stress + CTS (seeds pretreated with CTS germinated under water stress condition).

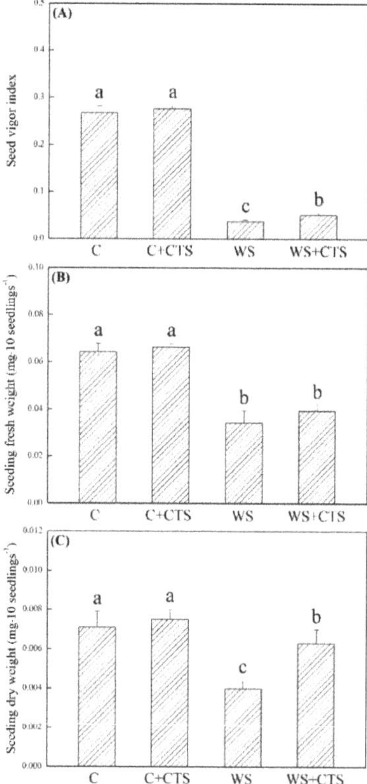

Figure 2. Seed priming with chitosan affected (**A**) seed vigor index, (**B**) seedling fresh weight, and (**C**) seedling dry weight after 7 days of germination under water stress. Vertical bars above columns indicate ± SE of mean, and different letters above columns indicate significant difference ($p < 0.05$). C, control (seeds pretreated with H_2O germinated under normal condition); C+CTS, control + CTS (seeds pretreated with CTS germinated under normal condition); WS, water stress (seeds pretreated with H_2O germinated under water stress condition); WS+CTS, water stress + CTS (seeds pretreated with CTS germinated under water stress condition).

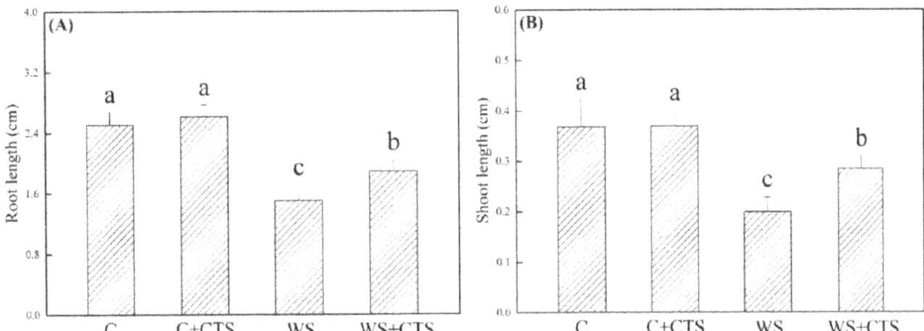

Figure 3. Seed priming with chitosan affected (**A**) seedling root length and (**B**) seedling shoot length after 7 days of germination under water stress. Vertical bars above columns indicate ± SE of mean, and different letters above columns indicate significant difference ($p < 0.05$). C, control (seeds pretreated with H_2O germinated under normal condition); C+CTS, control + CTS (seeds pretreated with CTS germinated under normal condition); WS, water stress (seeds pretreated with H_2O germinated under water stress condition); WS+CTS, water stress + CTS (seeds pretreated with CTS germinated under water stress condition).

3.2. Seed Priming with CTS Affected Oxidative Damage and Antioxidant Defense under Water Stress

ROS (O_2^- and H_2O_2) and MDA significantly accumulated in seedlings after 7 days of germination under water stress (Figure 4A–D). A 30%, 32%, or 16% lower $O_2^{\cdot-}$, H_2O_2, or MDA content was detected in the WS+CTS treatment as compared to the WS treatment under water stress, respectively (Figure 4A–C). As compared to normal condition, SOD activity did not significantly change in the WS treatment but significantly increased in the WS+CTS treatment (Figure 5A). Water stress inhibited the POD activity but improved the CAT activity in both of WS and WS+CTS treatments (Figure 5B,C). Seedlings germinated from seeds priming with the CTS showed significantly higher POD and CAT activities than seedlings without CTS priming (Figure 5B,C). APX activity significantly decreased under water stress, and no significant difference in APX activity was observed between the WS and WS+CTS (Figure 5D).

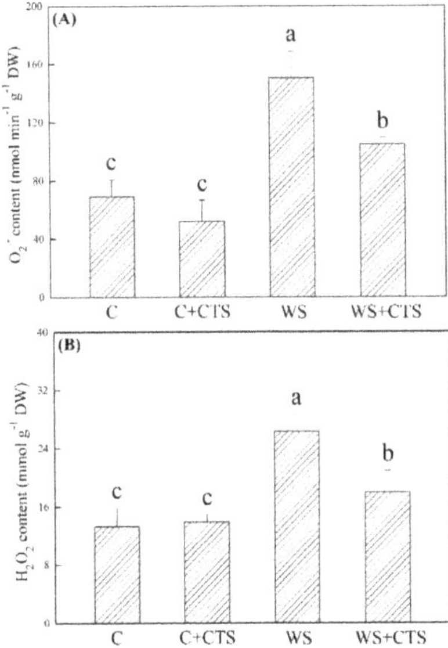

Figure 4. *Cont.*

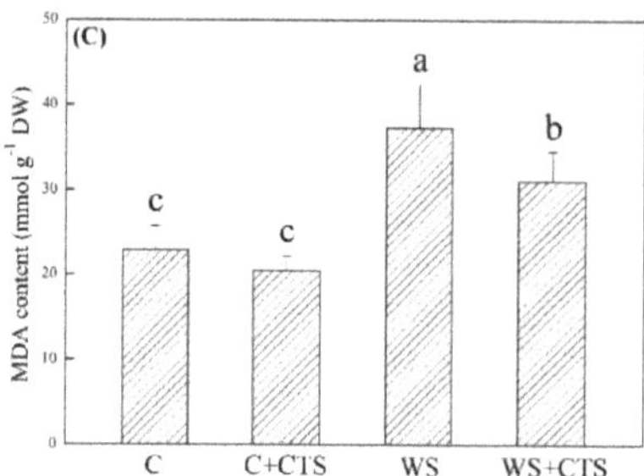

Figure 4. Seeds priming with chitosan affected (**A**) superoxide anion (O_2^-) content, (**B**) hydrogen peroxide (H_2O_2) content, and (**C**) malondialdehyde (MDA) content in seedlings after 7 days of germination under water stress. Vertical bars above columns indicate $\pm$ SE of mean, and different letters above columns indicate significant difference ($p < 0.05$). C, control (seeds pretreated with H_2O germinated under normal condition); C+CTS, control + CTS (seeds pretreated with CTS germinated under normal condition); WS, water stress (seeds pretreated with H_2O germinated under water stress condition); WS+CTS, water stress + CTS (seeds pretreated with CTS germinated under water stress condition).

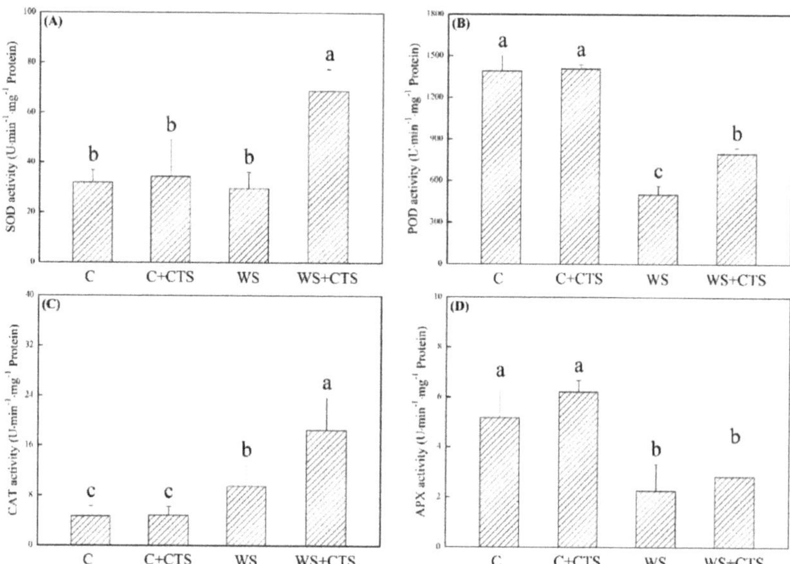

Figure 5. Seeds priming with chitosan affected (**A**) superoxide dismutase (SOD) activity, (**B**) peroxidase (POD) activity, (**C**) catalase (CAT) activity, and (**D**) ascorbate peroxidase (APX) activity in seedlings after 7 days of germination under water stress. Vertical bars above columns indicate $\pm$ SE of mean, and different letters above columns indicate significant difference ($p < 0.05$). C, control (seeds pretreated with H_2O germinated under normal condition); C+CTS, control + CTS (seeds pretreated with CTS germinated under normal condition); WS, water stress (seeds pretreated with H_2O germinated under water stress condition); WS+CTS, water stress + CTS (seeds pretreated with CTS germinated under water stress condition).

3.3. Seeds Priming with CTS Affected Genes Expression Levels Involved in Dehydration-Responsive Pathway under Water Stress

Relative expression levels of genes encoding dehydration-responsive element-binding proteins, including *DREB2*, *DREB3*, *DREB4*, and *DREB5*, are shown in Figure 6A–D. *DERB2* expression level was not affected significantly by water stress in the WS treatment, whereas it was significantly increased in the WS+CTS treatment (Figure 6A). *DERB3* expression level was inhibited significantly by water stress in both of the WS and WS+CTS, and there was no significant difference in the *DERB3* expression level between the WS and WS+CTS (Figure 6B). Water stress induced more pronounced increases in the *DREB4* and *DREB5* expression in the WS+CTS than that in the WS (Figure 6C,D). As compared to normal condition, water stress inhibited *Y2K* and *Y2SK* expression levels in the WS but up-regulated the *Y2K* and *Y2SK* expression levels in the WS+CTS (Figure 7A,B). The CTS priming significantly up-regulated the *SK2* expression level in seedling under normal condition and water stress (Figure 7C).

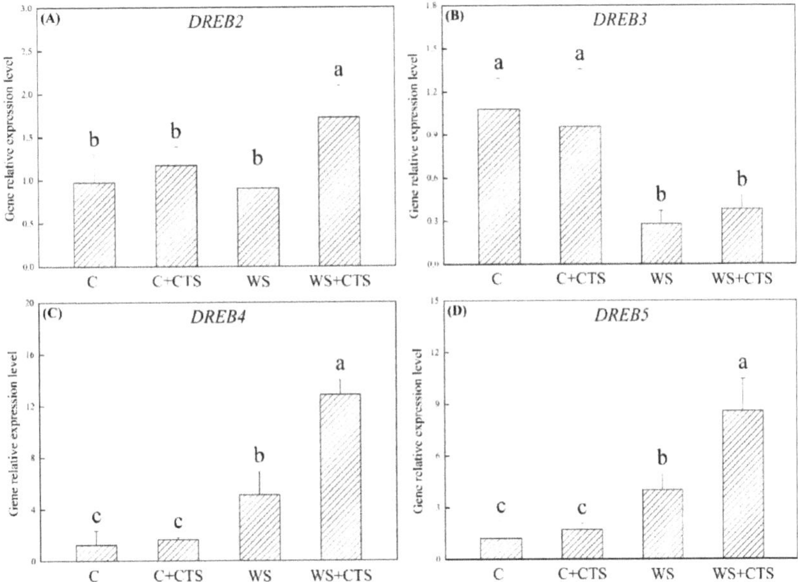

Figure 6. Seeds priming with chitosan affected genes expression levels of (**A**) *DREB2*, (**B**) *DERB3*, (**C**) *DREB4*, and (**D**) *DREB5* encoding different types of dehydration responsive element-binding proteins in seedlings after 7 days of germination under water stress. Vertical bars above columns indicate $\pm$ SE of mean, and different letters above columns indicate significant difference ($p < 0.05$). C, control (seeds pretreated with H_2O germinated under normal condition); C+CTS, control + CTS (seeds pretreated with CTS germinated under normal condition); WS, water stress (seeds pretreated with H_2O germinated under water stress condition); WS+CTS, water stress + CTS (seeds pretreated with CTS germinated under water stress condition).

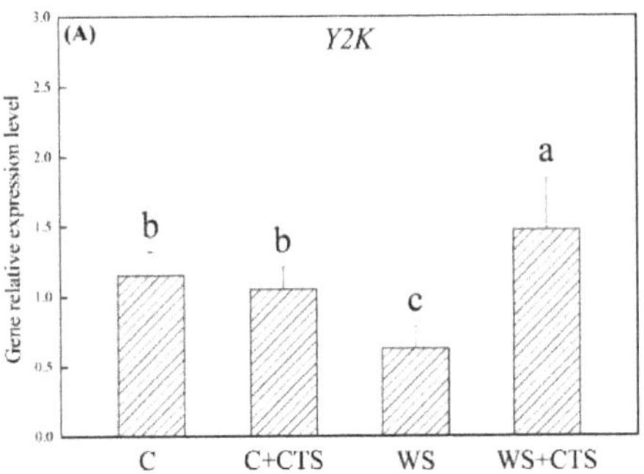

Figure 7. *Cont.*

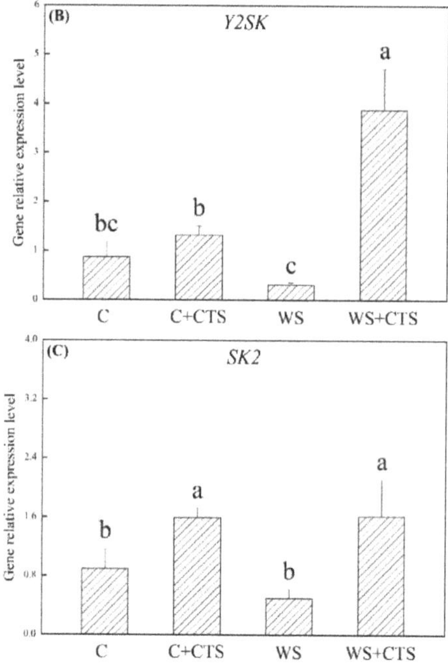

Figure 7. Seeds priming with chitosan affected genes expression levels of (**A**) *Y2K*, (**B**) *Y2SK*, and (**C**) *SK2* encoding different types of dehydrins in seedlings after 7 days of germination under water stress. Vertical bars above columns indicate ± SE of mean, and different letters above columns indicate significant difference ($p < 0.05$). C, control (seeds pretreated with H_2O germinated under normal condition); C+CTS, control + CTS (seeds pretreated with CTS germinated under normal condition); WS, water stress (seeds pretreated with H_2O germinated under water stress condition); WS+CTS, water stress + CTS (seeds pretreated with CTS germinated under water stress condition).

4. Discussion

Water stress decreases turf quality and also increases maintenance cost, especially in arid and semi-arid regions worldwide [34]. White clover is a leguminous ground cover plant that is widely used for urban landscaping and conservation of water and soil [35]. As compared to other leguminous species such as alfalfa, white clover is more susceptible to water deficit at the germination stage. Seed priming with bioactive compound has become an important agronomic strategy for improving seed vigor and germination under normal and stress conditions [36]. It has been proven that white clover seed priming with a low concentration of NaCl solution could significantly mitigate adverse effects induced by water stress, including declines in GP, GV, SVI, and radicle length [37]. Recent research also showed that white clover seeds soaking in an appropriate dose of diethyl aminoethyl hexanoate solution before being geminated under water stress effectively improved germination rate, root length, and shoot length of seedlings [23]. In addition, seed coating with CTS has been reported to significantly promote GP and seedling growth of hybrid rice under salt stress [38]. Our study demonstrated that seed priming with the CTS showed better GP, GV, GI, dry weight, root length, and shoot length of seedlings than those seeds primed with H_2O under water stress. Current findings indicated that the CTS could be used as a beneficial elicitor to improve seed germination under stressful conditions.

Drought-induced high amounts of ROS in cells caused lipid peroxidation, protein degradation, and membrane leakage, resulting in retarded growth, premature senescence, and even death [39]. The overaccumulation of ROS (O_2^- and H_2O_2) and the aggravation of membrane lipid peroxidation were observed in our current study when white clover seeds germinated under water-limited conditions. Previous study has found that zinc priming ameliorated adverse effects of drought stress associated with enhancement in total antioxidant capacity and reduction in membrane lipid peroxidation during seed germination [19]. In addition, the regulatory role of CTS in activating the antioxidant defense system to scavenge free radicals has also been reported in response to water stress. For example, the CTS coating could mitigate drought-induced oxidative damage by activating SOD, CAT, and POD activities in wheat seedlings [40]. A combination of seed priming and foliar application of CTS improved shoot and root growth as well as antioxidant enzyme activities, including POD and APX in rice seedling under drought stress [41]. Seeds pretreated by exogenous CTS increased drought tolerance in alfalfa through enhancing the antioxidant defense system [42]. White clover seed priming with CTS significantly alleviated oxidative damage induced by water stress through improving ROS-scavenging enzyme activities, including SOD, POD, and CAT, which indicated the beneficial function of CTS in maintaining ROS homeostasis for better adaptation to a water-deficit environment during early seedling establishment.

The DREBs family is considered one of the most critical classes of TFs in relation to drought tolerance in plants [43]. DREBs regulate stress-defensive genes expression by binding to the DRE/C-repeat core component of these downstream genes under different abiotic stresses [44]. It has been found that transgenic tobacco overexpressing an *RcDREB 5-A* showed better growth and higher biomass than non-transgenic lines in response to drought stress [45]. Similarly, up-regulation of *PcDREB2A* could significantly improve drought tolerance of *Arabidopsis* [46]. On the contrary, RNAi-tomato plants exhibited a significantly lower expression level of *SlDREB2* and higher lipid membrane peroxidation than the wild-type under drought stress [47]. During seed germination, significant increases in expression levels of different types of *DREBs* are also propitious to achieve stress tolerance. For instance, drought tolerance of transgenic *Arabidopsis* overexpressing an *AmDREB2* was improved significantly at the seed-germination stage [48]. A *SgDREB2* overexpression in *Arabidopsis* increased the seed-germination rate, seedlings survival rate, and antioxidant enzyme activities, including SOD and APX, under drought stress, suggesting that *SgDREB2* regulated drought tolerance involved in antioxidant defense [49]. Exogenous CTS priming significantly up-regulated expression levels of *DREB2*, *DREB4*, and *DREB5* during white clover seed germination, which indicated that the potential role of CTS in regulating adaptability to water stress could be associated with the DREB-responsive pathway.

DHNs are diverse classes of stress-responsive proteins that are regulated by the DREBs [50]. DHNs quickly accumulate during seed germination or when plants suffer dehydration due to their positive functions as chaperones, ROS scavengers, and osmoprotectants in cells [51]. Previous study has proven that white clover seed priming with DA-6 significantly mitigated adverse effects of water stress on seed germination and seedling establishment in relation to significant accumulation of DHN and upregulation of *Y2K*, *Y2SK*, and *SK2* genes encoding different types of DHNs [23]. Enhanced *AnDHN* or *CaDHN3* expression in *Arabidopsis* increased seed germination and initial root length under drought stress and also promoted antioxidant capacity to alleviate drought-induced ROS accumulation in seedlings [52,53]. However, silencing of the *CaDHN3* in pepper (*Capsicum annuum* L.) plants significantly decreased drought tolerance, as demonstrated by more accumulation of ROS and MDA contents than the wild-type [53]. A recent study of Decena et al. found that *DHNs* expression among 32 *Brachypodium* grass ecotypes was highly correlated with drought-responsive traits, such as plant biomass and water-use efficiency, and drought-tolerant ecotypes often had higher expression levels of *DHNs* in response to drought stress [54]. Water stress could also induce more or higher *DHNs* expression in drought-tolerant Kentucky bluegrass (*Poa pratensis*) germplasms [55]. Our findings indicated that seed priming with CTS activated the expression of *Y2K-*, *Y2SK-*, and *SK2*-encoding DHNs, which could be a key factor affecting water-stress tolerance in white clover.

5. Conclusions

Water stress significantly decreased seed germination characteristics and hindered seedling establishment. Seed priming with CTS could be used as a simple, effective, economical, and environmentally friendly technique to improve germination and seedling establishment under water-deficit conditions. Stress-induced overaccumulation of ROS damaged cell membrane, leading to membrane lipid peroxidation, but this symptom could be significantly alleviated by the CTS priming through activating SOD, POD, and CAT activities. In addition, seeds pretreated with CTS exhibited significantly higher expression levels of *DREB2*, *DREB4*, *DREB5*, *Y2K*, *Y2SK*, and *SK2* than those seeds without the CTS priming. Current findings indicated that the CTS-induced tolerance to water stress could be associated with the enhancement in dehydration-responsive pathway during seed germination.

Author Contributions: Conceptualization, Z.L.; methodology, Z.L., B.C. and Y.Z. (Yan Zhang); formal analysis, Y.L., Y.Z. (Yue Zhao) and Z.L.; investigation, Y.L., B.C., M.T. and Y.Z. (Yue Zhao); writing—original draft preparation, Y.L. and Y.Z. (Yue Zhao); writing—review and editing, Y.Z. (Yan Zhang), Y.L. and Z.L.; funding acquisition, Z.L. and Y.L. All authors have read and agreed to the published version of the manuscript.

Funding: This research was supported by the Ya'an City and Sichuan Agricultural University Cooperation Project (20SXHZ0022) and the Science and Technology Innovation Project (2021064).

Institutional Review Board Statement: Not applicable.

Informed Consent Statement: Not applicable.

Data Availability Statement: All data presented in this study are available in the article.

Conflicts of Interest: All authors declare no conflict of interest.

References

1. Aroca, R. Plant responses to drought stress. In *From Morphological to Molecular Features*; Springer: Berlin/Heidelberg, Germany, 2012; pp. 1–5.
2. Ahmad, S.; Ahmad, R.; Ashraf, M.Y.; Ashraf, M.; Waraich, E.A. Sunflower (*Helianthus annuus* L.) response to drought stress at germination and seedling growth stages. *Pak. J. Bot.* **2009**, *41*, 647–654.
3. Sucre, B.; Suarez, N. Effect of salinity and PEG-induced water stress on water status, gas exchange, solute accumulation and leaf growth in Ipomoea pes-caprae. *Environ. Exp. Bot.* **2011**, *70*, 2. [CrossRef]
4. Farooq, M.; Hussain, M.; Wahid, A.; Siddique, K. Drought stress in plants: An overview. In *Plant Responses to Drought Stress*; Springer: Berlin/Heidelberg, Germany, 2012; pp. 1–33.
5. Hussain, Q.; Asim, M.; Zhang, R.; Khan, R.; Farooq, S.; Wu, J. Transcription factors interact with ABA through gene expression and signaling pathways to mitigate drought and salinity stress. *Biomolecules* **2021**, *11*, 1159. [CrossRef]
6. Gujjar, R.S.; Akhtar, M.; Singh, M. Transcription factors in abiotic stress tolerance. *Indian J. Plant Physiol.* **2014**, *19*, 306–316. [CrossRef]
7. Mizoi, J.; Shinozaki, K.; Yamaguchi-Shinozaki, K. AP2/ERF family transcription factors in plant abiotic stress responses. *Biochim. Biophys. Acta (BBA)-Gene Regul. Mech.* **2012**, *1819*, 86–96. [CrossRef] [PubMed]
8. Thapa, S.; Yadav, J.; Chakdar, H.; Saxena, A. Deciphering the mechanisms of microbe mediated drought stress alleviation in wheat. *Res. Sq. Prepr.* **2021**, 1–30.
9. Xie, X.; He, Z.; Chen, N.; Tang, Z.; Wang, Q.; Cai, Y. The roles of environmental factors in regulation of oxidative stress in plant. *BioMed Res. Int.* **2019**, *2019*, 9732325. [CrossRef] [PubMed]
10. Sharma, P.; Jha, A.B.; Dubey, R.S.; Pessarakli, M. Reactive oxygen species, oxidative damage, and antioxidative defense mechanism in plants under stressfull conditions. *J. Bot.* **2012**, *2012*, 217037.
11. Wang, W.B.; Kim, Y.H.; Lee, H.S.; Kim, K.Y.; Deng, X.P.; Kwak, S.S. Analysis of antioxidant enzyme activity during germination of alfalfa under salt and drought stresses. *Plant Physiol. Biochem.* **2009**, *47*, 570–577. [CrossRef] [PubMed]
12. Begum, N.; Hasanuzzaman, M.; Li, Y.; Akhtar, K.; Zhang, C.; Zhao, T. Seed germination behavior, growth, physiology and antioxidant metabolism of four contrasting cultivars under combined drought and salinity in soybean. *Antioxidants* **2022**, *11*, 498. [CrossRef]
13. Shi, Y.; Zhang, Y.; Yao, H.; Wu, J.; Sun, H.; Gong, H. Silicon improves seed germination and alleviates oxidative stress of bud seedlings in tomato under water deficit stress. *Plant Physiol. Biochem.* **2014**, *78*, 27–36. [CrossRef] [PubMed]
14. Kocięcka, J.; Liberacki, D. The potential of using chitosan on cereal crops in the face of climate change. *Plants* **2021**, *10*, 1160. [CrossRef] [PubMed]

15. Ali, E.; El-Shehawi, A.; Ibrahim, O.; Abdul-Hafeez, E.; Moussa, M.; Hassan, F. A vital role of chitosan nanoparticles in improvisation the drought stress tolerance in *Catharanthus roseus* (L.) through biochemical and gene expression modulation. *Plant Physiol. Biochem.* **2021**, *161*, 166–175. [CrossRef] [PubMed]

16. Sadak, M.; Bakhoum, G.; Tawfic, M. Chitosan and chitosan nanoparticle effect on growth, productivity and some biochemical aspects of *Lupinus termis* L. plant under drought conditions. *Egypt. J. Chem.* **2022**, *65*, 1–2. [CrossRef]

17. Nurliana, S.; Fachriza, S.; Hemelda, N.; Yuniati, R. *Chitosan Application for Maintaining the Growth of Lettuce (Lactuca sativa) under Drought Condition*; IOP Conference Series: Earth and Environmental Science, 2022; IOP Publishing: Bristol, UK, 2022; p. 012013.

18. Akhtar, G.; Faried, H.N.; Razzaq, K.; Ullah, S.; Wattoo, F.M.; Shehzad, M.A.; Sajjad, Y.; Ahsan, M.; Javed, T.; Dessoky, E.S. Chitosan-induced physiological and biochemical regulations confer drought tolerance in pot marigold (*Calendula officinalis* L.). *Agronomy* **2022**, *12*, 474. [CrossRef]

19. Farooq, M.; Almamari, S.A.D.; Rehman, A.; Al-Busaidi, W.M.; Wahid, A.; Al-Ghamdi, S.S. Morphological, physiological and biochemical aspects of zinc seed priming-induced drought tolerance in faba bean. *Sci. Hortic.* **2021**, *281*, 109894. [CrossRef]

20. Zhou, M.; Hassan, M.J.; Peng, Y.; Liu, L.; Liu, W.; Zhang, Y.; Li, Z. γ-Aminobutyric acid (GABA) priming improves seeds germination and stress tolerance associated with enhanced antioxidant metabolism, *DREB* expression, and dehydrin accumulation in white clover under drought stress. *Front. Plant Sci.* **2021**, *12*, 2675. [CrossRef]

21. Cheng, B.; Hassan, M.J.; Feng, G.; Zhao, J.; Liu, W.; Peng, Y.; Li, Z. Metabolites reprogramming and Na^+/K^+ transportation associated with putrescine-regulated white clover seed germination and seedling tolerance to salt toxicity. *Front. Plant Sci.* **2022**, *13*, 856007. [CrossRef]

22. Li, Z.; Peng, Y.; Zhang, X.-Q.; Ma, X.; Huang, L.-K.; Yan, Y.-H. Exogenous spermidine improves seed germination of white clover under water stress via involvement in starch metabolism, antioxidant defenses and relevant gene expression. *Molecules* **2014**, *19*, 18003–18024. [CrossRef]

23. Hassan, M.J.; Geng, W.; Zeng, W.; Raza, M.A.; Khan, I.; Iqbal, M.Z.; Peng, Y.; Zhu, Y.; Li, Z. Diethyl aminoethyl hexanoate priming ameliorates seed germination via involvement in hormonal changes, osmotic adjustment, and dehydrins accumulation in white clover under drought stress. *Front. Plant Sci.* **2021**, *12*, 709187. [CrossRef]

24. Lizárraga-Paulín, E.G.; Miranda-Castro, S.P.; Moreno-Martínez, E.; Lara-Sagahón, A.V.; Torres-Pacheco, I. Maize seed coatings and seedling sprayings with chitosan and hydrogen peroxide: Their influence on some phenological and biochemical behaviors. *J. Zhejiang Univ. Sci.* **2013**, *14*, 87–96. [CrossRef] [PubMed]

25. Zeng, D.; Luo, X.; Tu, R. Application of bioactive coatings based on chitosan for soybean seed protection. *Int. J. Carbohydr. Chem.* **2012**, *2012*, 1–5. [CrossRef]

26. Wang, X.; Yong, H.; Gao, L.; Li, L.; Jin, M.; Liu, J. Preparation and characterization of antioxidant and pH-sensitive films based on chitosan and black soybean seed coat extract. *Food Hydrocoll.* **2019**, *89*, 56–66. [CrossRef]

27. Elstner, E.F.; Heupel, A. Inhibition of nitrite formation from hydroxylammoniumchloride: A simple assay for superoxide dismutase. *Anal. Biochem.* **1976**, *70*, 616–620. [CrossRef]

28. Velikova, V.; Yordanov, I.; Edreva, A. Oxidative stress and some antioxidant systems in acid rain-treated bean plants: Protective role of exogenous polyamines. *Plant Sci.* **2000**, *151*, 59–66. [CrossRef]

29. Dhindsa, R.S.; Plumb-Dhindsa, P.; Thorpe, T.A. Leaf senescence: Correlated with increased levels of membrane permeability and lipid peroxidation, and decreased levels of superoxide dismutase and catalase. *J. Exp. Bot.* **1981**, *32*, 93–101. [CrossRef]

30. Giannopolitis, C.N.; Ries, S.K. Superoxide dismutases: I. Occurrence in higher plants. *Plant Physiol.* **1977**, *59*, 309–314. [CrossRef]

31. Chance, B.; Maehly, A. Assay of catalases and peroxidases. *Method Enzym.* **1955**, *2*, 764–775.

32. Nakano, Y.; Asada, K. Hydrogen peroxide is scavenged by ascorbate-specific peroxidase in spinach chloroplasts. *Plant Cell Physiol.* **1981**, *22*, 867–880.

33. Livak, K.J.; Schmittgen, T.D. Analysis of relative gene expression data using real-time quantitative PCR and the $2^{-\Delta\Delta CT}$ method. *Methods* **2001**, *25*, 402–408. [CrossRef]

34. Colmer, T.D.; Barton, L. A review of warm-season turfgrass evapotranspiration, responses to deficit irrigation, and drought resistance. *Crop Sci.* **2017**, *57* (Suppl. S1), S98–S110. [CrossRef]

35. McCurdy, J.D.; McElroy, J.S.; Guertal, E.A. White clover (*Trifolium repens*) establishment within dormant bermudagrass turf: Cultural considerations, establishment timing, seeding rate, and cool-season companion grass species. *HortScience* **2013**, *48*, 1556–1561. [CrossRef]

36. Paparella, S.; Araújo, S.; Rossi, G.; Wijayasinghe, M.; Carbonera, D.; Balestrazzi, A. Seed priming: State of the art and new perspectives. *Plant Cell Rep.* **2015**, *34*, 1281–1293. [CrossRef] [PubMed]

37. Cao, Y.; Liang, L.; Cheng, B.; Dong, Y.; Wei, J.; Tian, X.; Peng, Y.; Li, Z. Pretreatment with NaCl promotes the seed germination of white clover by affecting endogenous phytohormones, metabolic regulation, and dehydrin-encoded genes expression under water stress. *Int. J. Mol. Sci.* **2018**, *19*, 3570. [CrossRef] [PubMed]

38. Songlin, R.; Qingzhong, X. Effects of chitosan coating on seed germination and salt-tolerance of seedling in hybrid rice (*Oryza sativa* L.). *Zuo Wu Xue Bao* **2002**, *28*, 803–808.

39. Hasanuzzaman, M.; Bhuyan, M.B.; Zulfiqar, F.; Raza, A.; Mohsin, S.M.; Mahmud, J.A.; Fujita, M.; Fotopoulos, V. Reactive oxygen species and antioxidant defense in plants under abiotic stress: Revisiting the crucial role of a universal defense regulator. *Antioxidants* **2020**, *9*, 681. [CrossRef] [PubMed]

40. Zeng, D.; Luo, X. Physiological effects of chitosan coating on wheat growth and activities of protective enzyme with drought tolerance. *Open J. Soil Sci.* **2012**, *2*, 282. [CrossRef]
41. Moolphuerk, N.; Lawson, T.; Pattanagul, W. Chitosan mitigates the adverse effects and improves photosynthetic activity in rice (*Oryza sativa* L.) seedlings under drought condition. *J. Crop Improv.* **2021**, *36*, 638–655. [CrossRef]
42. Mustafa, G.; Shehzad, M.A.; Tahir, M.; Nawaz, F.; Akhtar, G.; Bashir, M.A.; Ghaffar, A. Pretreatment with chitosan arbitrates physiological processes and antioxidant defense system to increase drought tolerance in alfalfa (*Medicago sativa* L.). *J. Soil Sci. Plant Nutr.* **2022**, *22*, 2169–2186. [CrossRef]
43. Manna, M.; Thakur, T.; Chirom, O.; Mandlik, R.; Deshmukh, R.; Salvi, P. Transcription factors as key molecular target to strengthen the drought stress tolerance in plants. *Physiol. Plant.* **2021**, *172*, 847–868. [CrossRef]
44. Lata, C.; Prasad, M. Role of DREBs in regulation of abiotic stress responses in plants. *J. Exp. Bot.* **2011**, *62*, 4731–4748. [CrossRef]
45. do Rego, T.F.C.; Santos, M.P.; Cabral, G.B.; de Moura Cipriano, T.; de Sousa, N.L.; de Souza Neto, O.A.; Aragão, F.J.L. Expression of a DREB 5-A subgroup transcription factor gene from *Ricinus communis* (*RcDREB1*) enhanced growth, drought tolerance and pollen viability in tobacco. *Plant Cell Tissue Organ Cult. (PCTOC)* **2021**, *146*, 493–504. [CrossRef]
46. Hu, H.; Wang, X.; Wu, Z.; Chen, M.; Chai, T.; Wang, H. Overexpression of the *Polygonum cuspidatum PcDREB2A* gene encoding a DRE-Binding transcription factor enhances the drought tolerance of transgenic *Arabidopsis thaliana*. *J. Plant Biol.* **2021**, 1–11. [CrossRef]
47. Tao, L.; Yu, G.; Chen, H.; Wang, B.; Jiang, L.; Han, X.; Lin, G.; Cheng, X.-G. *SlDREB2* gene specifically recognizing to the universal DRE elements is a transcriptional activator improving drought tolerance in tomato. *Sci. Hortic.* **2022**, *295*, 110887. [CrossRef]
48. Wang, X.; Ma, L.; Zhang, H.; Zhang, Y.; Xue, M.; Ren, M.; Tang, K.; Guo, H.; Wang, M. Constitutive expression of two A-5 subgroup DREB transcription factors from desert shrub *Ammopiptanthus mongolicus* confers different stress tolerances in transgenic *Arabidopsis*. *Plant Cell Tissue Organ Cult. (PCTOC)* **2021**, *147*, 351–363. [CrossRef]
49. Han, Y.; Xiang, L.; Song, Z.; Lu, S. Overexpression of *SgDREB2C* from *Stylosanthes guianensis* leads to increased drought tolerance in transgenic *Arabidopsis*. *Int. J. Mol. Sci.* **2022**, *23*, 3520. [CrossRef] [PubMed]
50. Riyazuddin, R.; Nisha, N.; Singh, K.; Verma, R.; Gupta, R. Involvement of dehydrin proteins in mitigating the negative effects of drought stress in plants. *Plant Cell Rep.* **2021**, *41*, 519–533. [CrossRef]
51. Tiwari, P.; Chakrabarty, D. Dehydrin in the past four decades: From chaperones to transcription co-regulators in regulating abiotic stress response. *Curr. Res. Biotechnol.* **2021**, *3*, 249–259. [CrossRef]
52. Sun, Y.; Liu, L.; Sun, S.; Han, W.; Irfan, M.; Zhang, X.; Zhang, L.; Chen, L. AnDHN, a dehydrin protein from *Ammopiptanthus nanus*, mitigates the negative effects of drought stress in plants. *Front. Plant Sci.* **2021**, *12*, 3067. [CrossRef]
53. Meng, Y.C.; Zhang, H.F.; Pan, X.X.; Chen, N.; Hu, H.F.; Haq, S.U.; Khan, A.; Chen, R.G. *CaDHN3*, a pepper (*Capsicum annuum* L.) dehydrin gene enhances the tolerance against salt and drought stresses by reducing ROS accumulation. *Int. J. Mol. Sci.* **2021**, *22*, 3205. [CrossRef] [PubMed]
54. Decena, M.A.; Gálvez-Rojas, S.; Agostini, F.; Sancho, R.; Contreras-Moreira, B.; Des Marais, D.L.; Hernandez, P.; Catalán, P. Comparative genomics, evolution, and drought-induced expression of dehydrin genes in model *Brachypodium* grasses. *Plants* **2021**, *10*, 2664. [CrossRef] [PubMed]
55. Bushman, B.S.; Robbins, M.D.; Thorsted, K.; Robins, J.G.; Warnke, S.E.; Martin, R.; Harris-Shultz, K. Transcript responses to drought in Kentucky bluegrass (*Poa pratensis* L.) germplasm varying in their tolerance to drought stress. *Environ. Exp. Bot.* **2021**, *190*, 104571. [CrossRef]

Article

Transcriptome Analysis Reveals the Stress Tolerance to and Accumulation Mechanisms of Cadmium in *Paspalum vaginatum* Swartz

Lei Xu [†], Yuying Zheng [†], Qing Yu, Jun Liu, Zhimin Yang and Yu Chen *

College of Agro-Grassland Science, Nanjing Agricultural University, Nanjing 210095, China
* Correspondence: cyu801027@njau.edu.cn
† These authors contributed equally to this work.

Abstract: Cadmium (Cd) is a non-essential heavy metal and high concentrations in plants causes toxicity of their edible parts and acts as a carcinogen to humans and animals. *Paspalum vaginatum* is widely cultivating as turfgrass due to its higher abiotic stress tolerance ability. However, there is no clear evidence to elucidate the mechanism for heavy metal tolerance, including Cd. In this study, an RNA sequencing technique was employed to investigate the key genes associated with Cd stress tolerance and accumulation in *P. vaginatum*. The results revealed that antioxidant enzyme activities catalase (CAT), peroxidase (POD), superoxide dismutase (SOD), and glutathione S-transferase GST) were significantly higher at 24 h than in other treatments. A total of 6820 (4457/2363, up-/down-regulated), 14,038 (9894/4144, up-/down-regulated) and 17,327 (7956/9371, up-/down-regulated) differentially expressed genes (DEGs) between the Cd1 vs. Cd0, Cd4 vs. Cd0, and Cd24 vs. Cd0, respectively, were identified. The GO analysis and the KEGG pathway enrichment analysis showed that DEGs participated in many significant pathways in response to Cd stress. The response to abiotic stimulus, the metal transport mechanism, glutathione metabolism, and the consistency of transcription factor activity were among the most enriched pathways. The validation of gene expression by qRT-PCR results showed that heavy metal transporters and signaling response genes were significantly enriched with increasing sampling intervals, presenting consistency to the transcriptome data. Furthermore, over-expression of *PvSnRK2.7* can positively regulate Cd-tolerance in *Arabidopsis*. In conclusion, our results provided a novel molecular mechanism of the Cd stress tolerance of *P. vaginatum* and will lay the foundation for target breeding of Cd tolerance in turfgrass.

Keywords: Cadmium; *Paspalum vaginatum*; RNA-seq; qRT-PCR; *PvSnRK2.7*

Citation: Xu, L.; Zheng, Y.; Yu, Q.; Liu, J.; Yang, Z.; Chen, Y. Transcriptome Analysis Reveals the Stress Tolerance to and Accumulation Mechanisms of Cadmium in *Paspalum vaginatum Swartz*. *Plants* **2022**, *11*, 2078. https://doi.org/10.3390/plants11162078

Academic Editors: Zhou Li and David Jespersen

Received: 28 May 2022
Accepted: 30 July 2022
Published: 9 August 2022

Publisher's Note: MDPI stays neutral with regard to jurisdictional claims in published maps and institutional affiliations.

1. Introduction

Heavy metal pollution is a major environmental concern that severely damages plant, animal, and human health [1]. Cadmium (Cd) is one of the non-essential heavy metals and one of the most toxic pollutants, deleterious to the environment and agriculture [2]. The higher accumulation of Cd in the edible parts of plants that enter the human and animal food chain thus leads to severe health risks [3]. Inorganic fertilizer application and spraying of synthetic fungicides are the major sources of Cd contamination in the food chain [4]. In addition, a higher accumulation of Cd in plants reduces seed germination, early seedling growth, plant biomass, physiological, and biochemical processes of the plant such as photosynthetic gas exchange parameters, soluble sugar, and soluble protein and chlorophyll synthesis in plants [5]. Additionally, Cd concentration in plants activates reactive oxygen species such as hydrogen peroxide (H_2O_2), singlet oxygen (O_2), organic hydroperoxide (ROOH), and oxygen-derived free radicals, e.g., hydroxyl (HO), peroxyl (RO_2), superoxide anion (O^{2-}), and alkoxyl (RO^-) radicals, which disturb antioxidant defense mechanisms and induce chromosomal aberrations, gene mutations, and DNA damage [1,2,5,6]. To avoid plant cell damage by ROS, plants enhance various enzymatic and non-enzymatic antioxidant activities to avoid plasma membrane and other cell membrane damage by oxidative molecules [7]. Studies reported that a higher accumulation of glutathione reductase, catalase, ascorbate peroxidase, superoxide dismutase, and peroxidase enhances tolerance to plant Cd stress [8–10]. In the phenylpropanoid pathway, 10 enzymes are involved in catalysis including phenylalanine ammonialyase (PAL), peroxidase (POD), 4-coumarate CoA ligase (4CL), caffeic acid 3-O-methyl transferase (COMT), cinnamyl alcohol dehydrogenase (CAD), and cafeoy1-CoA3-O-methy1transferase (CCoAOMT) [11,12].

There are several mechanisms reported in the accumulation and translocation of Cd in the aerial part of plants. Cadmium accumulation in the plant system depends on physiological processes including binding on root cell walls, and sequestration in root vacuoles and xylem tissues [13]. The subcellular distribution of Cd showed that Cd mainly accumulated in the cell wall, followed by a soluble fraction in organelles and membrane in root, bark, and leaf tissues [14,15]. Cell walls serve as the first stress barrier consisting of pectin components. The higher amount of negatively charged groups in the pectin components positively interact with Cd, increasing the accumulation of Cd in the cell wall [16]. The soluble fraction was higher in the root system compared to the cell wall, organelle, and membrane fractions of bark and leaves [15]. The vacuole is the major cell organelle in plant cells and plays an important role in the retention of Cd. The metal-phytochelatin complex is sequestered in the vacuole and the mechanism is regulated by two cassette transporters (*AtABCC1* and *AtABCC2*) [17]. In addition, higher Cd accumulation and tolerance depend on the activation of several key genes associated with metal transporters, chelator proteins, antioxidant enzymes, defending genes, and transcription factors in plants [18]. Cadmium can move from the soil and accumulates in the edible parts of plants mediated by various transporters. Many transporter families play an important role in Ca^{2+}, Fe^{2+}, Mn^{2+}, and Zn^{2+} absorption into plant cells and this may be involved in the transport of Cd. Nonessential elements (e.g., Cd) compete with other essential elements (Ca^{2+}, Fe^{2+}, Mn^{2+}, and Zn^{2+}) in the same membrane transport channels [18]. These transporter families include heavy metal ATPase (HMA), also known as P-type ATPases that absorb and transport both essential (Cu^{2+}, Zn^{2+}) and non-essential heavy metal ions (Cd^{2+} and Pb^{2+}) [19]. Similarly, natural resistance-associated macrophage proteins (NRAMPs) uptake the Cd^{2+} ions [20], and iron-regulated transporter (*IRT1*) (ZRT/IRT-like proteins) are one of the ZIP family members of Cd^{2+} transport [21]. Moreover, cation exchangers (CAX) help to transport the Cd-chelate complex crossing tonoplast [13]. Furthermore, the SNF1-related protein kinase 2 subfamily protein (SnRK) is involved in environmental Cd stress signaling [22,23]. In addition, transcription factors (TFs) including MYB, WRKY, C2H2, bZIP, AP2, ERF, and DREB also play a significant role in metal stress tolerance in various plants by regulation of functional gene expression [24–26].

High throughput sequencing techniques provided fundamental knowledge of genes associated with biotic and abiotic stress in the plant system [24]. Many molecular approaches have been developed, among them, the RNA sequencing (RNA-seq) strategy, that plays an essential role and helps to identify expected and unexpected gene expression and regulatory networks including the pathogenesis mechanism of avocado [27], dwarfing regulation in Seashore paspalum [12], storage root development in sweet potato [28], drought-resistant mechanism of sorghum [29], and Cd stress in *Nicotiana tabacum* and *Nicotiana rustica* [13]. Seashore paspalum (*Paspalum vaginatum*), a halophytic warm-season turfgrass widely cultivated in salinity-affected areas and is applied for phytoremediation in Cd contaminated areas due to its excellent tolerance to salinity and Cd [12,30–32]. However, the molecular mechanism of Cd tolerance of Seashore paspalum remains unclear. In this study, we uncovered the molecular pathways of seashore paspalum response to Cd stress conditions using comprehensive RNA seq analysis. This finding will help to develop candidate Cd-tolerant genes for molecular breeding of Cd-tolerant turfgrass and other grasses.

2. Results

2.1. Antioxidant Activities in Paspalum vaginatum under Cd Stress

The enzymatic and non-enzymatic antioxidant enzyme activities were investigated under cadmium (Cd) stress conditions with different sampling intervals. All antioxidant activities were higher in the Cd1 sampling time compared to the Cd0 sampling time, but no significant difference was noticed between these timing intervals (Figure 1). Peroxidase (POD) and superoxide dismutase (SOD) activities were higher at 4 h sampling intervals compared to the 0 h and 1 h sampling intervals, but there was no significant difference in SOD activity compared to the 1 h sampling time (Figure 1A–C). Similarly, glutathione S-transferase (GST) activity was also significantly higher in the 24 h time point compared to 4, 1, and 0 h. However, GST was higher at the 4 h sampling intervals compared to 0 and 1 h sampling intervals, but there was no significant difference between the 4 h and 1 h sampling intervals (Figure 1D). The results showed that the activities of catalase (CAT), POD, and SOD were significantly higher at the 24 h sampling interval compared to other sampling times.

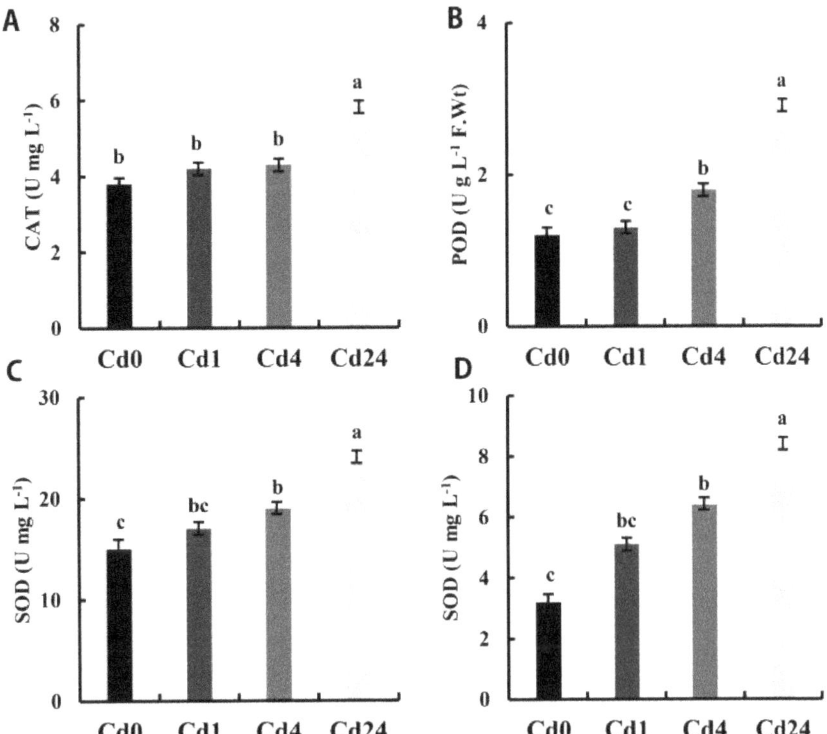

Figure 1. Enzymatic and non-enzymatic antioxidant activities in leaves of *Paspalum vaginatum* under stress from Cd. (**A**) Catalase (CAT); (**B**) peroxidase (POD); (**C**) superoxide dismutase (SOD); (**D**) glutathione S-transferase (GST). Different letters in the column indicate significant differences at $p < 0.05$, according to Tukey's test; Vertical bars indicate standard errors of each mean value (n = 3).

2.2. Transcriptomic Analysis after Cd Treatment

To understand the in-depth underlying molecular mechanisms of Cd stress tolerance of, RNA-sequencing of the Cd treated leaves was performed. A total of 12 cDNA libraries prepared from the leaves at 0, 1, 4, and 24 h after Cd treatment with three biological replicates from each timing sampled were used to identify the genes responsible for heavy metal tolerance in turfgrass. After assessing the quality and filtering the data, 84.91 gigabytes (Gb) were obtained from the samples. Each sample had an average of approximately 7.08 Gb of clean data. Of the 12 libraries, the GC% of the sequenced data varied from 51.19 to 52.29%, while more than 90.57% of the reads had an average quality score exceeding 30. The obtained high-quality sequencing results illustrated the suitability of subsequent analysis. The expression profile of DEGs was identified with false discovery rate (FDR)/$p < 0.05$ and fold change > 2 for comparison of Cd treatment with different sampling intervals and compared with different sampling intervals based on the FPKM value. Based on RNA-sequence data, DEGs were classified according to up-and down-regulation of genes as well as different time points after Cd treatment of seedlings.

A total of 6820 (4457 up-regulated, 2363 down-regulated) genes were expressed differentially between Cd1 and Cd0, while this value was 14,038 (9894 up-regulated, 4144 down-regulated) and 17,327 (7956 up-regulated, 9371 down-regulated) in Cd4 vs. Cd0 and Cd24 vs. Cd0, respectively. Similarly, 10,980 (7468 up-regulated, 3512 down-regulated) genes expressed differentially between Cd4 and Cd1, while 17,304 (6573 up-regulated, 10,731 down-regulated) and 11,121 (2658 up-regulated, 8463 down-regulated) genes expressed differentially between Cd24 vs. Cd1 and Cd24 vs. Cd4, respectively (Figure 2A). Furthermore, Venn diagram analysis displayed that among significantly regulated DEGs, 2170 genes were commonly regulated in all libraries, of which 796 DEGs were inclusively up-regulated and 1374 were down-regulated among all treatments, respectively. These results also showed that DEGs were higher in Cd24 vs. Cd0 (3227/5741, up-/down-regulated) followed by Cd4 vs. Cd0 (3743/1011, up-/down-regulated) and Cd1 vs. Cd0 (1857/376, up-/down-regulated) (Figure 2B,C).

2.3. Functions and Processes Influenced by the Cd Treatment

To further understand the modulated expression of the biological function of DEGs in *P. vaginatum* under Cd stress, DEG enrichment by GO analysis was performed, and the functions of the up/down-regulated DEGs were categorized. The GO enrichment of down-regulated DEGs showed that the most significantly enriched GO terms involved in the biological process were translation, protein metabolic process, generation of precursor metabolites, and energy biosynthetic processes. Of the cellular components, the cytosol, ribosome, and cytoplasm were highly enriched. In addition, molecular function category GO enrichment analysis showed that they are mostly involved in structural molecule activity, RNA binding, translation activity, and hydrolase activity (Figure 3A). Furthermore, to understand the biological functions and signal transduction pathways of these DEGs, we applied the KEGG pathway database to know the DEG-associated pathways under Cd stress. The highly enriched pathways were Valine, leucine, and isoleucine degradation (ko00280, 28), DNA repair and recombination proteins (ko03400, 175), genetic information processing (ko09120, 193), glycerophospholipid metabolism (ko00564, 25), and transcription machinery (ko03021, 182) (Figure 3B, Supplementary data Table S1). For up-regulated DEGs, these enriched genes were annotated into three fundamental GO categories, namely, biological, cellular components, and molecular function. Of these GO terms, the importantly enriched GO terms involved in the biological process were response to chemicals, response to abiotic stimulus, and response to various stresses. Likely, the plasma membrane, membrane, vacuole, and peroxisome were higher enriched GO terms in the cellular component category. In addition, molecular function category GO enrichment analysis showed that they are mostly associated with DNA-binding transcription factor activity, transporter activity, and binding activity (Figure 3C). The highest enriched KEGG pathways were Transfer RNA biogenesis (ko03016, 15), C5-Branched dibasic acid metabolism (ko00660, 9), and Valine, leucine, and isoleucine degradation (ko00280, 11) (Figure 3D, Supplementary data Table S2). Importantly, the obtained GO analysis and KEGG pathway enrichment analysis showed that the DEGs participated in many significant pathways in response to Cd stress. The response to abiotic stimulus, the metal transport mechanism, glutathione metabolism, and the transcription factor activity were among the most enriched pathways.

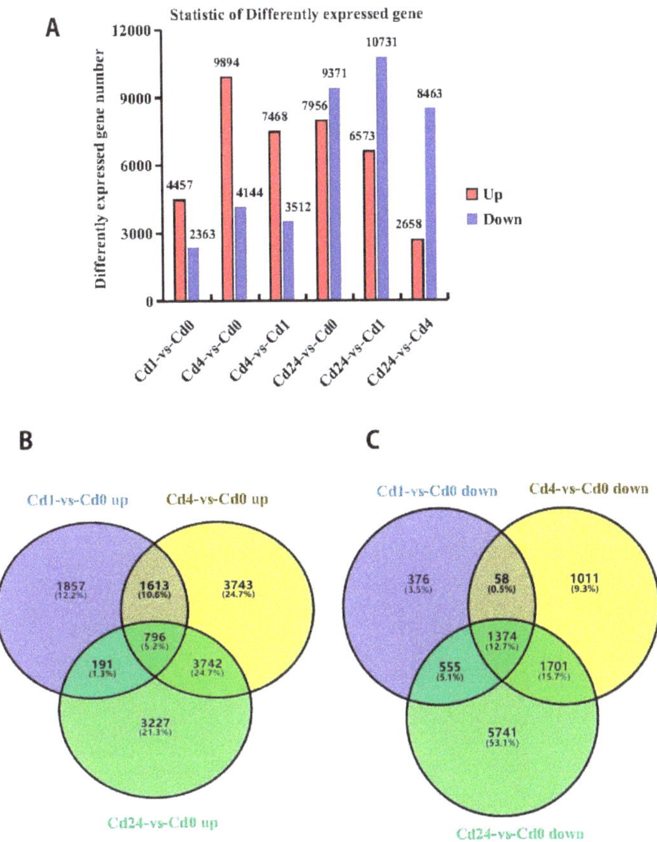

Figure 2. Analysis of gene expression by cadmium (Cd) stress tolerance of *Paspalum vaginatum*. (**A**) Compiled data of differentially expressed genes (DEGs) at different sampling times in *Paspalum vaginatum* under cd stress; (**B**) Venn diagrams of up-regulated DEGs for Cd1 vs. Cd0, Cd4 vs. Cd0 and Cd24 vs. Cd0. (**C**) Venn diagrams of down regulated DEGs for cd1 vs. Cd0, Cd4 vs. Cd0, and Cd24 vs. Cd0.

2.4. Cd Treatment Effects on Genes Involved in Glutathione Metabolism, Metal Transport, and Transcription Factors (TFs)

In the present study, we found that among the DEGs across the treatment groups, glutathione metabolism and phenylpropanoid biosynthesis pathways were significantly enriched under Cd stress conditions. In the glutathione metabolic pathway, there were 58, 111, and 82 genes up-regulated in the Cd1 vs. Cd0, Cd4 vs. Cd0, and Cd24 vs. Cd0 treatment samples, respectively. Among them, 11 genes were comparatively highly up-regulated between treatments. The up-regulated genes belonging to *glutathione synthase* (*GS*) (1 *GS*), *glutathione S-transferase* (*GST*) (7 *GST*), *glutathione peroxidase* (*GPX*) (2 *GPX*) and *glutathione hydrolase* (*GGT*) (1 *GGT*), which are the key enzymes of glutathione metabolism (Figure 4A).

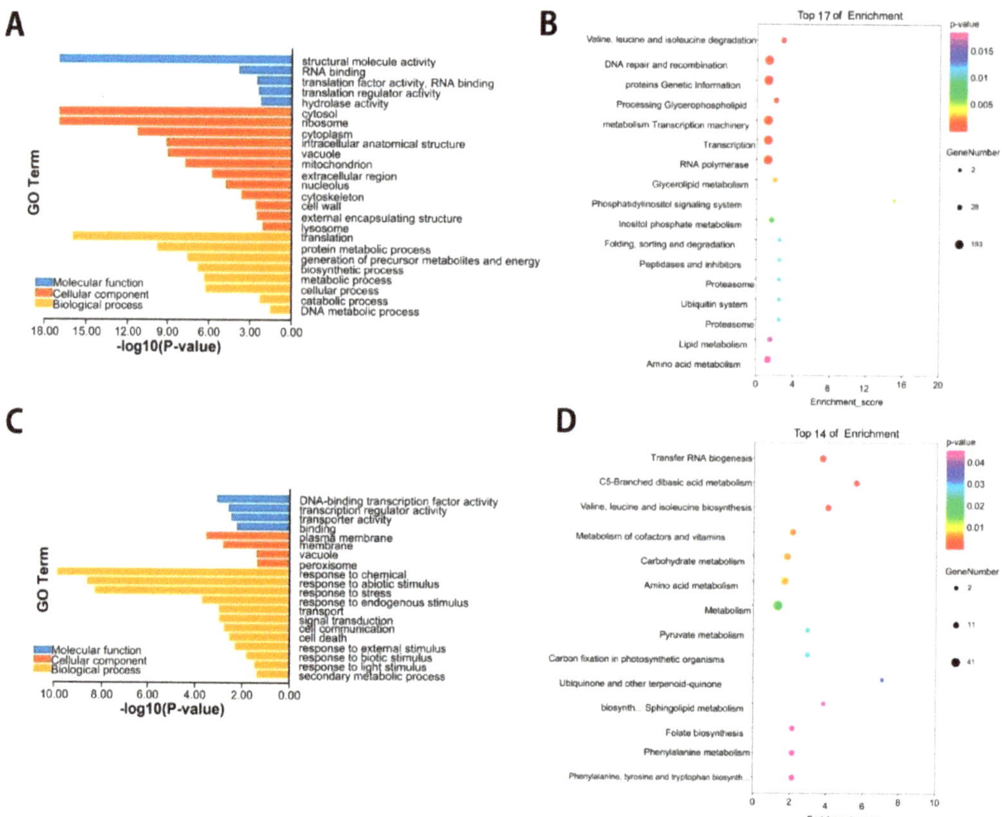

Figure 3. Gene ontology (GO) enrichment and KEGG enrichment analysis of DEGs. (**A**) Go terms of down-regulated DEGs in *Paspalum vaginatum* under Cd stress. (**B**) Graphs present 17 KEGG pathways with the highest transcriptional variations out of the down-regulated DEGs in *Paspalum vaginatum*. (**C**) Go terms of up-regulated DEGs in *Paspalum vaginatum* under Cd stress. (**D**) Graphs present 14 KEGG pathways with the highest transcriptional variations, out of the up-regulated DEGs in *Paspalum vaginatum*.

Furthermore, a total of 54 metal transport genes were identified from the differentially expressed genes, and the number of DEGs among the treatments were significantly varied. The results showed that the treatment with Cd24 vs. Cd0 had more metal transporter-related DEGs than Cd4 vs. Cd0 and Cd1 vs. Cd0 treatments (Figure 4B. *NRAMP*, *HMA*, *ZIP*, and *CAX* were the significant enriched metal transporter genes detected in this study. In addition, five *HMA* genes (4/1, up-/down-regulated) were commonly enriched between the Cd1 vs. Cd0, Cd4 vs. Cd0, and Cd24 vs. Cd0 treatments (Figure 4C). Furthermore, *NRAMP* (2), *HMA* (13) (12/1, up-/down-regulated), *ZIP* (7) (5/2, up-/down-regulated), and *CAX* (1) genes were significantly enriched in Cd24 vs. Cd0 sampled plants. Likewise, *NRAMP* (2), *HMA* (10), *ZIP* (5), and *CAX* (1) genes were enriched in Cd4 vs. Cd0 treatments, *NRAMP* (0), *HMA* (5), *ZIP* (0), and *CAX* (0) genes were highly enriched in Cd1 vs. Cd0 (Figure 4B). Additionally, we also identified 6 *SnRK2* genes (6/1, up-/down-regulated) that were significantly enriched in the treatment of Cd24 vs. Cd0, and 3 *SnRK2* genes were positively enriched in Cd4 vs. Cd0, whereas there were no DEGs regarding *SnRK2* observed in the treatment of Cd1 vs. Cd0. These results indicate that higher metal transport-associated gene expression occurred with an increase of sampling time after Cd treatment.

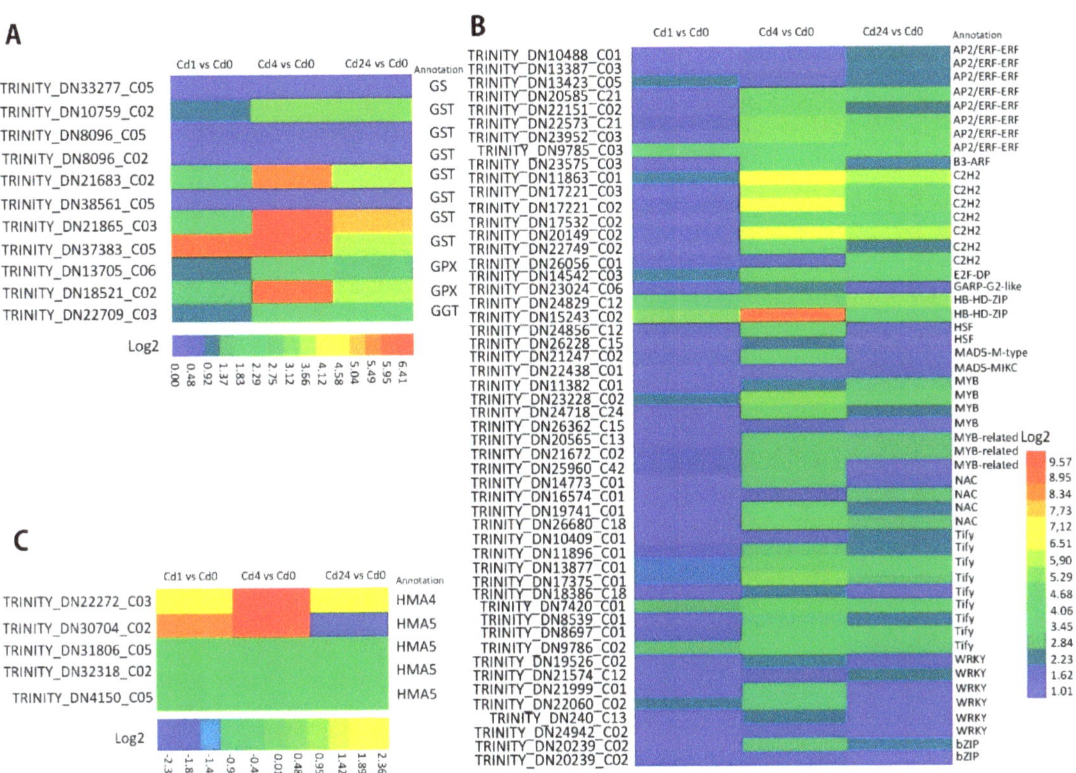

Figure 4. Heat map of DEGs involved in glutathione metabolism, metal transporter, and transcription factors. (**A**) Heat map of DEGs involved in glutathione metabolism; (**B**) Heat map of DEGs involved in metal transport; (**C**) Heat map of DEGs involved in transcription factors.

The role of several Cd stress responsive transcription factors such as, MYB, AP2/ERF, WRKY, bHLH, etc., have been identified in different plants. In this study, several DEGs belonging to different transcription families (at least 65) such as ZIP, bHLH, MYB, AP2/ERF-ERF, Tify, C2H2, and WRKY were identified in *Paspalum vaginatum* under Cd stress. A total of 141 (up-/down-regulated), 425 (up-/down-regulated), and 426 (up-/down-regulated) differentially expressed genes associated with TF families were identified from the sample of Cd1 vs. Cd0, Cd4 vs. Cd0, and Cd24 vs. Cd0, respectively, under Cd stress Similarly, 366 DEGs encoding TFs were up-regulated in Cd4 vs. Cd0 followed by Cd24 vs. Cd0 (320 DEGs) and Cd1 vs. Cd0 (122 DEGs) treatments. In total, we also detected 52 DEGs that were commonly regulated encoding TFs between the Cd1 vs. Cd0, Cd4 vs. Cd0, and Cd24 vs. Cd0 treatments. As shown in Figure 4B, 13 TF families were identified from the predicted DEGs. Among these identified TFs, *AP2/ERF-ERF* and *Tify* are the largest TF families and each TF family has eight up-regulated DEGs followed by *C2H2* containing seven up-regulated DEGs. In addition, six up-regulated DEGs were found in the *WRKY* family, followed by four up-regulated DEGs identified both in the *MYB* and *NAC* families, three up-regulated DEGs in the *MYB-related* family; the *HB-HD-ZIP*, HSF, and *bZIP* family each contained two up-regulated DEGs, and finally, the *B3-ARF*, *E2F-DP*, *GARP-G2-like*, *MADS-M-type*, and *MADS-MIKC* TF families each contained only one up-regulated DEG. The results indicated that Cd stress in the plants induced higher expression of *AP2/ERF-ERF*, *Tify*, *C2H2*, and *WRKY* TF families.

2.5. Gene Expression Validation by qRT-PCR

For the validation of the expression trends of DEGs from RNA-seq data, we selected 4 highly expressed DEGs for qRT-PCR analysis and these genes were highly associated with metal transporter genes. The transcript abundance results showed that *CAX2*, *NRAMP2*, *HMA5*, and *SnKR2.7* were significantly induced by Cd stress, and the transcript abundance pattern of these genes were up-regulated at the time intervals. The highest gene expression was observed in 24 h after Cd stress (Figure 5).

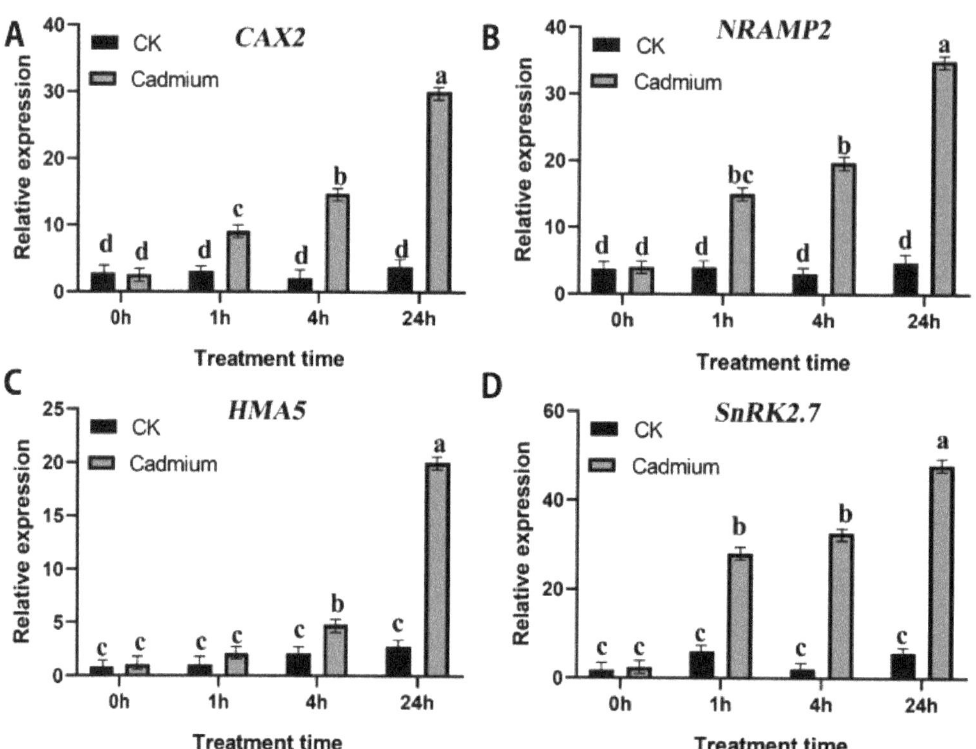

Figure 5. Validation of the expression pattern of 4 selected DEGs in the RNA-Seq by qRT-PCR in *Paspalum vaginatum* root after Cd treatments. (**A**) *CAX2*. (**B**) *NRAMP2*. (**C**) *HMA5*. (**D**) *SnKR2.7*. The lowercase letters above columns represent significant differences between over-expressed and control lines under different treatments ($p < 0.05$).

2.6. Generation of PvSnRK2.7 Transgenic A. thaliana and Cd-Tolerance Analysis

To investigate the role of PvSnRK2.7 in Cd tolerance, two PvSnRK2.7 overexpression Arabidopsis lines (Line1 and Line2) were generated. Then, the effect of *PvSnRK2.7* overexpression on root growth was analyzed by observing Arabidopsis seedlings grown on 1/2 MS media with or without Cd stress (Figure 6). Genomic DNA and RNA (reverse transcription to obtain cDNA) were extracted from the leaves of T_2 generation seedlings. *PvSnRK2.7* gene primers were designed for PCR and RT-PCR identification. The roots of all the plants were apparently similar under normal growth conditions, but the root growth of the WT plants were shorter under 80 μM $CdCl_2$ than those under 0 μM $CdCl_2$. Moreover, WT and two transgenic lines displayed similar root lengths under control conditions, whereas the root length of the WT plant was significantly shorter than transgenic lines under Cd stress. In addition, the fresh weight of the roots of transgenic plant roots were significantly higher than WT plants under normal as well as Cd stress conditions. These results indicated that overexpression of *PvSnRK2.7* promotes root growth under normal and Cd stress conditions.

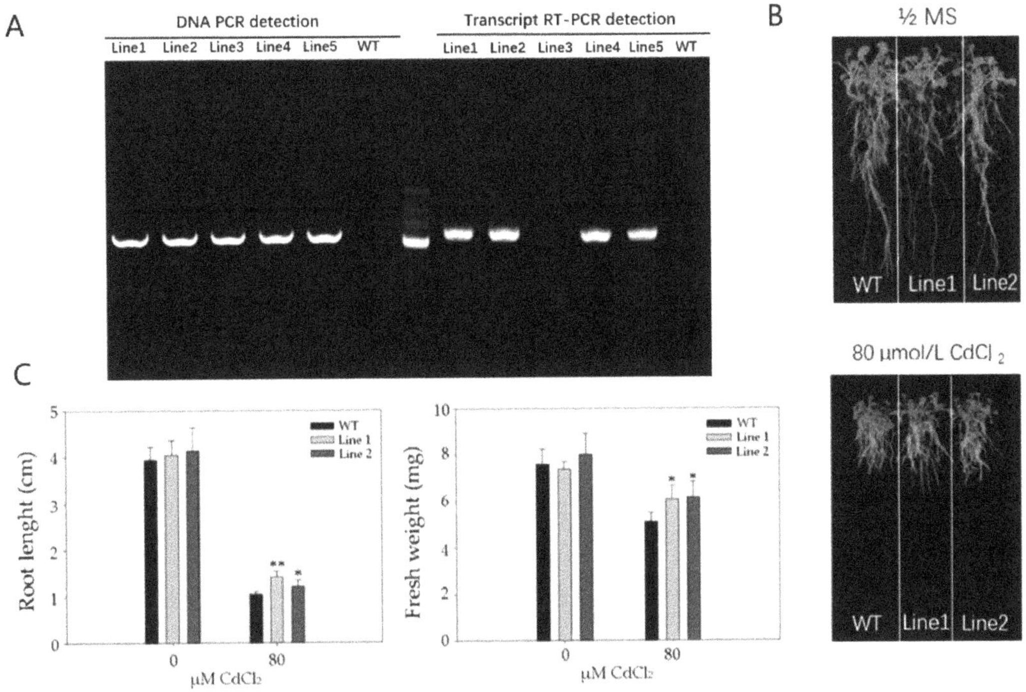

Figure 6. Cd resistant phenotypes of the *PvSnRK2.7* over-expression lines (Line-1 and Line-2). WT and PvSnRK2.7-OE seedlings were grown on 1/2 MS nutrient solution containing 0 or 80 µM CdCl$_2$ for 14 d. (**A**) *PvSnRK2.7* gene primers were designed for PCR and RT-PCR identification, then phenotypes were photographed (**B**), the root length and fresh weight were measured (**C**). The mean value and standard error were obtained from 3 biological replicates, and the significance difference level $p \leq 0.05$ (*), $p \leq 0.01$ (**).

2.7. Antioxidant Enzyme Activities Are Increased in PvSnRK2.7 Overexpressing Plants

Several studies referred to the protective role of antioxidant enzyme activities in plants under Cd stress. To investigate the antioxidant enzyme activities after Cd exposure, we measured antioxidant enzyme activities level in PvSnRK2.7-Arabidopsis and control plants. In the absence of Cd, similar levels of superoxide and hydrogen peroxide were produced in both PvSnRK2.7 transgenic and control plants. In the presence of 80 µM Cd, PvSnRK2.7 transgenic plants exhibited a significantly higher induction in the CAT, POD, SOD, and GST activities, compared with control plants. Increases in CAT, POD, and SOD activities were rather small. Only GST activity clearly increased in both transgenic lines after 14 d of treatment (Figure 7).

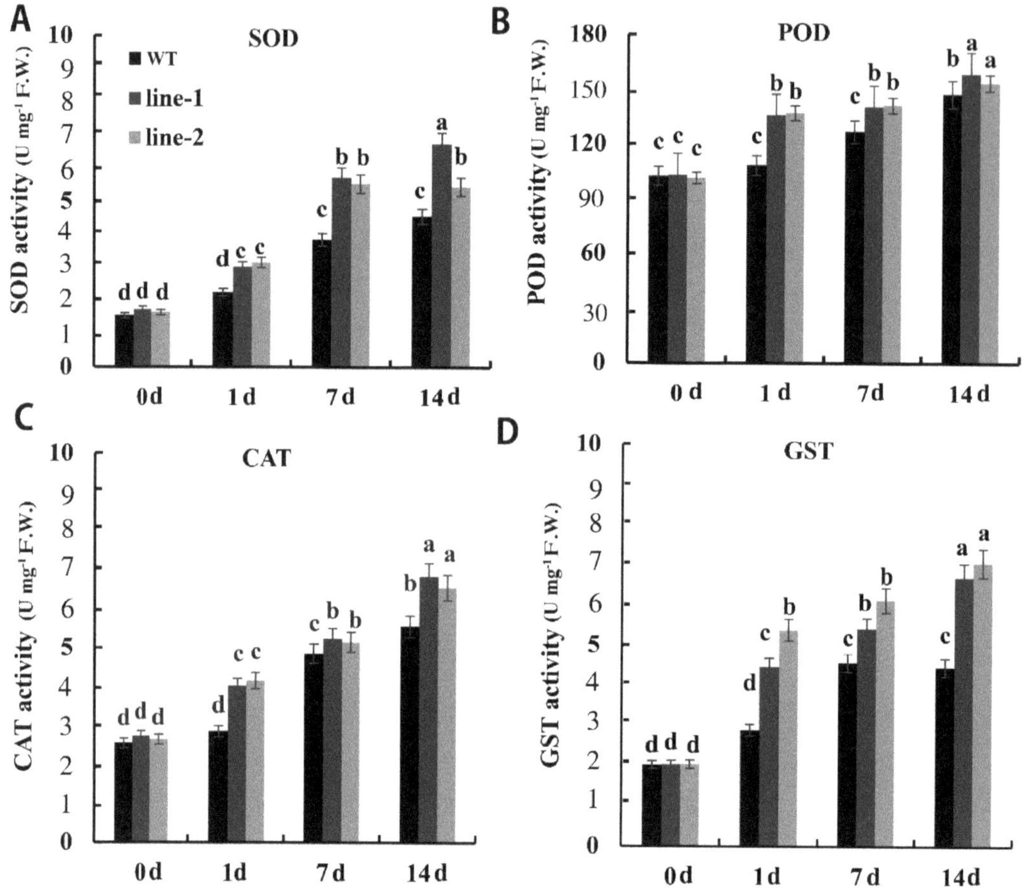

Figure 7. Oxidative stress in *PvSnRK2.7*-overexpressing and control Arabidopsis. (**A**) Superoxide, (**B**) hydrogen peroxide, (**C**) malondialdehyde and (**D**) glutathione S-transferases contents in *PvSnRK2.7*-expressed and control plants. Three-week-old seedlings grown on 1/2 MS media with and without 80 μM $CdCl_2$ were used for experiments. Data indicate the mean ± SE of three independent biological experiments. The lowercase letters above columns represent significant differences between over-expressed and control lines under different treatments ($p < 0.05$).

3. Discussion

Plants have evolved various effective strategies to manage heavy metal toxicity, including Cd, but the molecular mechanism of plants exposed to Cd stress and accumulation in plant systems is still poorly understood [10]. Primarily under stress conditions, the physiological process of plant cells was altered, thus increasing reactive oxygen species, leading to a series of defense reactions. To mitigate the cell damage caused by ROS, the plant activates both the enzymatic and non-antioxidant enzyme system, which promotes plant stress tolerance by removing or regulating excess ROS production and avoiding oxidative cell damage [24]. Under metal stress conditions, the plant activates various antioxidant enzymes, among which SOD, POD, CAT, GSH, and GST are the main antioxidant enzymes playing a crucial role in stress tolerance [33]. In this study, we observed that the enzymatic antioxidants (CAT, POD, and SOD) and non-enzymatic antioxidant (glutathione S-transferase/GST) activity was higher in different sampling intervals under Cd stress conditions (Figure 1). However, a significant level of antioxidant activity was observed in the 24 h sampling points followed by others (0, 1, and 4 h). Similar results were observed in other research investigations, and they concluded that Cd stress increased POD, CAT, SOD, and GST activities, and proline content in Populus × canadensis and tartary buckwheat [34]. SOD is the first antioxidant enzyme that plays a significant role in ROS metabolism and increase H_2O_2 production by dismuting of $O_2\bullet-/HO_2\bullet-$. Subsequently, higher accumulation of H_2O_2 is reduced by CAT and POD, which helps to reduce oxidative cell damage caused by excess H_2O_2 [35]. Previous studies also have reported that GST, GPX, and reduced glutathione (GSH) regulate ROS production under various stress conditions including drought, fungal infection, and Cd stress [29,36,37].

To elucidate the Cd-stress induced inherent mechanism at the transcriptomic level, we extensively conducted the present study in *Paspalum vaginatum*. GSH metabolism is significantly involved in metal tolerance and accumulation [38]. Here, we identified several glutathione (GSH) metabolism related enriched DEGs such as GS, GST, GPX, and GGT genes in Cd stressed plants (Figure 4). Similar to our findings, we also observed that the GSH and GST genes were up-regulated in *Populus × canadensis* under Cd stress. In addition, GSH was specifically allocated to phytochelatin synthesis (PC), which chelates heavy metal ions, including Cd^{2+} [39]. The higher expression of GS, GST, and GPX in this study suggested that they may be significantly involved in both plant defense and metal chelation of Cd. Lignin acts as a physical barrier that helps immobilize Cd into the secondary cell wall and prevents the absorption of Cd into the protoplast [11]. In the present study, we identified five genes; namely, peroxidase, cytochrome P450, 4CL, O-methyltransferase (OMT), and beta-glucosidase in phenylpropanoid biosynthesis which are associated with the enrichment of lignin synthesis in *Paspalum vaginatum* under Cd stress [13]. We observed that the 4CL genes were significantly enriched in *Nicotiana tabacum* and *Nicotiana rustica* under Cd conditions. We also noted higher accumulation of Cd in *N. tabacum* xylem sap compared to *N. rustica* and this might be related to higher expression of 4CL and casparian strip genes in the *N. rustica* plant species thus helping to reduce the accumulation of Cd in *N. rustica*. Similarly, The higher lignin content increases the plant stability and robustness against mechanical, chemical, and environmental damage such as wounding, heavy metals, and drought [40], which is strongly supported by our present findings.

The vacuole is the largest cell organelle in mature plant cells, and metal toxic substances (Cd and arsenic) are stored in this organelle [38]. Ion transporters help with uptake and translocation of essential elements in plants such as ABC families, HMA, NRAMP, and ZIP [13,26]. In this study, we identified 23 transporter genes belonging to HMA, NRAMP, ZIP, and CAX, and they were differentially expressed at different sampling intervals. The maximum number of DEGs were observed in Cd24 vs. Cd0 followed by Cd4 vs. Cd0 and Cd1 vs. Cd0; indicating that different gene expression regulation occurred based on various time intervals between treatments. In addition, we also identified 13 HMA proteins encoding genes and 12 HMA genes (HMA2/HMA4/HMA5) that were significantly enriched in *Paspalum vaginatum* under Cd stress and the qRT-PCR results (HMA5) confirmed this upregulation (Figure 4). Higher expression of HMA2/HMA4 in *N. tabacum* indicates a stronger translocation of Cd from the root to the shoot of the plants [24], which strongly supports our present findings. The ZIP protein is the best nonspecific transporter and helps the uptake of Cd as Cd+ into the plant system [24,41]. In the current study, we identified seven genes involved in ZIP transport, among which four genes were highly enriched in *Paspalum vaginatum* exposed to Cd stress; suggesting a higher accumulation of Cd^{2+} in plant leaves. CAX and NRAMP ion transporters also play a significant role in the translocation and compartmentalization of Cd into the vacuole of plant cell vacuole cells [38]. In this study, higher expression CAX2 and NRAMP2 genes was observed in *Paspalum vaginatum* under Cd stress (Figure 5). The enriching activities of SaCAX2h in *Sedum alfredii* was increased Cd accumulation in tobacco plants [42], and NRAMP1, NRAMP3, and NRAMP4 are involved in Cd^{2+} accumulation in Arabidopsis and rice plants, while the NRAMP2 genes can transport Mn^{2+} and Fe^{2+} ions [38,43,44], our findings were consistent with previous results.

Transcription factors (TFs) play a significant role in stress tolerance in plants [24]. In the present study, we identified 65 TF families including bZIP, bHLH, MYB, NAC, HSF, Tify, AP2/ERF-ERF, and C_2H_2, and, among them, AP2/ERF-ERF, Tify, and C_2H_2 are the largest TF families that were significantly enriched under Cd stress (Figure 4). Our findings were strongly supported by previous research that identified a total of 59 TF families (AP2-ERBP, C_2H_2, NAC, MYB, and MYB-related are the most prominent families) that are highly expressed in two barley genotypes under copper and cobalt stress conditions [26]. In addition, Tify, previously known as ZIM, is also one of the important members of TF families playing a significant role in the cross-talk between phytohormones and their signaling pathways, which are involved in biotic and abiotic stresses [45]. In this study, we observed a higher number of genes in the Tify TF family genes under Cd stress (Figure 4). Similar results were also found by the Wang et al.'s [46] research team and they identified 25 TIFY genes in the Populus trichocarpa genome. Moreover, C_2H_2 (Cys2/His2-type) zinc finger proteins are key transcriptional regulators in response to various abiotic stresses including extreme temperatures, salinity, drought, oxidative stress, excessive light, and silique shattering [47]. The higher number of C_2H_2 TF families associated with gene expression was identified under Cd stress (Figure 4) and analogous results were observed by Lwalaba et al. (2021) [26] in barley under cobalt and copper stress. Furthermore, we identified some other common TF families such as WRKY, MYB, NAC, MYB-related, HB-HD-ZIP, HSF, B3-ARF, E2F-DP, GARP-G2-like, MADS-M-type, and MADS-MIKC that were up-regulated in plants under Cd stress. Previous studies have reported that WRKY, MYB, and NAC are the predominant transcription factors playing a significant role in plant metal stress tolerance mechanisms [24,26].

Sucrose non-fermenting (SNF1) related protein kinase 1 subfamily protein (SnRK) plays a crucial role in ABA-dependent and ABA-independent environmental stress-signaling responses [48,49]. In this study, we identified that SnRK2.7 was up-regulated in *Paspalum vaginatum* under Cd stress and plants overexpressing SnRK2.7 could enhance tolerance to Cd (Figures 5 and 6). SnRK2 is a relatively large gene family and is significantly expressed in both cells and seedlings of *Arabidopsis thaliana* [50] and *Oryza sativa* [51] under different stress conditions. Similarly, [52] reported that ABA signaling pathways participated in Cd detoxification in the low-Cd-accumulating genotype (LAJK) of *B. chinensis* by ABA-induced antioxidant pathway. Based on these study results, we concluded that the enriched expression of the SnRK2 genes in *Paspalum vaginatum* under Cd stress conditions indicate the active involvement of ABA signaling in the regulation of heavy metal stress regulation in plants. Finally, we concluded that antioxidant activities and their encoding gene expression, stress-responsive genes, metal transporter genes, stress signaling response genes, and transcription factor regulation-associated genes were highly enriched with increasing sampling time points compared with earlier sampling time points and thus all help to increase Cd tolerance of the plant.

4. Materials and Methods

4.1. Plant Materials and Cd Treatment

Paspalum vaginatum SW (Seashore paspalum) were grown in 11-cm diameter and 21-cm deep plastic pots filled with 1:1:1 ratio of sand, soil, and perlite. The experimental pots were placed in turfgrass germplasm (Nanjing Agricultural University, Jiangsu province, China) and kept at optimal temperature and relative humidity. An adequate level of water and fertilizer was supplied to the plants. The plants were treated with 300 μmol/L of Cd^{2+}, supplemented with $CdCl_2 \cdot 5H_2O$. After Cd treatment, *Paspalum vaginatum* leaves were harvested at 0, 1, 4 and 24 h time intervals, immediately freeze-dried, and stored at $-80\,°C$ for transcriptome analysis, gene expression validation, and antioxidant analysis.

4.2. Antioxidants Activities Estimation

Leaf samples were collected at 0, 1, 4, and 24 h after Cd treatment and then processed for the analysis of antioxidant enzyme activities as described [53]. Catalase (CAT) activity was determined by following the consumption of H_2O_2 at 240 nm for 3 min [54], peroxidase (POD) activity was measured by the change in absorbance of 470 nm due to that determined using 4-methyl catechol as substrate according to the method described [55]. Superoxide dismutase (SOD) activity was determined by measuring the inhibition in the photoreduction of nitroblue tetrazolium (NBT). Glutathione S-transferase (GST) activities were determined according to the manufacturer's instructions (Jiancheng Bioengineering Institute, Nanjing, China). The antioxidant enzyme activities were statistically analyzed with three independent replicates.

4.3. Transcriptome Assembly and Genes Annotation

The raw data were filtered, and then clean data were obtained by removing adapter sequences, poly-N, and low-quality reads. Simultaneously, GC content levels, Q20, and Q30 of the clean data were evaluated. The functions of the unigenes were annotated using a series of databases, including BLASTx, against NCBI non-redundant protein (Nr), NCBI nucleotide collection (Nt), and Swiss-Prot databases. The reference genome was directly downloaded from the genome website (https://www.ncbi.nlm.nih.gov/genome/?term= Paspalum+vaginatum (accessed on 14 July 2022)). After filtering, the reference genome was built using hisat2-build and paired-end clean reads were aligned to the reference genome using hisat2 v2.0.12 software (Johns Hopkins University, Baltimore, MD, USA). Functional annotations of putative unigenes were grouped using the Kyoto Encyclopedia of Genes and Genomes (KEGG) Gene Ontology (GO).

4.4. Gene Expression Level Quantification and Differential Expression Analysis

Gene expression levels were calculated using the fragments per kilobase of transcript sequence per million base pairs sequenced (FPKM) method. Differentially expressed genes (DEGs) were identified via differentially DEseq package [54] The DEGs were identified with false discovery rate (FDR)/fold change > 2 and $p < 0.05$ were deemed to be DEGs. Gene Ontology (GO) enrichment analysis of DEGs was carried out by the GO seq R package. GO terms with corrected $p < 0.05$ were significantly enriched.

4.5. RNA Extraction, cDNA Library, and qRT-PCR for Validation

To evaluate the reliability of differentially expressed genes under Cd stress conditions revealed by RNA sequencing, four genes were selected from the important pathways and validated by quantitative real-time RT-PCR (qPCR). The leaf samples were collected from Cd-treated plants at 0, 1, 4, and 24 h intervals, and freeze-dried in liquid nitrogen. Total RNA was extracted from the ground leaves according to the manufacturer's instructions (EASYspin RNA Plant Mini Kit, Aidlab, China). Extracted total RNA was reverse-transcribed to cDNA in a total volume of 10 µL using the TransScript II cDNA synthesis SuperMix kit (Transgen Biotech, China). The primer sequences were designed based on the corresponding gene sequence by NCBI (https://www.ncbi.nlm.nih.gov/tools/primer-blast/, accessed on 14 July 2022). Gene expression analysis was performed using a qRT-PCR CFX96 thermocycler (Bio-Rad) with GAPDH serving as an internal control. The primers of target genes and GAPDH were diluted in a ChamQ SYBR qPCR master mix (Vazyme Biotech Co., Ltd., Nanjing, China), and the mix was transferred to a 96-well plate. Reactions were performed and relative expression values were calculated according to the $2^{-\Delta\Delta Ct}$ [55].

4.6. Statistical Analysis

GraphPad Prism 5 (GraphPad software, San Diego, CA, USA) was used to compute the data and statistical analyses. All the records were statistically analyzed with appropriate replications, the values are expressed as the mean value with standard error means (SE±), and the results were statistically analyzed using one-way ANOVA (analysis of variance) with a Tukey's correction ($p < 0.05$).

4.7. Agrobacterium-Mediated Transformation of Arabidopsis

The monoclonal colonies containing the target gene were inoculated into 1 mL culture medium (50 mg/L Kan and 50 mg/L Rif), and shaken at 28 °C for 2–3 days, then 100 µL of solution was extracted and inoculated into 200 mL of fresh medium (OD600 = 0.2) allowing growth until OD600 = 0.8 under the same conditions. After centrifugation at 4 °C, the thalli were collected in a 50 mL tube. Next, the thalli were resuspended with pre-cooled 5% sucrose solution until no obvious thalli ware found. Before dipping, 1–2 drops of Silwet-L77 surfactant were added. Apical parts of the Arabidopsis (growing to 2 months old) were dipped in the dipping dye for about 5 s and shaken slightly.

4.8. Positive Lines Screening of Transgenic Arabidopsis

The primary seeds were harvested after *Agrobacterium*-mediated transformation was recorded as the T_0 generation. The T_0 generation seeds were sown on 20 mg/mL Basta resistance screening 1/2 MS culture medium. After 7 days, the normal surviving seedlings were transplanted into the soil and cultured until maturity, and the harvested seeds were recorded as the T_1 generation. According to the above methods, until the T2 generation seeds were obtained, gDNA and RNA (reverse transcription to obtain cDNA) were extracted from the leaves of the T_2 generation seedlings. Precise gene primers were designed for PCR and RT-PCR identification.

4.9. The Treatment of Cadmium on Culture Medium

The transgenic *Arabidopsis* seeds of the T_2 generation were disinfected (20% 84 disinfectant for 20 min, washed 6 times with sterile water, and vernalized at 4 °C for 2 days) and then seeded on 1/2 MS medium. After germination, seedlings with consistent growth were selected and transferred to a new square medium containing different concentrations of cadmium. Seedlings grew under light (photoperiod 16 h/8 h) until the obvious phenotype was observed, then pictures were collected and root length and fresh weight were measured. Three biological replicates were performed for each treatment, collected at 0, 1, 7, and 14 days after Cd treatment, and then processed for the analysis of antioxidant enzyme activities as described in Section 4.2.

5. Conclusions

The transcriptomic profile obtained from the results of the RNA seq analysis revealed a distinctly different gene expression pattern between the sampling intervals after Cd treatment in *Paspalum vaginatum*. A higher number of DEGs were observed in Cd24 vs. Cd0 (17,327 DEGs), followed by Cd4 vs. Cd0 (14,038 DEGs) and Cd1 vs. Cd0 (6820). The DEG results indicated that higher Cd accumulation and stress response occurred with increasing sampling times. The schematic diagram represents the transcriptional changes of genes responsible for Cd stress tolerance and accumulation in *Paspalum vaginatum*. In this study many genes encoding lignin biosynthesis (POD, 4CL, and OMT), non-enzymatic antioxidants (GST, GS, and GPX), signaling response (SnRK2), and transcription factors (AP2/ERF-ERF, Tify, C2H2, and WRKY) were highly up-regulated in the plants under Cd treatment. Similarly, heavy metal transporter genes (HMA, ZIP, CAX2, and NRAMP2) were highly expressed among the treatments (Cd1 vs. Cd0, Cd4 vs. Cd0, and Cd24 vs. Cd0) and their expression increased with an increase of sampling time indicating that higher Cd accumulation occurred based on the incubation intervals. Based on transcriptome analysis, *Paspalum vaginatum* can be used as a heavy metal remover from polluted soil, however, further experimental studies are required to conclude this statement.

Supplementary Materials: The following supporting information can be downloaded at: https://www.mdpi.com/article/10.3390/plants11162078/s1, Table S1: Graphs present 17 KEGG pathways with the highest transcriptional variations, out of the down-regulated DEGs in *Paspalum vaginatum*; Table S2: Graphs present 14 KEGG pathways with the highest transcriptional variations, out of the up-regulated DEGs in *Paspalum vaginatum*.

Author Contributions: Y.C., Z.Y. and J.L. conceived the study and designed the experiments. L.X. and Y.Z. performed the experiments. Y.C. and L.X. analyzed the data with suggestions by Z.Y. and J.L., Y.C., Q.Y. and L.X. wrote the manuscript. All authors have read and agreed to the published version of the manuscript.

Funding: This work was supported by the program of National Natural Science Foundation of China NSFC (31872953, 31672193).

Data Availability Statement: The datasets used in this study can be found in the NCBI SRA database under the accession number SUB11901595.

Conflicts of Interest: The authors declare that they have no conflict of interest.

References

1. Altaf, M.A.; Shahid, R.; Ren, M.X.; Khan, L.U.; Altaf, M.M.; Jahan, M.S.; Nawaz, M.A.; Naz, S.; Shahid, S.; Lal, M.K. Protective Mechanisms of Melatonin Against Vanadium Phytotoxicity in Tomato Seedlings: Insights into Nutritional Status, Photosynthesis, Root Architecture System, and Antioxidant Machinery. *J. Plant Growth Regul.* **2021**, *10*, 1–17. [CrossRef]
2. Shanmugaraj, B.M.; Malla, A. Chapter 1-cadmium stress and toxicity in plants: An Overview. In *Cadmium Toxicity and Tolerance in Plants*; Hasanuzzaman, M., Prasad, M.N.V., Eds.; Academic Press: Cambridge, MA, USA, 2019; pp. 1–17. [CrossRef]
3. El Rasafi, T.; Oukarroum, A.; Haddioui, A.; Song, H.; Kwon, E.E.; Bolan, N.; Tack, F.M.G.; Sebastian, A.; Prasad, M.N.V.; Rinklebe, J. Cadmium stress in plants: A critical review of the effects, mechanisms, and tolerance strategies. *Crit. Rev. Environ. Sci. Technol.* **2020**, *52*, 675–726. [CrossRef]
4. Carne, G.; Leconte, S.; Sirot, V.; Breysse, N.; Badot, P.M.; Bispo, A.; Deportes, I.Z.; Dumat, C.; Crepet, A. Mass balance approach to assess the impact of cadmium decrease in mineral phosphate fertilizers on health risk: The case-study of french agricultural soils. *Sci. Total Environ.* **2021**, *760*, 143374. [CrossRef] [PubMed]
5. Zhao, H.Y.; Guan, J.L.; Liang, Q.; Zhang, X.Y.; Hu, H.L.; Zhang, J. Effects of cadmium stress on growth and physiological characteristics of sassafras seedlings. *Sci. Rep.* **2021**, *11*, 9913. [CrossRef] [PubMed]
6. Dai, S.; Wang, B.Q.; Song, Y.; Xie, Z.M.; Li, C.; Li, S.; Huang, Y.; Jiang, M. Astaxanthin and its gold nanoparticles mitigate cadmium toxicity in rice by inhibiting cadmium translocation and uptake. *Sci. Total Environ.* **2021**, *786*, 147496. [CrossRef] [PubMed]
7. Cai, Z.D.; Xian, P.Q.; Wang, H.; Lin, R.B.; Lian, T.X.; Cheng, Y.B.; Ma, Q.B.; Nian, H. Transcription factor GmWRKY142 confers cadmium resistance by up-regulating the cadmium tolerance 1-like genes. *Front. Plant Sci.* **2020**, *11*, 724. [CrossRef] [PubMed]
8. Anjum, N.A.; Aref, I.M.; Duarte, A.C.; Pereira, E.; Ahmad, I. Glutathione and proline can coordinately make plants withstand the joint attack of metal(loid) and salinity stresses. *Front. Plant Sci.* **2014**, *5*, 662. [CrossRef]
9. Ali, B.; Gill, R.A.; Yang, S.; Gill, M.B.; Farooq, M.A.; Liu, D.; Daud, M.K.; Ali, S.; Zhou, W.J. Regulation of cadmium-induced proteomic and metabolic changes by 5-aminolevulinic acid in leaves of *Brassica napus* L. *PLoS ONE* **2015**, *10*, e0123328. [CrossRef]
10. Zhang, L.; Ding, H.; Jiang, H.L.; Wang, H.S.; Chen, K.X.; Duan, J.J.; Feng, S.J.; Wu, G. Regulation of cadmium tolerance and accumulation by miR156 in *Arabidopsis*. *Chemosphere* **2020**, *242*, 125168. [CrossRef]
11. He, M.; Jahan, M.S.; Wang, Y.; Sun, J.; Shu, S.; Guo, S. Compost amendments based on vinegar residue promote tomato growth and sup-press bacterial wilt caused by *Ralstonia Solanacearum*. *Pathogens* **2020**, *9*, 227. [CrossRef]
12. Zhang, Y.; Liu, J.; Yu, J.; Zhang, H.; Yang, Z. Relationship between the phenylpropanoid pathway and dwarfism of paspalum seashore based on RNA-Seq and iTRAQ. *Int. J. Mol. Sci.* **2021**, *22*, 9568. [CrossRef] [PubMed]
13. Huang, W.X.; Zhang, D.M.; Cao, Y.Q.; Dang, B.J.; Jia, W.; Xu, Z.C.; Han, D. Differential cadmium translocation and accumulation between *Nicotiana tabacum* L. and *Nicotiana rustica* L. by transcriptome combined with chemical form analyses. *Ecotoxicol. Environ. Saf.* **2021**, *208*, 111412. [CrossRef]
14. Zhou, J.; Wan, H.; He, J.; Lyu, D.; Li, H. Integration of cadmium accumulation, subcellular distribution, and physiological responses to understand cadmium tolerance in apple rootstocks. *Front. Plant Sci.* **2017**, *8*, 966. [CrossRef] [PubMed]
15. Zhang, H.; Guo, Q.; Yang, J.; Shen, J.; Chen, T.; Zhu, G.; Chen, H.; Shao, C. Subcellular cadmium distribution and antioxidant enzymatic activities in the leaves of two castor (*Ricinus communis* L.) cultivars exhibit differences in Cd accumulation. *Ecotoxicol. Environ. Saf.* **2015**, *120*, 184–192. [CrossRef]
16. Ai, T.N.; Naing, A.H.; Yun, B.W.; Lim, S.H.; Kim, C.K. Overexpression of RsMYB1 enhances anthocyanin accumulation and heavy metal stress tolerance in transgenic Petunia. *Front. Plant Sci.* **2018**, *9*, 1388. [CrossRef] [PubMed]
17. Hasan, M.M.; Alharbi, B.M.; Alhaithloul, H.A.S.; Abdulmajeed, A.M.; Alghanem, S.M.; Al-Mushhin, A.A.M.; Jahan, M.S.; Corpas, F.J.; Fang, X.-W.; Soliman, M.H. Spermine-Mediated Toleranceto Selenium Toxicity in Wheat (*Triticum aestivum* L.) Depends onEndogenous Nitric Oxide Synthesis. *Antioxidants* **2021**, *10*, 1835. [CrossRef]

18. Romè, C.; Huang, X.Y.; Danku, J.; Salt, D.E.; Sebastiani, L. Expression of specific genes involved in Cd uptake, translocation, vacuolar compartmentalisation and recycling in *Populus alba* Villafranca clone. *J. Plant Physiol.* **2016**, *202*, 83–91. [CrossRef]
19. Li, D.D.; Xu, X.M.; Hu, X.Q.; Liu, Q.G.; Wang, Z.C.; Zhang, H.Z.; Wang, H.; Wei, M.; Wang, H.Z.; Liu, H.M.; et al. Genome-wide analysis and heavy metal-induced expression profiling of the HMA gene family in *Populus trichocarpa*. *Front. Plant Sci.* **2015**, *6*, 1149. [CrossRef]
20. Ullah, I.; Wang, Y.; Eide, D.J.; Dunwell, J.M. Evolution, and functional analysis of natural resistance-associated macrophage proteins (NRAMPs) from theobroma cacao and their role in cadmium accumulation. *Sci. Rep.* **2018**, *8*, 14412. [CrossRef]
21. Ye, P.; Wang, M.H.; Zhang, T.; Liu, X.Y.; Jiang, H.; Sun, Y.P.; Cheng, X.Y.; Yan, Q. Enhanced cadmium accumulation and tolerance in transgenic hairy roots of *Solanum nigrum* L. expressing iron-regulated transporter gene IRT1. *Life* **2020**, *10*, 324. [CrossRef]
22. Bucholc, M.; Ciesielski, A.; Goch, G.; Anielska-Mazur, A.; Kulik, A.; Krzywinska, E.; Dobrowolska, G. SNF1-related protein kinases 2 are negatively regulated by a plant-specific calcium sensor. *J. Biol. Chem.* **2011**, *286*, 3429–3441. [CrossRef] [PubMed]
23. Wang, Y.J.; Yan, H.F.; Qiu, Z.F.; Hu, B.; Zeng, B.S.; Zhong, C.L.; Fan, C.J. Comprehensive analysis of SnRK gene family and their responses to salt stress in *Eucalyptus grandis*. *Int. J. Mol. Sci.* **2019**, *20*, 2786. [CrossRef] [PubMed]
24. Li, X.; Mao, X.H.; Xu, Y.J.; Li, Y.; Zhao, N.; Yao, J.X.; Dong, Y.F.; Tigabu, M.; Zhao, X.Y.; Li, S.W. Comparative transcriptomic analysis reveals the coordinated mechanisms of *Populus × canadensis* 'Neva' leaves in response to cadmium stress. *Ecotoxicol. Environ. Saf.* **2021**, *216*, 112179. [CrossRef] [PubMed]
25. Jalmi, S.K.; Bhagat, P.K.; Verma, D.; Noryang, S.; Tayyeba, S.; Singh, K.; Sharma, D.; Sinha, A.K. Traversing the links between heavy metal stress and plant signaling. *Front. Plant Sci.* **2018**, *9*, 12. [CrossRef] [PubMed]
26. Lwalaba, J.L.W.; Zvobgo, G.; Gai, Y.P.; Issaka, J.H.; Mwamba, T.M.; Louis, L.T.; Fu, L.B.; Nazir, M.M.; Kirika, B.A.; Tshibangu, A.K.; et al. Transcriptome analysis reveals the tolerant mechanisms to cobalt and copper in barley. *Ecotoxicol. Environ. Saf.* **2021**, *209*, 111761. [CrossRef]
27. Zumaquero, A.; Kanematsu, S.; Nakayashiki, H.; Matas, A.; Martinez-Ferri, E.; Barcelo-Munoz, A.; Pliego-Alfaro, F.; Lopez-Herrera, C.; Cazorla, F.M.; Pliego, C. Transcriptome analysis of the fungal pathogen *Rosellinia necatrix* during infection of a susceptible avocado rootstock identifies potential mechanisms of pathogenesis. *BMC Genom.* **2019**, *20*, 1016. [CrossRef]
28. Dong, T.T.; Zhu, M.K.; Yu, J.W.; Han, R.P.; Tang, C.; Xu, T.; Liu, J.R.; Li, Z.Y. RNA-Seq and iTRAQ reveal multiple pathways involved in storage root formation and development in sweet potato (*Ipomoea batatas* L.). *BMC Plant Biol.* **2019**, *19*, 136. [CrossRef]
29. Varoquaux, N.; Cole, B.; Gao, C.; Pierroz, G.; Baker, C.R.; Patel, D.; Madera, M.; Jeffers, T.; Hollingsworth, J.; Sievert, J.; et al. Transcriptomic analysis of field-droughted sorghum from seedling to maturity reveals biotic and metabolic responses. *Proc. Natl. Acad. Sci. USA* **2019**, *116*, 27124–27132. [CrossRef]
30. Ntoulas, N.; Varsamos, I. Performance of two seashore paspalum (*Paspalum vaginatum* Sw.) varieties growing in shallow green roof substrate depths and irrigated with seawater. *Agronomy* **2021**, *11*, 250. [CrossRef]
31. Wu, P.P.; Cogill, S.; Qiu, Y.J.; Li, Z.G.; Zhou, M.; Hu, Q.; Chang, Z.H.; Noorai, R.E.; Xia, X.X.; Saski, C.; et al. Comparative transcriptome profiling provides insights into plant salt tolerance in seashore paspalum (*Paspalum vaginatum*). *BMC Genom.* **2020**, *21*, 131. [CrossRef]
32. Fabbri, L.T.; PLoSchuk, E.L.; López, M.V.; Insausti, P.; Rua, G.H. Freeze tolerance differs between two ecotypes of *Paspalum vaginatum* (Poaceae). *Acta Bot. Bras.* **2016**, *30*, 152–156. [CrossRef]
33. Li, F.T.; Qi, J.M.; Zhang, G.Y.; Lin, L.H.; Fang, P.P.; Tao, A.F.; Xu, J.T. Effect of cadmium stress on the growth, antioxidative enzymes and lipid peroxidation in two kenaf (*Hibiscus cannabinus* L.) plant seedlings. *J. Integr. Agric.* **2013**, *12*, 610–620. [CrossRef]
34. Lu, Y.; Wang, Q.F.; Li, J.; Xiong, J.; Zhou, L.N.; He, S.L.; Zhang, J.Q.; Chen, Z.A.; He, S.G.; Liu, H. Effects of exogenous sulfur on alleviating cadmium stress in *Tartary buckwheat*. *Sci. Rep.* **2019**, *9*, 7397. [CrossRef] [PubMed]
35. Hasanuzzaman, M.; Bhuyan, M.H.M.; Zulfiqar, F.; Raza, A.; Mohsin, S.M.; Al Mahmud, J.; Fujita, M.; Fotopoulos, V. Reactive oxygen species and antioxidant defense in plants under abiotic stress: Revisiting the crucial role of a universal defense regulator. *Antixidants* **2020**, *9*, 681. [CrossRef]
36. Baek, S.A.; Han, T.; Ahn, S.K.; Kang, H.; Im, K.H. Effects of heavy metals on plant growths and pigment contents in *Arabidopsis thaliana*. *Plant Pathol. J.* **2012**, *28*, 446–452. [CrossRef]
37. Ederli, L.; Reale, L.; Ferranti, F.; Pasqualini, S. Responses induced by high concentration of cadmium in *Phragmites australis* roots. *Physiol. Plant.* **2004**, *121*, 66–74. [CrossRef]
38. Zhang, J.; Martinoia, E.; Lee, Y. Vacuolar transporters for cadmium and arsenic in plants and their applications in phytoremediation and crop development. *Plant Cell Physiol.* **2018**, *59*, 1317–1325. [CrossRef]
39. Jozefczak, M.; Keunen, E.; Schat, H.; Bliek, M.; Hernandez, L.E.; Carleer, R.; Remans, T.; Bohler, S.; Vangronsveld, J.; Cuypers, A. Differential response of *Arabidopsis* leaves and roots to cadmium: Glutathione-related chelating capacity vs. antioxidant capacity. *Plant Physiol. Biochem.* **2014**, *83*, 1–9. [CrossRef]
40. Moura, J.C.M.S.; Bonine, C.; Viana, J.; Dornelas, M.C.; Mazzafera, P. Abiotic and biotic stresses and changes in the lignin content and composition in plants. *J. Int. Plant Biol.* **2010**, *52*, 360–376. [CrossRef]
41. Ismael, M.A.; Elyamine, A.M.; Moussa, M.G.; Cai, M.M.; Zhao, X.H.; Hu, C.X. Cadmium in plants: Uptake, toxicity, and its interactions with selenium fertilizers. *Metallomics* **2019**, *11*, 255–277. [CrossRef]
42. Zhang, J.; Zhang, M.; Tian, S.; Lu, L.; Shohag, M.J.; Yang, X. Metallothionein 2 (SaMT2) from Sedum alfredii Hance confers increased Cd tolerance and accumulation in yeast and tobacco. *PLoS ONE* **2014**, *17*, e102750. [CrossRef] [PubMed]

43. Thomine, S.; Wang, R.; Ward, J.M.; Crawford, N.M.; Schroeder, J.I. Cadmium and iron transport by members of a plant metal transporter family in *Arabidopsis* with homology to nramp genes. *Proc. Natl. Acad. Sci. USA* **2000**, *97*, 4991–4996. [CrossRef] [PubMed]
44. Takahashi, R.; Ishimaru, Y.; Senoura, T.; Shimo, H.; Ishikawa, S.; Arao, T.; Nakanishi, H.; Nishizawa, N.K. The OsNRAMP1 iron transporter is involved in Cd accumulation in rice. *J. Exp. Bot.* **2011**, *62*, 4843–4850. [CrossRef] [PubMed]
45. Singh, P.; Mukhopadhyay, K. Comprehensive molecular dissection of TIFY transcription factors reveal their dynamic responses to biotic and abiotic stress in wheat (*Triticum aestivum* L.). *Sci. Rep.* **2021**, *11*, 9739. [CrossRef] [PubMed]
46. Wang, H.; Leng, X.; Xu, X.; Li, C. Comprehensive analysis of the TIFY gene family and its expression profiles under phytohormone treatment and abiotic stresses in roots of *Populus trichocarpa*. *Forests* **2020**, *11*, 315. [CrossRef]
47. Wang, K.; Ding, Y.; Cai, C.; Chen, Z.; Zhu, C. The role of C2H2 zinc finger proteins in plant responses to abiotic stresses. *Physiol. Plant.* **2019**, *165*, 690–700. [CrossRef]
48. Holappa, L.D.; Ronald, P.C.; Kramer, E.M. Evolutionary analysis of snf1-related protein kinase2 (SnRK2) and calcium sensor (SCS) gene lineages, and dimerization of rice homologs, suggest deep biochemical conservation across angiosperms. *Front. Plant Sci.* **2017**, *8*, 395. [CrossRef]
49. Todaka, D.; Shinozaki, K.; Yamaguchi-Shinozaki, K. Recent advances in the dissection of drought-stress regulatory networks and strategies for development of drought-tolerant transgenic rice plants. *Front Plant* **2015**, *6*, 84. [CrossRef]
50. Boudsocq, M.; Barbier-Brygoo, H.; Laurière, C. Identification of nine sucrose nonfermenting 1-related protein kinases 2 activated by hyperosmotic and saline stresses in *Arabidopsis thaliana*. *J. Biol. Chem.* **2004**, *279*, 41758–41766. [CrossRef]
51. Coello, P.; Hirano, E.; Hey, S.J.; Muttucumaru, N.; Martinez-Barajas, E.; Parry, M.A.J.; Halford, N.G. Evidence that abscisic acid promotes degradation of SNF1-related protein kinase (SnRK) 1 in wheat and activation of a putative calcium-dependent SnRK2. *J. Exp. Bot.* **2012**, *63*, 913–924. [CrossRef]
52. Zhou, Q.; Guo, J.J.; He, C.T.; Shen, C.; Huang, Y.Y.; Chen, J.X.; Guo, J.H.; Yuan, J.G.; Yang, Z.Y. Comparative transcriptome analysis between low-and high-cadmium-accumulating genotypes of pakchoi (*Brassica chinensis* L.) in response to cadmium stress. *Environ. Sci. Technol.* **2016**, *50*, 6485–6494. [CrossRef] [PubMed]
53. Li, X.; Wei, J.P.; Scott, E.R.; Liu, J.W.; Guo, S.; Li, Y.; Zhang, L.; Han, W.Y. Exogenous melatonin alleviates cold stress by promoting antioxidant defense and redox homeostasis in *Camellia sinensis* L. *Molecules* **2018**, *23*, 165. [CrossRef] [PubMed]
54. Anders, S.; Huber, W. Differential Expression of RNA-Seq Data at the Gene Level-the DESeq Package. Available online: https://citeseerx.ist.psu.edu/viewdoc/download?doi=10.1.1.359.7464&rep=rep1&type=pdf (accessed on 14 July 2022).
55. Livak, K.J.; Schmittgen, T.D. Analysis of relative gene expression data using real-time quantitative PCR and the $2^{-\Delta\Delta CT}$ method. *Methods* **2001**, *25*, 402–408. [CrossRef] [PubMed]

Article

Assessing Heat Tolerance in Creeping Bentgrass Lines Based on Physiological Responses

Qianqian Fan and David Jespersen *

Department of Crop and Soil Sciences, University of Georgia, Griffin Campus, 1109 Experiment Street, Griffin, GA 30223, USA
* Correspondence: djesper@uga.edu

Abstract: Heat stress is a major concern for the growth of cool-season creeping bentgrass (*Agrostis stolonifera* L.). Nonetheless, there is a lack in a clear and systematic understanding of thermotolerance mechanisms for this species. This study aimed to assess heat tolerance in experimental lines and cultivars to determine important physiological and biochemical traits responsible for improved tolerance, including the use of OJIP fluorescence. Ten creeping bentgrass lines were exposed to either control (20/15 °C day/night) or high temperature (38/33 °C day/night) conditions for 35 d via growth chambers at Griffin, GA. Principal component analysis and clustering analysis were performed to rank stress performance and divide lines into different groups according to their tolerance similarities, respectively. At the end of the trial, S11 729-10 and BTC032 were in the most thermotolerant group, followed by a group containing BTC011, AU Victory and Penncross. Crenshaw belonged to the most heat-sensitive group while S11 675-02 and Pure Eclipse were in the second most heat-sensitive group. The exceptional thermotolerance in S11 729-10 and BTC032 was associated with their abilities to maintain cell membrane stability and protein metabolism, plus minimize oxidative damages. Additionally, among various light-harvesting steps, energy trapping, dissipation and electron transport from Q_A to PQ were more heat-sensitive than electron transport from Q_A to final PSI acceptors. Along with the strong correlations between multiple OJIP parameters and other traits, it reveals that OJIP fluorescence could be a valuable tool for dissection of photosynthetic processes and identification of the critical steps responsible for photosynthetic declines, enabling a more targeted heat-stress screening. Our results indicated that variability in the level of heat tolerance and associated mechanisms in creeping bentgrass germplasm could be utilized to develop new cultivars with improved thermotolerance.

Keywords: heat stress; creeping bentgrass; OJIP fluorescence

Citation: Fan, Q.; Jespersen, D. Assessing Heat Tolerance in Creeping Bentgrass Lines Based on Physiological Responses. *Plants* **2023**, *12*, 41. https://doi.org/10.3390/plants12010041

Academic Editor: Francisco Javier Cano Martin

Received: 30 November 2022
Revised: 16 December 2022
Accepted: 20 December 2022
Published: 22 December 2022

1. Introduction

As an important cool-season turfgrass native to Eurasia and North Africa, creeping bentgrass (*Agrostis stolonifera* L.) is widely used in high value turf areas across temperate regions of the globe, such as golf courses, due to its ability to tolerate low mowing heights and quick recovery from traffic and golf ball marks [1]. Although highly prized for its turf quality, creeping bentgrass has only low to moderate tolerance to high temperatures [2]. This makes heat stress a major concern in many transitional or sub-tropical areas such as the Southeastern China and the Southeastern United States where there are typically long hot summers combined with high temperatures, with damages being further exacerbated with more frequent and intense heat wave events as a function of climate change [3,4]. Many golf courses have been converted from creeping bentgrass to warm-season species, particularly bermudagrass (*Cynodon* sp.), due to a lack of heat tolerance in recent years [5].

High temperature can result in a number of physiological and biochemical injuries to plants, primarily including oxidative stress, photosynthesis inhibition and change in protein metabolism. Oxidative stress results from excess accumulation of reactive oxygen species (ROS) which are a group of free radicals, such as singlet oxygen ($^1O_2$)and hydroxyl radical (OH^-) [6]. They can attack a range of essential cellular components, like proteins, carbohydrates, and lipids in particular to cause leakage of cellular contents, eventually leading to lipid peroxidation and decreased integrity of cell membranes [7]. Photosynthesis inhibition occurs when elevated temperature brings about damages to photosynthetic machinery, including chlorophyll breakdown and reduced photosystem II (PSII) activity [7]. In addition, change in protein metabolism is another common stress symptom. Heat stress generally causes decreased protein abundance, which has been stated in various cool-season turfgrasses including creeping bentgrass [7–9]. It impacts many important cellular activities including photosynthesis and oxidative stress. The D1 protein plays a key role in PSII repair. Damaged D1 protein undergoes proteolysis to be removed and then a newly synthesized D1 is assembled into PSII, thus recovering PSII activity [10]. One study on wheat (*Triticum aestivum*) suggested that faster turnover of D1 protein contributed to better PSII photochemical efficiency [11]. In the case of oxidative stress, new proteins are synthesized to defend against ROS, and oxidized proteins go on to degradation to be removed, which otherwise would accumulate, causing damage and cell death [12]. These negative effects ultimately result in declines in carbon stores, reduced growth, loss of green color, thinning of the turf canopy and eventual plant death. To this end, development of heat-tolerant creeping bentgrass cultivars is desperately needed.

A few defense pathways have been clarified to be common strategies responding to heat stress in turfgrasses, like enhanced ROS detoxification, greater maintenance of photosynthesis ability, or altered protein metabolism [7]. Nonetheless, it should be noted that the specific changes may differ between species, or even cultivars and genotypes, which plays a pivotal role in the wide divergence in thermotolerance [9,13,14]. Creeping bentgrass shows considerable intraspecific diversity among different lines for its tolerance to heat. Previous research has identified differences in germplasm for important stress-related traits and there is potential to develop new cultivars with improved ability to withstand high temperatures [15]. However, despite progress made, comparisons of specific mechanisms, and physiological and biochemical parameters of heat tolerance among creeping bentgrass germplasm are still limited and need to be explored further.

As noted previously, PSII inhibition is a typical heat-stress induced symptom. The chlorophyll fluorescence parameter, photochemical efficiency (TRo/ABS or Fv/Fm), reflects the quantum efficiency of energy trapping by PSII and has been widely used as a reliable and sensitive tool for stress detection in different plant species including creeping bentgrass [16–19]. A relatively new development in fluorescence methodology is OJIP fluorescence [20]. It monitors rise of fluorescence intensity to a maximum at various states [21]. The O state is dark-adapted state when all the reaction centers, quinone A (Q_A), quinone B (Q_B) and plastoquinone (PQ) are oxidized. Upon exposure to saturating light, electrons will migrate into the PQ pool via Q_A and Q_B. When the majority of electrons have reduced Q_A, the J state is reached (at ~2 ms). When Q_B molecules are also reduced, the I state is reached (at 30 ms). Lastly, the P state is reached when maximum fluorescence intensity is obtained with a concurrent peak reduction in PQ pool, regardless of exposure time. By studying the OJIP curve, multiple photosynthetic component processes unavailable through traditional fluorescence methodologies can be quantified, such as energy trapping by PSII photochemistry, energy dissipation in PSII antennae, as well as electron transport between PSII and photosystem I (PSI), thereby providing a deeper insight into the function of photosynthetic components that might impair plant performance due to unfavorable environmental conditions [22]. To date, the use of OJIP fluorescence in abiotic stress studies have been documented in quite a few species, like tomato [23], cotton [24,25] and soybean [26]. However, despite its wide-spread application in stress physiology, related reports in creeping bentgrass, to our knowledge, are non-existent.

A number of creeping bentgrass materials were previously screened for summer performance. Within this germplasm collection, several experimental lines were identified with exceptional level of thermotolerance and outperformed commercial cultivars currently available on the market [27]. Nevertheless, the specific physiological or biochemical responses involved in their enhanced tolerance to heat have not yet been clearly revealed. A more complete understanding of the mechanisms conferring improved thermotolerance is essential for the efficient development of elite cultivars. Hence, this project aimed to evaluate heat tolerance in various creeping bentgrass lines to confirm the exceptional performance under heat stress in these promising experimental lines, as compared to commercial cultivars that form a range of thermotolerance. A number of physiological and biochemical measurements, including OJIP fluorescence, were taken to explore the responses enabling superior lines to outperform others. Integration of multiple stress-related traits will shed further light on heat stress survival strategies in creeping bentgrass, and determine useful traits associated with stress tolerance which can be utilized to develop new cultivars with improved thermotolerance.

2. Results

Since significant effects of temperature, line, date, and their interactions were detected for most parameters (Table 1) and the focus is mainly on exploring variations among lines under stress, differences among lines were analyzed for a given day under individual treatment.

Table 1. ANOVA results for heat stress trials of creeping bentgrass.

Parameter [†]	p Value						
	Date (D)	Temperature (T)	Line (L)	D × T	D × L	L × T	D × T × L
TQ	<0.0001	0.0004	<0.0001	<0.0001	<0.0001	<0.0001	<0.0001
Percent green cover	<0.0001	0.0025	<0.0001	<0.0001	0.0002	<0.0001	<0.0001
EL	<0.0001	0.0064	<0.0001	<0.0001	0.0400	<0.0001	0.1994
Total chlorophyll content	<0.0001	0.0004	0.0044	<0.0001	0.4223	<0.0001	0.7669
MDA content	<0.0001	0.0004	0.0016	<0.0001	0.5540	0.0006	0.7585
Total protein content	<0.0001	0.0316	0.0002	<0.0001	0.7554	0.0001	0.0102
ABS/CSm	<0.0001	0.0008	0.0002	<0.0001	0.8551	0.0001	0.6623
TRo/CSm	<0.0001	0.0004	<0.0001	<0.0001	0.7140	<0.0001	0.4652
ETo/CSm	<0.0001	0.0005	0.0002	<0.0001	0.8837	<0.0001	0.7059
TRo/ABS	<0.0001	0.0010	<0.0001	<0.0001	<0.0001	<0.0001	<0.0001
DIo/ABS	<0.0001	0.001	<0.0001	<0.0001	<0.0001	<0.0001	<0.0001
ETo/ABS	<0.0001	0.0025	<0.0001	<0.0001	0.1282	<0.0001	0.0608
REo/ABS	<0.0001	0.0572	0.4620	0.0834	1.0000	0.1614	1.0000
ESC	\	0.0097	0.3803	\	\	0.9539	\

[†] TQ, turf quality; EL, electrolyte leakage; MDA, malondialdehyde; ABS/CSm, the energy flux absorbed by the antenna of photosystem II (PSII) per cross section; TRo/CSm, the excitation energy flux trapped by open PSII reaction centers per cross section, leading to the reduction of quinone A (Q_A); ETo/CSm, the energy flux associated with electron transport from Q_A to PQ per cross section; TRo/ABS, quantum efficiency of energy trapping by PSII; DIo/ABS, quantum efficiency of energy dissipation in PSII antenna; ETo/ABS, quantum efficiency of electron transport from Q_A to plastoquinone; REo/ABS, quantum efficiency of electron transport from Q_A to final photosystem I acceptors; ESC, ethanol soluble carbohydrate.

For TQ, there were no significant differences for control plants over the duration of the trial, with all lines maintaining values greater than 8.5, representing a lack of stress (Figure 1; Table S1). Conversely, TQ scores declined throughout the trial for heat-stressed plants, to a greater extent in Crenshaw, Pure Eclipse as well as S11 675-02 than others, presenting variations in thermotolerance. These were in accordance with the significant effects of temperature and temperature × line interaction (Table 1). TQ scores were not significantly different among lines until heat progressed beyond 14 d, with differences being more pronounced over time. At the end of the trial, Crenshaw had the worst performance with an average score of 1.9 but not significantly differed from Pure Eclipse and S11 675-02. The two top performers, S11 729-10 and BTC032, had values of 5.8 and 5.6, respectively, without significant differences relative to BTC011, AU Victory, Penn A4 and Penncross. As with TQ, change in percent green cover followed a similar pattern (Figure 2; Table S2).

All lines experienced significant drops in percent cover by 35 d of stress and the declines were greater in Crenshaw, Pure Eclipse and S11 675-02 than other lines. At 35 d, these three lines were also the poorest performers, whereas BTC032 was the top performer but was not significantly differed from S11 729-10, BTC011, AU Victory and Penncross.

Regarding photosynthetic attributes, consistent values of total chlorophyll content, TRo/ABS and DIo/ABS were maintained in most lines with little variation among lines on most sampling dates under control conditions (Figures 3–5; Tables S3–S5). For ETo/ABS, no significant difference was found between 0 d and 35 d under control conditions despite variations over time, with differences among lines detected within sampling dates (Figure 6; Table S6). Intriguingly, Dio/ABS significantly rose while the other three parameters fell as a result of heat stress. At the end of week 5, the top statistical group contained S11 729-10, AU Victory, BTC032 and BTC011 for Tro/ABS as well as Dio/ABS measurements, S11 729-10, Penncross and AU Victory for chlorophyll content, and AU Victory, BTC032 and S11 729-10 for Eto/ABS. Crenshaw consistently presented the lowest values regarding Tro/ABS, total chlorophyll content and Eto/ABS but was not significantly different compared to Pure Eclipse for Tro/ABS, or to S11 675-02 and Pure Eclipse for chlorophyll levels or Eto/ABS at 35 d. Regarding Dio/ABS, the highest values were also found in Crenshaw at 35 d although it showed no significant difference compared to Pure Eclipse. As for Reo/ABS, only the main effect of date was significant.

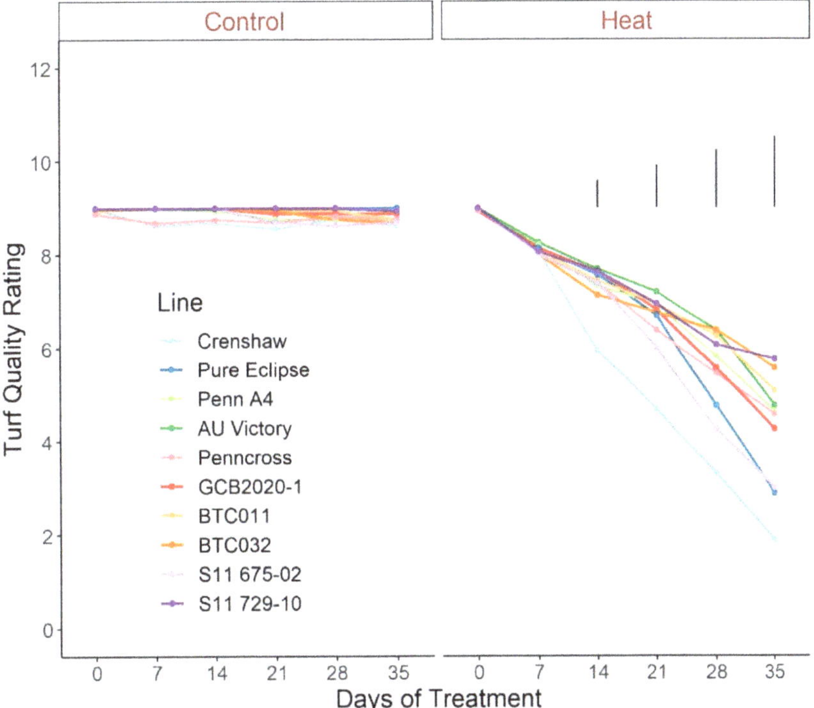

Figure 1. Change in visual turf quality rating for creeping bentgrass lines over time under control (20/15 °C day/night) and heat stress (38/33 °C day/night) conditions. Bars represent LSD values at $p = 0.05$ on days when significant differences among lines were found.

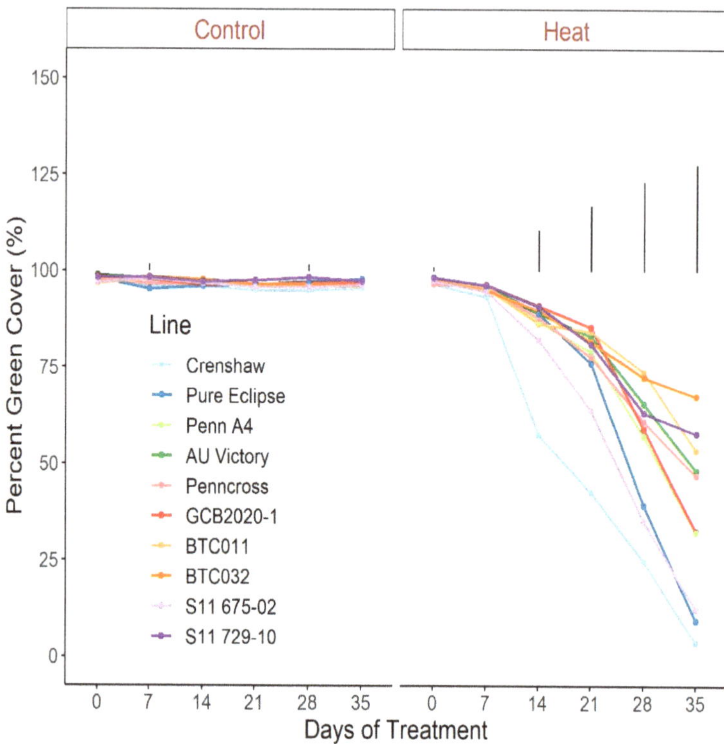

Percent Green Cover (%)

Days of Treatment

Figure 2. Change in percent green cover for creeping bentgrass lines over time under control (20/15 °C day/night) and heat stress (38/33 °C day/night) conditions. Bars represent LSD values at $p = 0.05$ on days when significant differences among lines were found.

As for the phenomenological energy fluxes involved in light-harvesting processes, significant differences were not found between 0 d and 35 d under control conditions for all parameters (ABS/CSm, Tro/CSm and Eto/CSm) although variations existed over time potentially as a consequence of chamber acclimation effects or natural genotypic variations, with differences detected within certain sampling dates (Figures 7–9; Tables S7–S9). In contrast with control, heat stress caused significant reductions in every line for all three parameters, with pronounced separations being observed from 14 d onwards. At 35 d, Crenshaw, Pure Eclipse and S11 675-02 were the three poorest performers whose values were significantly lower than S11 729-10, Penncross and BTC032 in terms of ABS/CSm, and all the 10 emaining lines except for GCB2020-1 regarding Tro/CSm and Eto/CSm.

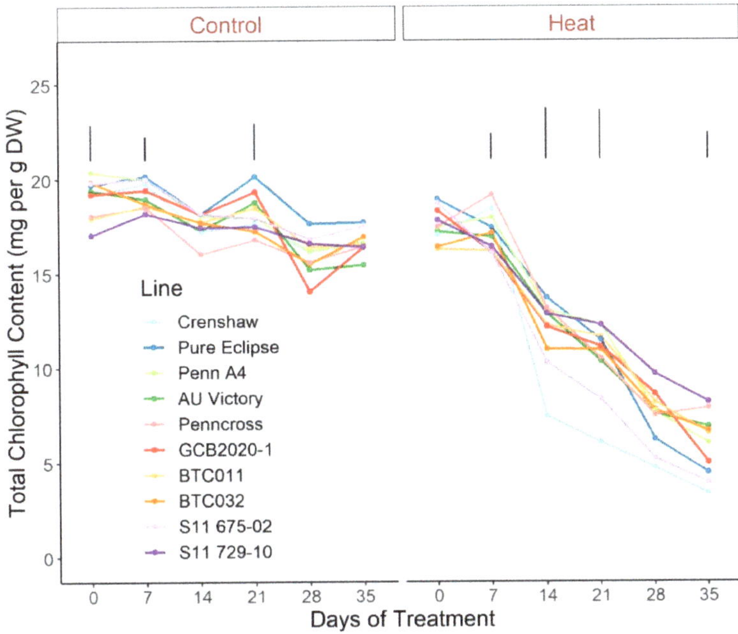

Figure 3. Change in total chlorophyll content for creeping bentgrass lines over time under control (20/15 °C day/night) and heat stress (38/33 °C day/night) conditions. Bars represent LSD values at $p = 0.05$ on days when significant differences among lines were found. DW, dry weight.

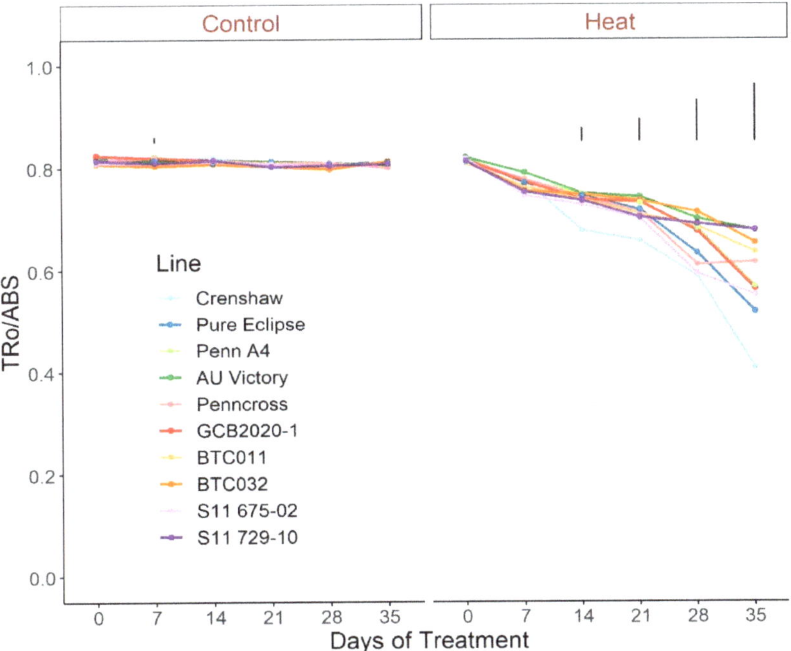

Figure 4. Change in Tro/ABS for creeping bentgrass lines over time under control (20/15 °C day/night) and heat stress (38/33 °C day/night) conditions. Bars represent LSD values at $p = 0.05$ on days when significant differences among lines were found.

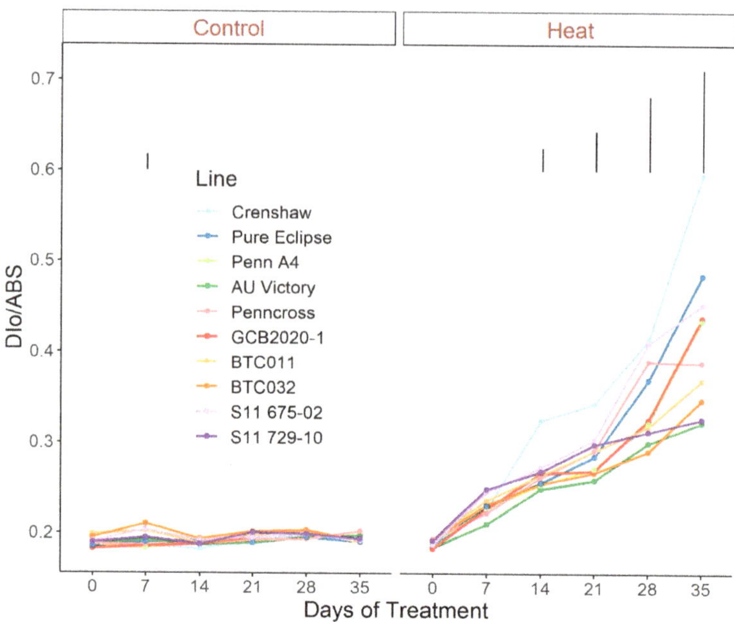

Figure 5. Change in Dio/ABS for creeping bentgrass lines over time under control (20/15 °C day/night) and heat stress (38/33 °C day/night) conditions. Bars represent LSD values at p = 0.05 on days when significant differences among lines were found.

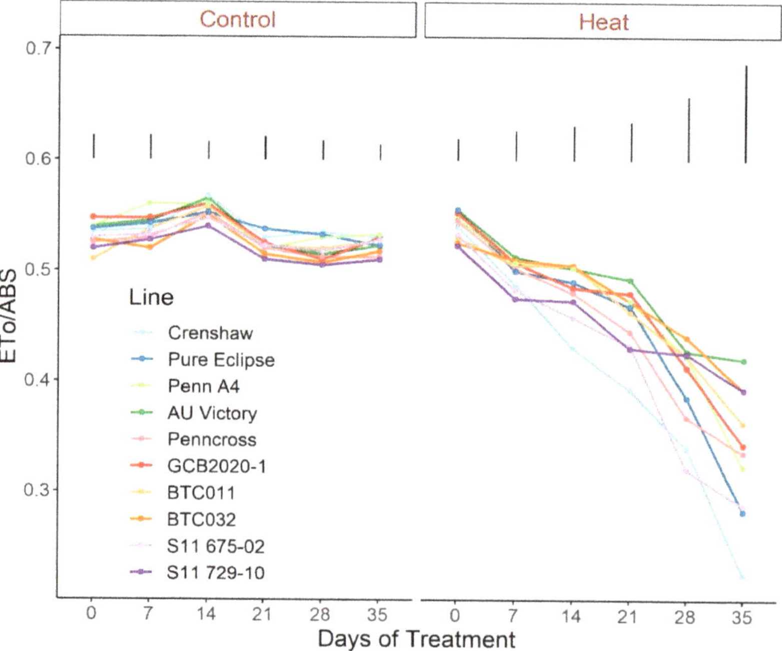

Figure 6. Change in Eto/ABS for creeping bentgrass lines over time under control (20/15 °C day/night) and heat stress (38/33 °C day/night) conditions. Bars represent LSD values at p = 0.05 on days when significant differences among lines were found.

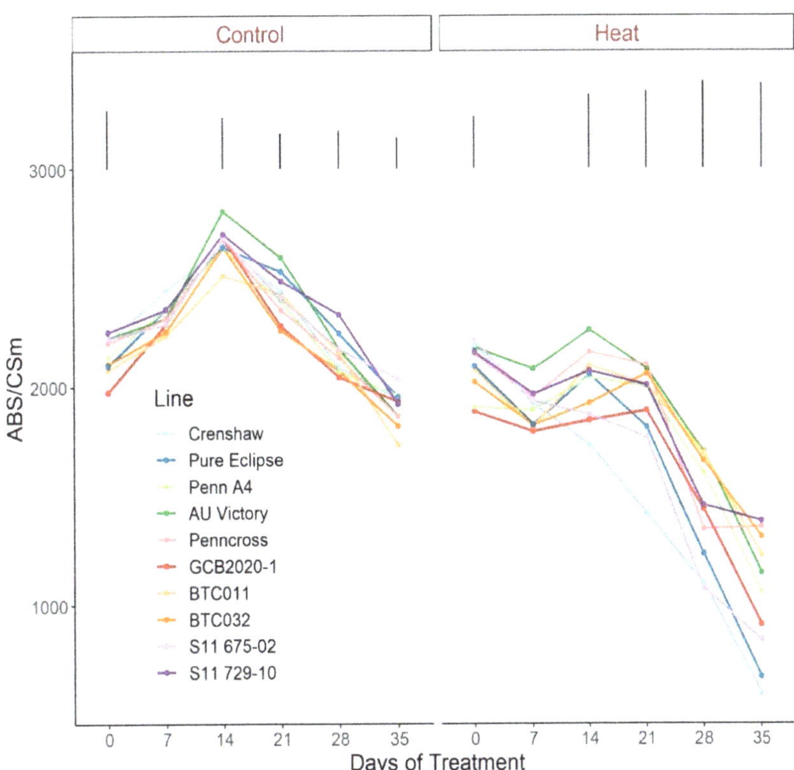

Figure 7. Change of ABS/CSm for creeping bentgrass lines over time under control (20/15 °C day/night) and heat stress (38/33 °C day/night) conditions. Bars represent LSD values at $p = 0.05$ on days when significant differences mong lines were found.

For EL under control conditions, no significant difference was seen between 0 d and 35 d despite some variance over time (Figure 10; Table S10). Contrastingly, values went up dramatically in response to heat stress in all lines with the exception of S11 729-10. Similar to most parameters mentioned above, variation among lines became apparent after two-weeks of stress with divergence increasing over time until 35 d. At 35 d of treatment, EL of S11 729-10 was the lowest but was not statistically different from BTC032, AU Victory, or BTC011, whereas Crenshaw had the highest value and was in the same statistical group as Pure Eclipse and S11 675-02.

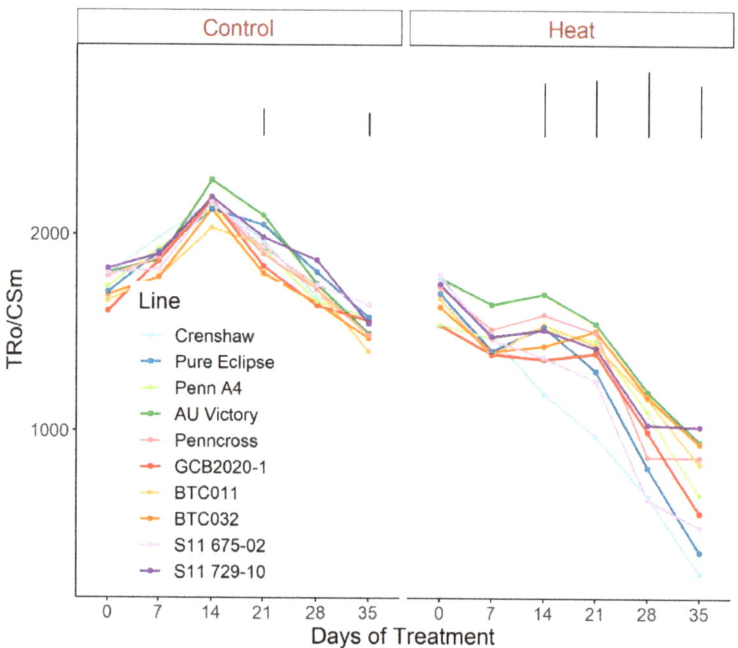

Figure 8. Change in Tro/CSm for creeping bentgrass lines over time under control (20/15 °C day/night) and heat stress (38/33 °C day/night) conditions. Bars represent LSD values at $p = 0.05$ on days when significant differences among lines were found.

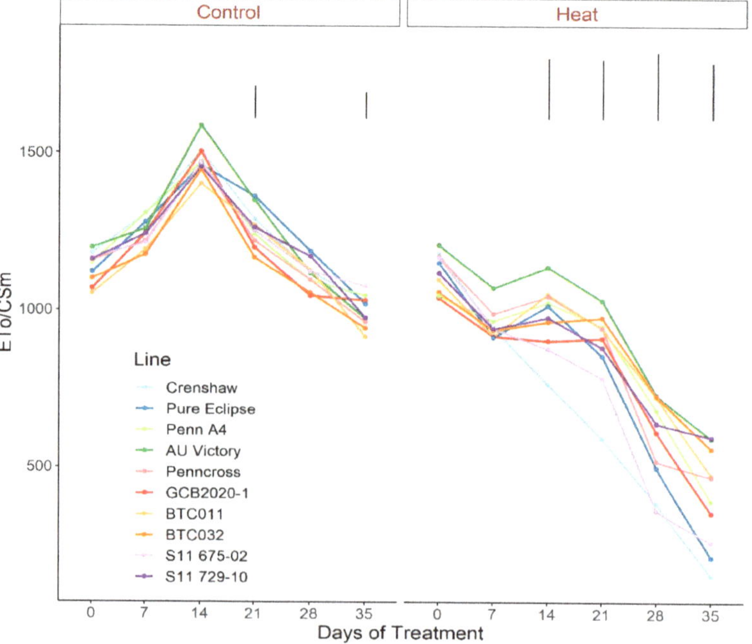

Figure 9. Change in Eto/CSm for creeping bentgrass lines over time under control (20/15 °C day/night) and heat stress (38/33 °C day/night) conditions. Bars represent LSD values at $p = 0.05$ on days when significant differences among lines were found.

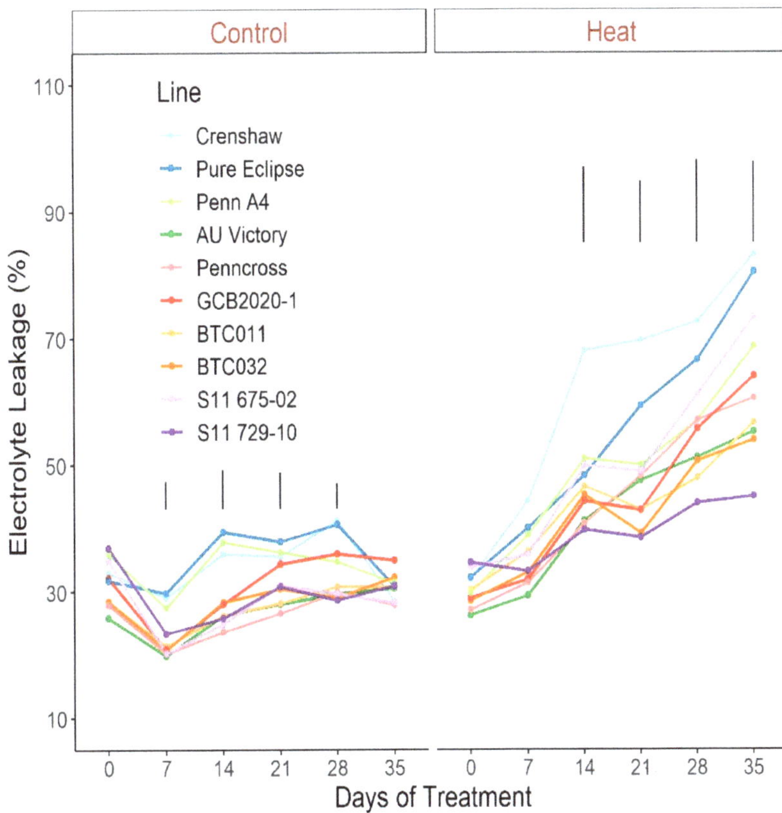

Figure 10. Change in electrolyte leakage for creeping bentgrass lines over time under control (20/15 °C day/night) and heat stress (38/33 °C day/night) conditions. Bars represent LSD values at $p = 0.05$ on days when significant differences among lines were found.

For MDA content, plants under control conditions maintained mostly consistent values over time whereas significant rises were detected in all lines under heat stress, with the exception of S11 729-10 and AU Victory (Figure 11; Table S11). Pronounced variation among lines was found in response to stress at 21 d and continued to diverge through the end of the experiment. At 35 d, S11 729-10, as the top performer, presented a significantly lower MDA content than Crenshaw, Pure Eclipse, GCB2020-1 and S11 675-02.

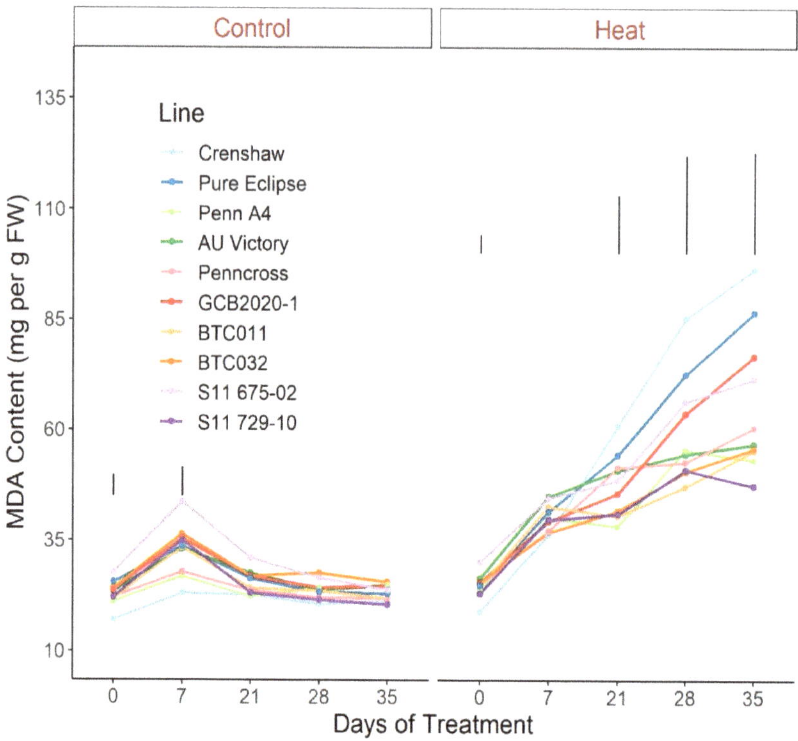

Figure 11. Change in MDA content for creeping bentgrass lines over time under control (20/15 °C day/night) and heat stress (38/33 °C day/night) conditions. Bars represent LSD values at $p = 0.05$ on days when significant differences among lines were found. FW, fresh weight.

For protein abundance, plants followed similar patterns of change over time when exposed to control conditions, with greater values generally found at 0 d and 7 d potentially due to variations in fresh weight (Figure 12; Table S12). On the contrary, heat stress caused an obvious separation among lines with prominent differences detected over the last two weeks of stress. Specifically, at 35 d, S11 729-10 was the top performer, showing greater contents than the other lines with the exception of BTC032 and BTC011, whereas Crenshaw's protein content was significantly lower than all others except Penn A4. Moreover, protein abundances presented no significant changes over the course of the five-week stress period for S11 729-10, BTC032 and BTC011 while the remaining lines all presented dramatic decreases.

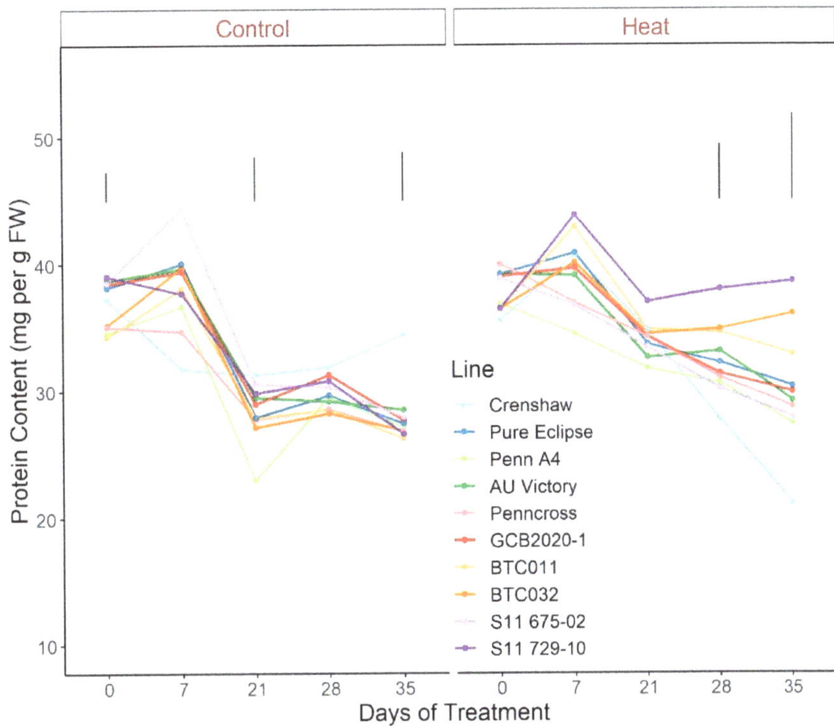

Figure 12. Change in total protein content for creeping bentgrass lines over time under control (20/15 °C day/night) and heat stress (38/33 °C day/night) conditions. Bars represent LSD values at $p = 0.05$ on days when significant differences among lines were found.

For ESC, lines exposed to elevated temperature presented increases compared to those under control conditions, with an average of 81.0 and 37. 0 mg g^{-1} dry weight under heat stress and control conditions, respectively (data not shown). No significance differences existed for the line or line × temperature interaction effects. In this study ESC was not a useful parameter for separating heat tolerance among lines.

Correlation analysis was conducted for all parameters except for ESC since there was limited data with no significant variation among lines. It revealed that all the parameters excluding protein content and REo/ABS were significantly and strongly correlated with each other, with the absolute values ranging from 0.70 to 0.99 (Figure 13). Conversely, the absolute values were not greater than 0.23 for the correlation coefficients between protein and other parameters, and not over 0.21 between REo/ABS and other parameters.

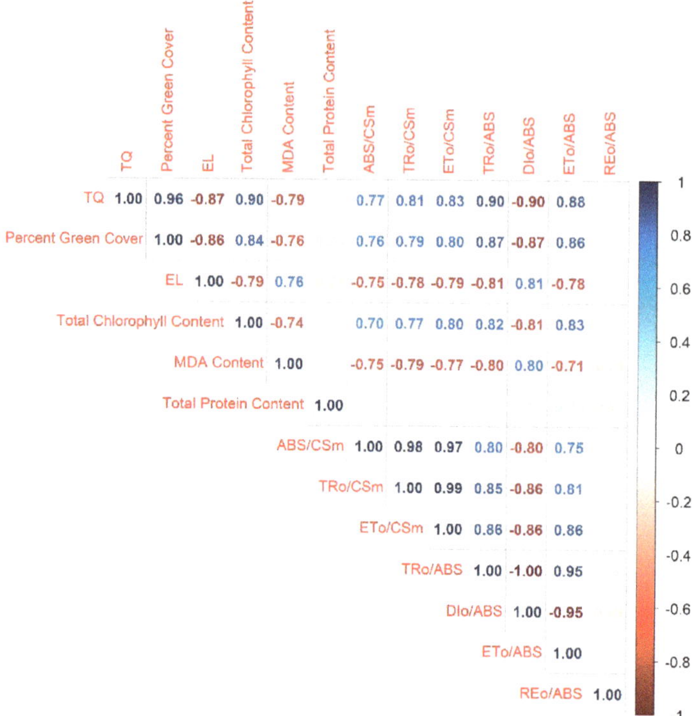

Figure 13. Correlation plot for different parameters of creeping bentgrass under control (20/15 °C day/night) and heat stress (38/33 °C day/night) conditions. Numbers indicate correlation coefficients. Color intensity is proportional to the correlation coefficients with blue indicating positive correlations and red representing negative correlations. Correlation coefficient values were left blank when not significant at $p = 0.05$.

To take all the measurements into account to rank stress performance of lines, principal component analysis was conducted. Analysis determined the contribution of each component to the overall variation among lines due to differences after five weeks of heat stress, and also revealed to what extent different parameters contributed to stress tolerance (Figure 14). The first principal component (PC1) explained 89.2% of variance while the second principal component accounted for only 4.6% of variance. Except for total protein content and REo/ABS accounting for 5.7% and 6.6% of PC1, respectively, the contribution to PC1 made by the remaining traits ranged from 7.3 to 8.4%. Furthermore, clustering analysis was performed to divide lines into different groups according to their similarities (within-group variation is minimized). Together with results of principal component analysis, it revealed that S11 729-10 and BTC032 were in the most thermotolerant group. The second most thermotolerant group contained BTC011, AU Victory and Penncross, followed by the group containing Penn A4 as well as GCB2020-1. Crenshaw belonged to the most heat-sensitive group while S11 675-02 and Pure Eclipse were in the second most heat-sensitive group.

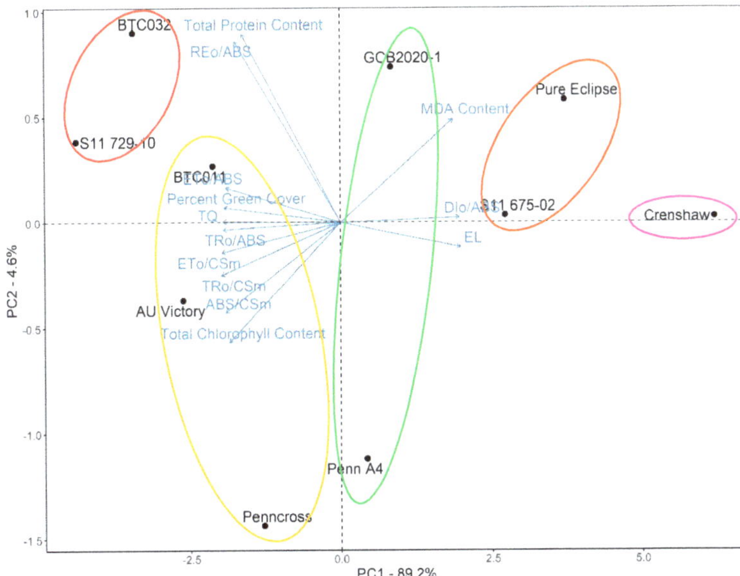

Figure 14. Principal component analysis for different parameters of creeping bentgrass at 35 d under heat stress (38/33 °C day/night) condition. Principal component 1 is represented on the *X* axis while principal component 2 is represented on the *Y* axis. Vectors indicate the direction and contribution of each parameter to the overall distribution of various lines. Circles of different colors indicate different clusters.

3. Materials and Methods

3.1. Growth and Treatment Conditions

A total of ten creeping bentgrass lines were used in this study, including five commercial cultivars ('Crenshaw', 'Pure Eclipse', 'Penn A4', 'Penncross' and 'AU Victory'), and five experimental lines which have shown to perform well during summer in preliminary studies in Georgia, namely, 'GCB2020-1', 'BTC011' and 'BTC032' (Paul Raymer, unpublished work, 2020), plus 'S11 675-02 'and 'S11 729-10' [27]. For each line, 6-cm-diameter plugs were established in plastic pots (10.5 cm long, 10.5 cm wide and 12.5 cm deep) filled with a mixture of 50% sand and 50% calcined clay (Turface; Profile Products LLC, Buffalo Grove, IL, USA) for ten weeks in greenhouse conditions [~23/~15 °C (light/dark period temperatures) and 70% relative humidity] before transferred to controlled environmental growth chambers (CG-72; Conviron, Winnipeg, MB, Canada). Plants were allowed one-week acclimation inside the growth chambers under conditions of 20/15 °C (day/night), 70% humidity and 14 h photoperiod with 600 µmol m^{-2} s^{-1} photosynthetically active radiation at the canopy level before the onset of different temperature treatments. Plants were maintained well-watered and fertilized weekly with a 24-8-16 (N-P-K) fertilizer (Scotts Miracle-Gro; Marysville, OH, USA) at the rate of 9.8 g N m^{-2} during establishment in the greenhouse as well as during the treatment period inside growth chambers. Applications of insecticide and fungicide were made as needed for disease control. Plants of each line were exposed to either heat stress (38/33 °C day/night) or control (20/15 °C day/night) conditions for 35 d after treatments began.

3.2. Measurements

3.2.1. Physiological Measurements

Measurements consisted of assessments of whole-plant responses along with physiological and biochemical factors. Overall turf performance was estimated using a visual turf quality (TQ) rating on a scale of 1–9 and percent green cover via digital image analysis. Turf quality was determined according to color, density and uniformity with 1 representing totally dead grass, 9 standing for completely healthy grass with lush green color, and 6 being the minimum acceptable quality [28]. Digital image analysis was conducted through images taken with a digital camera (Canon G9X; Canon, Tokyo, Japan) using a lightbox to ensure a uniform lighting, which were processed using ImageJ v.1.46 to obtain values of percent green cover [29].

Total chlorophyll content and OJIP fluorescence were used to reflect the health status of photosynthetic machinery. Plants were dark adapted overnight (10 h) in growth chambers prior to performing OJIP measurements via a chlorophyll fluorometer set with a 3500 μmol actinic light intensity (OSP 5+; Opti-sciences, Hudson, NH, USA). This study focused on several energy flux and quantum efficiency parameters to better understand the light-harvesting processes, which included the energy flux absorbed by the antenna of PSII per cross section (ABS/CSm), the excitation energy flux trapped by open PSII reaction centers per cross section leading to the reduction of Q_A (TRo/CSm), the energy flux associated with electron transport from Q_A to PQ per cross section (ETo/CSm), quantum efficiency of energy trapping by PSII (TRo/ABS), quantum efficiency of energy dissipation in PSII antenna (DIo/ABS),quantum efficiency of electron transport Q_A to PQ (ETo/ABS), and quantum efficiency of electron transport Q_A to final PSI acceptors (REo/ABS) [22]. Four measurements were taken at the midpoint on fully expanded leaves for each replicate. To obtain values of total chlorophyll content, 0.1 g fresh leaves were incubated in 5 mL dimethyl sulfoxide for 7 days to allow for chlorophyll extraction. Then, the absorbance of solutions at 665 and 649 nm were read using spectrophotometer (Evolution 300 UV-visible spectrophotometer; Thermo Scientific, Madison, WI, USA) and converted to chlorophyll content according to previously derived equations on a dry weight basis [30].

Electrolyte leakage (EL) serves as an indicator of cell membrane stability. Around 0.1 g fresh leaves were placed in a tube containing 35 mL deionized water. After agitating tubes on a shaker for 16 h, initial conductivity was recorded through a conductivity meter (Radiometer, Copenhagen, Denmark). Next, the samples were autoclaved at 120 °C for 20 min, followed by incubation for another 16 h on a shaker, after which the final conductivity was read. EL then was calculated as the percentage of initial conductivity over total conductivity [31].

3.2.2. Biochemical Measurements

Change in protein abundance was measured to represent change in protein metabolism, while malondialdehyde (MDA) content, a final product of lipid peroxidation, was quantified to indicate the extent of oxidative damage. Both analyses were performed through a microplate reader (Epoch 2 microplate reader, BioTek, Winooski, VT, USA). Approximately 50 mg fresh leaves were added into 1.1 mL 50 mM sodium phosphate buffer (pH 7.0 with 1 mM ethylenediaminetetraacetic acid). Supernatants were collected after homogenization and centrifugation at $15,000 \times g$, 4 °C for 20 min. Then, total protein content was quantified at 595 nm with Bradford dye reagent and a bovine serum albumin standard [32]. For the quantification of MDA content, 0.25 mL supernatant was mixed and reacted with 0.5 mL reaction solution (20% w/v trichloroacetic acid and 0.5% w/v thiobarbituric acid) at 95 °C, followed by absorbance measurement at the wavelengths of 532 and 600 nm. MDA content was acquired by subtracting background absorbance at 600 nm from absorbance at 532 nm and then divided by an extinction coefficient of 155 mM^{-1} cm^{-1} [33].

The content of ethanol soluble carbohydrates (ESC) were determined based on the anthrone method [34]. Approximately 30 mg dry leaf tissues were homogenized in 5 mL of 95% (v/v) ethanol and centrifuged at 3500 rpm, 4 °C for 10 min. The pellet was washed with 5 mL of 70% (v/v) ethanol twice. Then, all soluble portions were pooled, vortexed and centrifuged again to remove debris. Next, 100 μL of the ethanolic extract was added into 3 mL of anthrone-sulfuric acid reagent (200 mg anthrone dissolved in 100 mL 72% (v/v) H$_2$SO$_4$). After incubation in boiling water for 10 min, the absorbance of the resultant reaction mixture was recorded at 630 nm, and compared against glucose standards in the range of 20–100 µg mL^{-1}. All measurements were taken weekly except for contents of protein, MDA and ESC with the former two being measured every other week while the latter was analyzed once at the end of the trial.

3.3. Statistical Analysis

A completely randomized split-plot design was applied with temperature as the whole plot and line as the subplot, with each combination of temperature and line having four replications. During the trial, each temperature was repeated in four growth chambers. Inside each chamber, there were two pots for every line and the average of these two pots was used to represent an individual replicate.

Data were analyzed via ANOVA using a mixed model in JMP Pro 16.0.0 (SAS Institute Inc., Cary, NC, USA, 2021). Date, temperature, line, and their interactions were treated as fixed effects whereas experimental run and the whole plot were random effects. Before ANOVA, normal distribution of residuals and the homogeneity of variance were checked according to normal quantile-quantile plots and residuals versus fitted plots, respectively, to make sure data met ANOVA assumptions. Means were separated by Fisher's protected least significant difference (LSD) at the 0.05 probability level. Correlation analysis and K-means clustering analysis were performed using corrplot and cluster packages, respectively, while principal component analysis was conducted through plotly and ggfortify packages in RStudio (R 3.6.0, Boston, MA, USA, 2019).

4. Discussion

Although elevated temperature caused damages to all plants over the course of the 35 d stress period, a wide range of thermotolerance was observed among lines as evidenced by the differences in their visual characteristics. Specifically, S11 729-10, BTC032, BTC011 and AU Victory were the four top performers, outperforming others by maintaining greater overall quality as measured by TQ and green cover. Conversely, heat-sensitive lines, such as Crenshaw, S11 675-02 and Pure Eclipse, consistently performed poorly in terms of these two measured parameters, while the remaining lines were intermediate in their performances. Moreover, superior visual characteristics in the more heat-tolerant lines were attributed to their improved physiological as well as biochemical responses. These included greater abilities to withstand injuries to photosynthetic machinery as reflected in chlorophyll content and OJIP fluorescence traits (TRo/ABS, DIo/ABS, ETo/ABS, ABS/CSm, TRo/CSm and ETo/CSm), maintain cell membrane stability as evaluated by EL, minimize oxidative damage as measured by MDA content, and reduce change in protein metabolism as indicated by total protein content.

Maintaining chlorophyll levels and chlorophyll fluorescence traits is critically important for cool-season grass survival during heat stress. The former contributes to the absorption of light energy for use in photosynthesis while the latter estimates the health of PSII reaction centers which are the most thermally labile component of the electron transport chain, with constraints in either of them impairing photosynthetic capacity [35,36]. Within this study, despite heat-induced declines in chlorophyll content, TRo/ABS, ETo/ABS, ABS/CSm, TRo/CSm and ETo/CSm and increases in DIo/ABS over time, more heat-tolerant lines S11 719-10, AU Victory, BTC032 and BTC011 generally better maintained these characteristics, revealing less damage to their photosynthetic systems, which is in accordance with previous research [16,37,38]. Chlorophyll loss is one major characteristic of leaf senescence induced by heat stress damage [39]. The lesser decline in chlorophyll content in heat-tolerant plants could be a consequence of slower chlorophyll degradation resulting from relatively lower gene expression levels of chlorophyll-degrading enzymes, like chlorophyllase, pheophytinase and chlorophyll-degrading peroxidases [40,41].

The light energy absorbed by photosynthetic pigments is either used in PSII photochemistry or dissipated through heat and fluorescence [42]. Energy absorption decreased as measured by ABS/CSm, potentially as a result of heat-induced chlorophyll reduction or damage to photosynthetic complexes. With reduced energy absorption, the resulting energy trapped by PSII reaction centers, as measured by TRo/CSm, would also be expected to go down, reducing the efficiency of trapping and ultimately causing declines in the light-harvesting abilities of the leaf [23]. As noted previously, electrons migrate from PSII to PSI via Q_A and PQ during light harvesting. When the electron flow to Q_A declined (TRo/CSm), there would generally be a concomitant decline in energy flux from Q_A to PQ too, as evaluated by ETo/CSm [23,43]. Likewise, when OJIP traits were expressed as energy fluxes per absorbed photo flux, declines in TRo/ABS and ETo/ABS were detected as well. A decrease in the quantum efficiency of light photochemical reactions in PSII (TRo/ABS) resulted in a rise of energy dissipation as heat and fluorescence, as evidenced by increases in DIo/ABS, highlighting that stress-induced damage required the leaves to dissipate excess excitation energy instead of utilizing it for photosynthetic processes [44]. Intriguingly, in contrast with the significances observed for TRo/ABS, DIo/ABS and ETo/ABS, neither temperature effects nor line effects were significant for the quantum efficiency of electron transport from Q_A to final PSI acceptors, as measured by REo/ABS, despite heat-induced numerical declines over time among all lines (data not shown). This suggested that during light-harvesting processes, energy trapping, energy dissipation and electron transport from Q_A to PQ were more sensitive to temperature rise than electron transport from Q_A to final PSI acceptors [23,45]. Identifying these critical steps responsible for photosynthetic inhibitions would allow for a more targeted screening and improvement of plants with enhanced heat tolerance.

Previous studies pointed out that heat stress impaired a range of OJIP fluorescence traits in croftonweed (*Ageratina adenophora*) and peony (*P. lactiflora*) [43,44]. However, more heat-tolerant populations or cultivars typically better maintained absorbed energy flux, energy flux trapped by PSII, electron transport from Q_A to PQ, as well as quantum efficiencies of energy trapping, dissipation, and electron transport from Q_A to PQ. Similarly, another study on tall fescue (*Festuca arundinacea* Schreb.), stated that at the end of heat treatments, the heat-tolerant "TF71" presented significantly higher values in absorbed energy flux, energy flux trapped by PSII, and quantum efficiencies of energy trapping and electron transport from Q_A to PQ than the heat-sensitive "TF133", maintaining better photosynthetic capacity thus contributing to its enhanced adaptation to high temperature [38]. These are all in agreement with our findings. Additionally, the close associations among ABS/CSm, TRo/CSm, ETo/CSm, TRo/ABS, ETo/ABS and DIo/ABS were also supported by their significantly strong correlation coefficients among each other. Along with concurrent declines in ABS/CSm, TRo/CSm, ETo/CSm from 21 d onwards, and concurrent declines or increases in TRo/ABS, ETo/ABS and DIo/ABS from 7 d onwards, it could be inferred that injuries to photosynthetic components were wide spread in the chloroplast and these light-harvesting steps might be concomitantly damaged by heat stress. Furthermore, our study also detected strong correlations of physiological factors plus MDA content with all fluorescence traits except for REo/ABS. As a reliable parameter commonly used in heat stress screening, it was not surprising that TRo/ABS showed the strongest associations with other factors [14,17,46]. Nevertheless, it's noteworthy that DIo/ABS had the same correlation coefficients as TRo/ABS did, indicating its potential as another rapid and reliable measurement for thermotolerance evaluation in turfgrasses. The second strongest correlations with physiological traits plus MDA content were observed for ETo/ABS, which were close to those for TRo/ABS and DIo/ABS. Stronger relationships between whole-plant seedling vigor and ETo/ABS than other OJIP traits were reported in cotton previously and the authors proposed that ETo/ABS could be used as a surrogate for more time-consuming seedling vigor measurement [24,25]. To our knowledge, this is the first time that the application of OJIP fluorescence in abiotic stress response has been reported in creeping bentgrass. It could serve as a valuable tool for actual dissection of photosynthetic processes and help understand which steps of light-harvesting electron transport were more sensitive to heat stress [43,44]. Such information would provide deeper insights into heat-induced photosynthetic declines, allowing for a more targeted screening and improvement of heat tolerance in plants.

Accumulated ROS triggered by heat stress can attack lipids, resulting in decreased membrane integrity and lipid peroxidation, so typically the EL value increases with a concomitant increase in MDA content during plants' exposure to stress [16]. However, these responses could be specific at both species and cultivar levels with lower EL and MDA content representing improved thermotolerance [13,47]. This corroborates the results found in our study where, compared to heat-sensitive lines, lower values of EL and MDA content were detected in heat-tolerant lines at 35 d, particularly S11 729-10. S11 729-10 had relatively little increase in these two parameters during the entire period of stress, suggesting its superior ability to maintain cell membrane integrity and to minimize oxidative damage. Furthermore, although not measured in the current study, greater activities of antioxidant enzymes, for instance, superoxide dismutase, catalase, ascorbate peroxidase, and peroxidase, can result in lower MDA abundances in heat-tolerant lines by scavenging excess ROS to protect cells or macromolecules from severe oxidative damage [16,48]. This may also have contributed to better maintenance of chlorophyll content and fluorescence traits and lower EL, eventually leading to better overall quality. Strong correlations between MDA and many other physiological traits support that reduced oxidative damage may result in the maintenance of photosynthetic processes such as light harvesting, reduced cellular damage, and maintenance of overall turf quality.

Protein metabolism, a process controlled by the balance between protein synthesis and protein degradation, impacts many cellular activities such as the aforementioned photosynthesis and oxidative stress. Decrease in protein abundance is a typic stress-induced characteristic, which has been confirmed in a wide range of plant species besides creeping bentgrass, including but not limited to strawberry (*Fragaria* × ananassa cv. Camarosa) [49], tomato (*Lycopersicon esculentum* Mill.) and maize (*Zea mays* L.) [50]. In general, protein catabolism is accelerated to a greater degree compared to the biosynthesis process under unfavorable environmental conditions, taking major responsibility for reduced protein content in response to elevated temperature [51,52]. However, it was previously documented that heat-tolerant plants typically had lower declines in protein abundance [53,54]. This agrees with our findings among which, greater protein contents were seen in more heat-tolerant lines like S11 729-10 and BTC032 at the end of the experiment. The higher protein abundance in heat-tolerant plants could be a consequence of faster protein synthesis, slower protein degradation, or both. Proteins synthesized abundantly under heat are primarily heat shock proteins, functioning as chaperones by preventing other proteins from aggregation and refolding stress-damaged proteins [55], contributing to the maintenance of protein metabolism. Hence, a greater and earlier induction of heat shock proteins in the heat-tolerant plants could be one reason for their improved thermotolerance [53,56]. As an opposing process to protein synthesis, less protein catabolism or slower degradation may be due to reduced proteolysis activity carried out by the coordinated action of the ubiquitin-proteasome system and various proteases [57,58]. One previous study stated that less enhanced gene expression of cysteine protease and a slower rate of overall protein degradation were detected in heat-adapted *Agrostis scabra*, contributing to its higher protein thermostability, and thereby greater protein abundance compared creeping bentgrass [54]. Thus, in order to further understanding of protein regulation in response to heat stress in turfgrasses, more research is needed to understand proteolysis activity of both the ubiquitin-proteasome system and proteases.

As with the defense mechanisms discussed above, accumulation of sugars can be another important contributor to heat tolerance. It not only contributes to increased osmotic adjustment but also improves the integrity of cellular membranes, helping relieve plants from heat-induced damages, which has been documented in creeping bentgrass [59] and other plant species [60,61]. Moreover, ESC turned out to accumulate more in creeping bentgrass cultivars with better summer performance [14]. The authors proposed that elevated temperature resulted in rapid loss of water, causing dehydration, as manifested by the declines in leaf relative water content and osmotic potential, while accumulation of sugars could produce positive effects on water homeostasis, thus explaining the higher sugar contents along with better leaf water status and osmotic adjustment abilities found in the more heat-tolerant cultivars. These findings were not consistent with our research where no significance was shown among various lines in terms of ESC. The discrepancy between studies is likely due to differences in creeping bentgrass lines, environmental conditions such as stress duration and stress intensity, and measurement techniques.

5. Conclusions

In summary, a broad range of thermotolerance exists among creeping bentgrass lines At the end of the trial, the overall ranking for heat tolerance of lines was that S11 729-10 and BTC032 were in the most thermotolerant group while BTC011, AU Victory and Penncross were in the second most thermotolerant group; Crenshaw belonged to the most heat-sensitive group while S11 675-02 and Pure Eclipse were in the second most heat-sensitive group; The group containing Penn A4 and GCB2020-1 was intermediate in their tolerance ranking. The exceptional thermotolerance in S11 729-10 and BTC032 was mainly associated with their greater abilities to maintain integrity of cellular membranes, as well as protein metabolism, and the ability to minimize oxidative damages. In addition, among various light-harvesting steps, energy trapping, dissipation and electron transport from Q_A to PQ were more heat-sensitive than electron transport from Q_A to final PSI acceptors. Moreover, strong significant correlations were detected among multiple OJIP parameters and other stress-related factors. These suggest that OJIP fluorescence could be a valuable tool for dissection of photosynthetic processes and identification of the critical steps responsible for photosynthetic declines, enabling a more targeted screening and improvement of plants for enhanced heat tolerance. This is also the first time that the potential application of rapid OJIP assessments is addressed in creeping bentgrass. Additional research is needed to further reveal how heat-tolerant and heat-sensitive creeping bentgrass lines respond to high temperature from physiological, biochemical, and molecular mechanisms, particularly regarding proteolysis.

Supplementary Materials: The following supporting information can be downloaded at: https://www.mdpi.com/article/10.3390/plants12010041/s1, Table S1: Change in visual turf quality rating for creeping bentgrass lines over time under control (20/15 °C day/night) and heat stress (38/33 °C day/night) conditions; Table S2: Change in green cover (%) for creeping bentgrass lines over time under control (20/15 °C day/night) and heat stress (38/33 °C day/night) conditions; Table S3: Change in total chlorophyll content (mg per g DW) for creeping bentgrass lines over time under control (20/15 °C day/night) and heat stress (38/33 °C day/night) conditions; Table S4: Change in TRo/ABS for creeping bentgrass lines over time under control (20/15 °C day/night) and heat stress (38/33 °C day/night) conditions; Table S5: Change in ETo/ABS for creeping bentgrass lines over time under control (20/15 °C day/night) and heat stress (38/33 °C day/night) conditions; Table S6: Change in DIo/ABS for creeping bentgrass lines over time under control (20/15 °C day/night) and heat stress (38/33 °C day/night) conditions; Table S7: Change in ABS/CSm for creeping bentgrass lines over time under control (20/15 °C day/night) and heat stress (38/33 °C day/night) conditions; Table S8: Change in TRo/CSm for creeping bentgrass lines over time under control (20/15 °C day/night) and heat stress (38/33 °C day/night) conditions; Table S9: Change in ETo/CSm for creeping bentgrass lines over time under control (20/15 °C day/night) and heat stress (38/33 °C day/night) conditions; Table S10: Change in electrolyte leakage (%) for creeping bentgrass lines over time under control (20/15 °C day/night) and heat stress (38/33 °C day/night) conditions; Table S11: Change in MDA content (mg per g FW) for creeping bentgrass lines over time under control (20/15 °C day/night) and heat stress (38/33 °C day/night) conditions; Table S12: Change in protein content (mg per g FW) for creeping bentgrass lines over time under control (20/15 °C day/night) and heat stress (38/33 °C day/night) conditions.

Author Contributions: Formal analysis, investigation, project administration, visualization, writing—original draft preparation and editing, Q.F.; conceptualization, resources, supervision, writing—review and editing, D.J. All authors have read and agreed to the published version of the manuscript.

Funding: Support was provided by the University of Georgia Research Foundation.

Data Availability Statement: Data is contained within the article and supplementary material.

Acknowledgments: Thanks go to Paul Raymer and Stacy Bonos for providing experimental materials. Technical assistance from Somer Rowe and assistance on data analysis from Kaiwen Han is much appreciated.

Conflicts of Interest: The authors declare no conflict of interest.

References

1. Fry, J.; Huang, B. *Applied Turfgrass Science and Physiology*; John Wiley & Sons: Hoboken, NJ, USA, 2004.
2. Turgeon, A.J.N. *Turfgrass Management*, 5th ed.; Prentice Hall: Upper Saddle River, NJ, USA, 1999; Volume 4, p. 225.
3. Pachauri, R.K.; Meyer, L. *Climate Change 2014 Synthesis Report-Summary for Policymakers*; Intergovernmental Panel on Climate Change (IPCC): Geneva, Switzerland, 2014.
4. Zhou, Y.; Yin, M.; Liu, F. First Report of Summer Patch of Creeping Bentgrass Caused by Magnaporthiopsis poae in Southeastern China. *Plant Dis.* **2022**, *106*, 1527. [CrossRef]
5. O'Brien, P.; Hartwiger, C. A time to change: The ultradwarf bermudagrass putting green golf model is solid in the southern USA. *USGA Green Sect. Record. Febr.* **2011**, *25*, 8.
6. Mansoor, S.; Ali Wani, O.; Lone, J.K.; Manhas, S.; Kour, N.; Alam, P.; Ahmad, A.; Ahmad, P. Reactive Oxygen Species in Plants: From Source to Sink. *Antioxidants* **2022**, *11*, 225. [CrossRef]
7. Huang, B.R.; DaCosta, M.; Jiang, Y.W. Research Advances in Mechanisms of Turfgrass Tolerance to Abiotic Stresses: From Physiology to Molecular Biology. *Crit. Rev. Plant Sci.* **2014**, *33*, 141–189. [CrossRef]
8. Alam, M.N.; Yang, L.; Yi, X.; Wang, Q.; Robin, A.H.K. Role of melatonin in inducing the physiological and biochemical processes associated with heat stress tolerance in tall fescue (*Festuca arundinaceous*). *J. Plant Growth Regul.* **2022**, *41*, 2759–2768. [CrossRef]
9. Jespersen, D.; Xu, C.P.; Huang, B.R. Membrane Proteins Associated with Heat-Induced Leaf Senescence in a Cool-Season Grass Species. *Crop Sci.* **2015**, *55*, 837–850. [CrossRef]
10. Balfagón, D.; Sengupta, S.; Gómez-Cadenas, A.; Fritschi, F.B.; Azad, R.K.; Mittler, R.; Zandalinas, S.I. Jasmonic acid is required for plant acclimation to a combination of high light and heat stress. *Plant Physiol.* **2019**, *181*, 1668–1682. [CrossRef]
11. Li, H.; Xu, H.; Zhang, P.; Gao, M.; Wang, D.; Zhao, H. High temperature effects on D1 protein turnover in three wheat varieties with different heat susceptibility. *Plant Growth Regul.* **2017**, *81*, 1–9. [CrossRef]
12. Møller, I.M.; Kristensen, B.K. Protein oxidation in plant mitochondria as a stress indicator. *Photochem. Photobiol. Sci.* **2004**, *3*, 730–735. [CrossRef]
13. Du, H.; Wang, Z.; Huang, B. Differential responses of warm-season and cool-season turfgrass species to heat stress associated with antioxidant enzyme activity. *J. Am. Soc. Hortic. Sci.* **2009**, *134*, 417–422. [CrossRef]
14. Li, Z.; Zeng, W.; Cheng, B.; Xu, J.; Han, L.; Peng, Y. Turf Quality and Physiological Responses to Summer Stress in Four Creeping Bentgrass Cultivars in a Subtropical Zone. *Plants* **2022**, *11*, 665. [CrossRef]
15. Jespersen, D.; Meyer, W.; Huang, B. Physiological traits and genetic variations associated with drought and heat tolerance in creeping bentgrass. *Int. Turfgrass Soc. Res. J* **2013**, *12*, 459–464.
16. Liu, X.; Huang, B. Heat stress injury in relation to membrane lipid peroxidation in creeping bentgrass. *Crop Sci.* **2000**, *40*, 503–510. [CrossRef]
17. Zhou, R.; Yu, X.; Kjær, K.H.; Rosenqvist, E.; Ottosen, C.-O.; Wu, Z. Screening and validation of tomato genotypes under heat stress using Fv/Fm to reveal the physiological mechanism of heat tolerance. *Environ. Exp. Bot.* **2015**, *118*, 1–11. [CrossRef]
18. Pirdashti, H.; Sarvestani, Z.T.; Bahmanyar, M.A. Comparison of physiological responses among four contrast rice cultivars under drought stress conditions. *World Acad. Sci. Eng. Technol.* **2009**, *49*, 52–53.
19. Zhang, X.; Gao, Y.; Zhuang, L.; Hu, Q.; Huang, B. Phosphatidic acid and hydrogen peroxide coordinately enhance heat tolerance in tall fescue. *Plant Biol.* **2021**, *23*, 142–151. [CrossRef]
20. Strasser, R.J.; Srivastava, A.; Tsimilli-Michael, M. The fluorescence transient as a tool to characterize and screen photosynthetic samples. *Probing Photosynth. Mech. Regul. Adapt.* **2000**, *1*, 445–483.
21. Küpper, H.; Benedikty, Z.; Morina, F.; Andresen, E.; Mishra, A.; Trtílek, M. Analysis of OJIP chlorophyll fluorescence kinetics and QA reoxidation kinetics by direct fast imaging. *Plant Physiol.* **2019**, *179*, 369–381. [CrossRef]
22. Stirbet, A.; Lazár, D.; Kromdijk, J. Chlorophyll a fluorescence induction: Can just a one-second measurement be used to quantify abiotic stress responses? *Photosynthetica* **2018**, *56*, 86–104. [CrossRef]
23. Zushi, K.; Kajiwara, S.; Matsuzoe, N. Chlorophyll a fluorescence OJIP transient as a tool to characterize and evaluate response to heat and chilling stress in tomato leaf and fruit. *Sci. Hortic.* **2012**, *148*, 39–46. [CrossRef]

24. Snider, J.L.; Thangthong, N.; Pilon, C.; Virk, G.; Tishchenko, V. OJIP-fluorescence parameters as rapid indicators of cotton (*Gossypium hirsutum* L.) seedling vigor under contrasting growth temperature regimes. *Plant Physiol. Biochem.* **2018**, *132*, 249–257. [CrossRef] [PubMed]

25. Virk, G.; Snider, J.L.; Chee, P.; Jespersen, D.; Pilon, C.; Rains, G.; Roberts, P.; Kaur, N.; Ermanis, A.; Tishchenko, V. Extreme temperatures affect seedling growth and photosynthetic performance of advanced cotton genotypes. *Ind. Crops Prod.* **2021**, *172*, 114025. [CrossRef]

26. Strauss, A.; Krüger, G.; Strasser, R.; Van Heerden, P. Ranking of dark chilling tolerance in soybean genotypes probed by the chlorophyll a fluorescence transient OJIP. *Environ. Exp. Bot.* **2006**, *56*, 147–157. [CrossRef]

27. Jespersen, D.; Ma, X.; Bonos, S.A.; Belanger, F.C.; Raymer, P.; Huang, B.J.C.S. Association of SSR and candidate gene markers with genetic variations in summer heat and drought performance for creeping bentgrass. *Crop Sci.* **2018**, *58*, 2644–2656. [CrossRef]

28. Krans, J.V.; Morris, K. Determining a profile of protocols and standards used in the visual field assessment of turfgrasses: A survey of national turfgrass evaluation program-sponsored university scientists. *Appl. Turfgrass Sci.* **2007**, *4*, 1–16. [CrossRef]

29. Karcher, D.E.; Richardson, M.D. Digital image analysis in turfgrass research. *Turfgrass Biol. Use Manag.* **2013**, *56*, 1133–1149. [CrossRef]

30. Wellburn, A.R. The spectral determination of chlorophylls a and b, as well as total carotenoids, using various solvents with spectrophotometers of different resolution. *J. Plant Physiol.* **1994**, *144*, 307–313. [CrossRef]

31. Blum, A.; Ebercon, A.J.C.S. Cell membrane stability as a measure of drought and heat tolerance in wheat. *Crop Sci.* **1981**, *21*, 43–47. [CrossRef]

32. Bradford, M.M. A rapid and sensitive method for the quantitation of microgram quantities of protein utilizing the principle of protein-dye binding. *Anal. Biochem.* **1976**, *72*, 248–254. [CrossRef]

33. Hodges, D.M.; DeLong, J.M.; Forney, C.F.; Prange, R.K. Improving the thiobarbituric acid-reactive-substances assay for estimating lipid peroxidation in plant tissues containing anthocyanin and other interfering compounds. *Planta* **1999**, *207*, 604–611. [CrossRef]

34. Irigoyen, J.; Einerich, D.; Sánchez-Díaz, M. Water stress induced changes in concentrations of proline and total soluble sugars in nodulated alfalfa (*Medicago sativd*) plants. *Physiol. Plant.* **1992**, *84*, 55–60. [CrossRef]

35. Allakhverdiev, S.I.; Kreslavski, V.D.; Klimov, V.V.; Los, D.A.; Carpentier, R.; Mohanty, P. Heat stress: An overview of molecular responses in photosynthesis. *Photosynth. Res.* **2008**, *98*, 541–550. [CrossRef] [PubMed]

36. Xu, Y.; Huang, B. Heat-induced leaf senescence and hormonal changes for thermal bentgrass and turf-type bentgrass species differing in heat tolerance. *J. Am. Soc. Hortic. Sci.* **2007**, *132*, 185–192. [CrossRef]

37. Xu, C.P.; Huang, B.R. Comparative Analysis of Proteomic Responses to Single and Simultaneous Drought and Heat Stress for Two Kentucky Bluegrass Cultivars. *Crop Sci.* **2012**, *52*, 1246–1260. [CrossRef]

38. Hu, L.; Bi, A.; Hu, Z.; Amombo, E.; Li, H.; Fu, J. Antioxidant metabolism, photosystem II, and fatty acid composition of two tall fescue genotypes with different heat tolerance under high temperature stress. *Front. Plant Sci.* **2018**, *9*, 1242. [CrossRef] [PubMed]

39. Wahid, A.; Gelani, S.; Ashraf, M.; Foolad, M. Heat tolerance in plants: An overview. *Environ. Exp. Bot.* **2007**, *61*, 199–223. [CrossRef]

40. Jespersen, D.; Zhang, J.; Huang, B.R. Chlorophyll loss associated with heat-induced senescence in bentgrass. *Plant Sci.* **2016**, *249*, 1–12. [CrossRef] [PubMed]

41. Rossi, S.; Burgess, P.; Jespersen, D.; Huang, B.J.C.S. Heat-induced leaf senescence associated with Chlorophyll metabolism in Bentgrass lines differing in heat tolerance. *Crop Sci.* **2017**, *57* (Suppl. S1), S-169–S-178. [CrossRef]

42. Taiz, L.; Zeiger, E. *Plant Physiology*, 5th ed.; The Sinauer Associates: Sunderland, MA, USA, 2006.

43. Zhang, L.; Chang, Q.; Hou, X.; Wang, J.; Chen, S.; Zhang, Q.; Wang, Z.; Yin, Y.; Liu, J. The Effect of high-temperature stress on the physiological indexes, chloroplast ultrastructure, and photosystems of two herbaceous peony cultivars. *J. Plant Growth Regul.* **2022**, *41*, 1–16. [CrossRef]

44. Chen, S.; Yang, J.; Zhang, M.; Strasser, R.J.; Qiang, S. Classification and characteristics of heat tolerance in Ageratina adenophora populations using fast chlorophyll a fluorescence rise OJIP. *Environ. Exp. Bot.* **2016**, *122*, 126–140. [CrossRef]

45. Chen, L.-S.; Cheng, L. Photosystem 2 is more tolerant to high temperature in apple (*Malus domestica* Borkh.) leaves than in fruit peel. *Photosynthetica* **2009**, *47*, 112–120. [CrossRef]

46. Sharma, D.K.; Andersen, S.B.; Ottosen, C.O.; Rosenqvist, E. Wheat cultivars selected for high Fv/Fm under heat stress maintain high photosynthesis, total chlorophyll, stomatal conductance, transpiration and dry matter. *Physiol. Plant.* **2015**, *153*, 284–298. [CrossRef] [PubMed]

47. Huang, B.R.; Liu, X.Z.; Xu, Q.Z. Supraoptimal soil temperatures induced oxidative stress in leaves of creeping bentgrass cultivars differing in heat tolerance. *Crop Sci.* **2001**, *41*, 430–435. [CrossRef]

48. Li, F.F.; Zhan, D.; Xu, L.X.; Han, L.B.; Zhang, X.Z. Antioxidant and Hormone Responses to Heat Stress in Two Kentucky Bluegrass Cultivars Contrasting in Heat Tolerance. *J. Am. Soc. Hortic. Sci.* **2014**, *139*, 587–596. [CrossRef]

49. Gulen, H.; Eris, A. Effect of heat stress on peroxidase activity and total protein content in strawberry plants. *Plant Sci.* **2004**, *166*, 739–744. [CrossRef]

50. Ayub, M.; Ashraf, M.Y.; Kausar, A.; Saleem, S.; Anwar, S.; Altay, V.; Ozturk, M. Growth and physio-biochemical responses of maize (*Zea mays* L.) to drought and heat stresses. *Plant Biosyst.-Int. J. Deal. All Asp. Plant Biol.* **2021**, *155*, 535–542. [CrossRef]

51. Huffaker, R. Proteolytic activity during senescence of plants. *New Phytol.* **1990**, *116*, 199–231. [CrossRef]

52. Callis, J.J.T.P.C. Regulation of protein degradation. *Plant Cell.* **1995**, *7*, 845. [CrossRef]

53. He, Y.; Huang, B. Protein changes during heat stress in three Kentucky bluegrass cultivars differing in heat tolerance. *Crop Sci.* **2007**, *47*, 2513–2520. [CrossRef]
54. Huang, B.R.; Rachmilevitch, S.; Xu, J.C. Root carbon and protein metabolism associated with heat tolerance. *J. Exp. Bot.* **2012**, *63*, 3455–3465. [CrossRef]
55. Parsell, D.; Lindquist, S. The function of heat-shock proteins in stress tolerance: Degradation and reactivation of damaged proteins. *Annu. Rev. Genet.* **1993**, *27*, 437–496. [CrossRef] [PubMed]
56. Xu, Y.; Huang, B. Differential protein expression for geothermal Agrostis scabra and turf-type Agrostis stolonifera differing in heat tolerance. *Environ. Exp. Bot.* **2008**, *64*, 58–64. [CrossRef]
57. Roberts, I.N.; Caputo, C.; Criado, M.V.; Funk, C. Senescence-associated proteases in plants. *Physiol. Plant.* **2012**, *145*, 130–139. [CrossRef] [PubMed]
58. Ferguson, D.L.; Guikema, J.A.; Paulsen, G.M.J.P.P. Ubiquitin pool modulation and protein degradation in wheat roots during high temperature stress. *Plant Physiol.* **1990**, *92*, 740–746. [CrossRef]
59. Duff, D.T.; Beard, J.B. Supraoptimal Temperature Effects upon Agrostis palustris: Part II. Influence on Carbohydrate Levels, Photosynthetic Rate, and Respiration Rate. *Physiol. Plant.* **1974**, *32*, 18–22. [CrossRef]
60. Du, H.; Wang, Z.; Yu, W.; Liu, Y.; Huang, B. Differential metabolic responses of perennial grass *Cynodon transvaalensis×Cynodon dactylon* (C4) and *Poa Pratensis* (C3) to heat stress. *Physiol. Plant.* **2011**, *141*, 251–264. [CrossRef]
61. Wahid, A.; Close, T.J.B.P. Expression of dehydrins under heat stress and their relationship with water relations of sugarcane leaves. *Biol. Plant.* **2007**, *51*, 104–109. [CrossRef]

Article

Photosynthesis, Water Status and K⁺/Na⁺ Homeostasis of *Buchoe dactyloides* Responding to Salinity

Huan Guo [†], Yannong Cui [†], Zhen Li, Chunya Nie, Yuefei Xu * and Tianming Hu *

College of Grassland Agriculture, Northwest A&F University, Xianyang 712100, China; huan.guo@nwafu.edu.cn (H.G.); cuiyn@nwafu.edu.cn (Y.C.); lizhen135712@163.com (Z.L.); 17319697319@163.com (C.N.)
* Correspondence: xuyuefei@nwsuaf.edu.cn (Y.X.); hutianming@126.com (T.H.)
† These authors have contributed equally to this work.

Abstract: Soil salinization is one of the most serious abiotic stresses restricting plant growth. Buffalograss is a C_4 perennial turfgrass and forage with an excellent resistance to harsh environments. To clarify the adaptive mechanisms of buffalograss in response to salinity, we investigated the effects of NaCl treatments on photosynthesis, water status and K⁺/Na⁺ homeostasis of this species, then analyzed the expression of key genes involved in these processes using the qRT-PCR method. The results showed that NaCl treatments up to 200 mM had no obvious effects on plant growth, photosynthesis and leaf hydrate status, and even substantially stimulated root activity. Furthermore, buffalograss could retain a large amount of Na⁺ in roots to restrict Na⁺ overaccumulation in shoots, and increase leaf K⁺ concentration to maintain a high K⁺/Na⁺ ratio under NaCl stresses. After 50 and 200 mM NaCl treatments, the expressions of several genes related to chlorophyll synthesis, photosynthetic electron transport and CO_2 assimilation, as well as aquaporin genes (*BdPIPs* and *BdTIPs*) were upregulated. Notably, under NaCl treatments, the increased expression of *BdSOS1*, *BdHKT1* and *BdNHX1* in roots might have helped Na⁺ exclusion by root tips, retrieval from xylem sap and accumulation in root cells, respectively; the upregulation of *BdHAK5* and *BdSKOR* in roots likely enhanced K⁺ uptake and long-distance transport from roots to shoots, respectively. This work finds that buffalograss possesses a strong ability to sustain high photosynthetic capacity, water balance and leaf K⁺/Na⁺ homeostasis under salt stress, and lays a foundation for elucidating the molecular mechanism underlying the salt tolerance of buffalograss.

Keywords: *Buchloe dactyloides*; salt tolerance; photosynthesis; water status; ion homeostasis

Citation: Guo, H.; Cui, Y.; Li, Z.; Nie, C.; Xu, Y.; Hu, T. Photosynthesis, Water Status and K⁺/Na⁺ Homeostasis of *Buchoe dactyloides* Responding to Salinity. *Plants* **2023**, *12*, 2459. https://doi.org/10.3390/plants12132459

Academic Editor: James A. Bunce

Received: 27 May 2023
Revised: 17 June 2023
Accepted: 20 June 2023
Published: 27 June 2023

1. Introduction

Soil salinization is an increasingly serious threat to food security and ecological environments worldwide [1]. Salt stress restricts plant growth by affecting a series of physiological processes, such as photosynthetic processes, water uptake and nutrient (especially K⁺) acquisition [2,3]. Photosynthesis, the vital process of primary metabolism, provides energy and organic molecules for plant growth and development [4,5]. However, photosynthesis for numerous plant species is progressively reduced with increasing salinity as a consequence of lessened CO_2 availability, disturbed chloroplast light energy capture, hindered photosynthetic electron flow and carbon assimilation capacity [6,7]. Photosynthesis is a multi-step and complicated process that comprises several biological pathways [4], and it covers three stages in which many proteins are involved: primary reaction, photosynthetic electron transfer (including photosystem II, cytochrome b6/f and photosystem I complexes) and photophosphorylation (ATP-synthase system), as well as CO_2 assimilation (such as key enzymes in C_3 or C_4 photosynthetic pathways) [8–10]. Accordingly, the expression response of genes related to photosynthetic metabolic process is one of the most important reflections for plants coping with salt stress.

The root water uptake is restricted by a diminished water potential because of the presence of high concentrations of salt ions in soil solutions; thus, osmotic stress is also a serious challenge faced by higher plants exposed to saline environments [11,12]. Furthermore, studies have shown that the cell water content of most plant species greatly reduces after only a few hours of salt stress, and the rate of cell elongation and division also significantly slows [13–15]. As a consequence, osmotic stress is the primary factor limiting plant growth during the initial stage of salt stress. Acquiring sufficient amounts of water from the saline soil and keeping a higher shoot water content are essential for plants to maintain photosynthetic capacity and cell growth [16,17]. Aquaporins (AQPs) play a pivotal role in regulating water equilibrium inside and outside of the plant cells and water use efficiency (WUE) by specifically mediating the rapid transport of water across cell membranes [18,19]. Plasma membrane intrinsic proteins (*PIPs*) and tonoplast intrinsic proteins (*TIPs*), among the five subfamilies belonging to AQPs, are the most important proteins responsible for water transport because they govern transcellular and intracellular water movement, respectively [20,21], and therefore play vital roles in maintaining the water status of plants grown in saline environments.

The most direct damage of salt stress to plants is ionic toxicity. There are high amounts of salt ions in saline habitats, triggering severe toxicity and nutrient deficiency in plants [22,23]. Na^+ is the most dominant cation in saline soils, and the overaccumulation of Na^+ can directly destroy membrane systems, damage cellular organelles and impair photosynthesis [24]. Specifically, it competes with K^+ binding sites on cellular transporters and channels, causing K^+ deficiency in cells, which in turn affects protein biosynthesis and enzyme activation [25,26]. The ability of plants to resist Na^+ toxicity mainly depends on whether they can maintain a high K^+/Na^+ ratio in the cytoplasm [11,27]. Additionally, K^+/Na^+ homeostasis is an important aspect to maintain the intracellular environment under saline conditions, thereby influencing plant photosynthetic efficiency and water balance. A direct strategy for plants to cope with excessive Na^+ accumulation and a K^+ deficit is to regulate the expression of several genes encoding membrane transporters or channels associated with Na^+ and/or K^+ uptake, translocation or compartmentalization [28]. For example, the expression of plasma membrane Na^+/H^+ antiporter gene *SOS1* in root tips functions in the efflux of Na^+ from the roots, and the expression of the tonoplast Na^+/H^+ antiporter gene *NHX1* in roots/shoots helps with the sequestration of Na^+ in vacuoles to avoid disturbance to cell metabolisms [3,29,30]; the expression of the high-affinity K^+ transporter genes *HAKs* and inward-rectifier K^+ channel gene *AKT1* in roots facilitates the K^+ uptake [26,31,32]. Therefore, the genes related to Na^+ and K^+ transport are also closely associated with the salt tolerance of plants.

Buffalograss (*Buchloe dactyloides* (Nutt.) Engelm.), a C_4 perennial turfgrass and forage belonging to Poaceae, is native to northwestern America and is now widely distributed in arid and semi-arid regions around the world [33–35]. It is the preferred grass for the establishment of landscape green land, the governance of desert saline-alkali land and slope protection due to its outstanding tolerance to various abiotic stresses, rapid reproduction, dense growing habit as well as long lifetime [36–38]. Meanwhile, this species is also an attractive year-round forage grass because of its high palatability, rich nutrition and strong grazing resistance [39]. Researchers have found that buffalograss could well-tolerate a certain salinity (up to a 100 mM salt concentration) [33,40]. However, regarding the physiological mechanism of salt tolerance in this species, particularly the photosynthetic responses, hydration status and ion accumulation characteristics under salt treatments have not been intensively documented.

The objective of this study was to characterize the physiological responses of buffalograss to salinity by measuring parameters related to photosynthesis, water conservation and K^+/Na^+ homeostasis under treatments with different concentrations of NaCl. Furthermore, the expression patterns of some key genes involved in photosynthesis, encoding *PIP/TIP* aquaporins and K^+/Na^+ transporters/channels under NaCl treatments were analyzed.

2. Results

2.1. Buffalograss Exhibited a Strong Tolerance to Salinity

After treatment with 50 and 100 mM NaCl for 8 days, the seedlings remained vigorous in growth (Figure 1A). When the salinity increased to 200 mM, the leaf tips of seedlings appeared slightly wilting, and under higher salinity (400 mM NaCl), the leaves of plants were withered and grew weakly (Figure 1A). To further confirm the above observations, plant biomass, tiller number and leaf relative membrane permeability (RMP) were measured. The data showed that the addition of 50, 100 and 200 mM NaCl had no significant effects on plant biomass, tiller number and leaf RMP (except for the RMP exposed to 200 mM NaCl, which was higher than that for the control plants) (Figure 1B–E). It is worth noting that, under 50 and 100 mM NaCl treatments, the root fresh weight (FW) and dry weight (DW) significantly increased by 44 and 47%, as well as 32 and 51%, respectively, in comparison with those under the control condition (Figure 1B,C). A total of 400 mM NaCl treatment resulted in an obvious decrease in shoot biomass and tiller number, and a substantial increase in leaf RMP when compared with the control (Figure 1B–E).

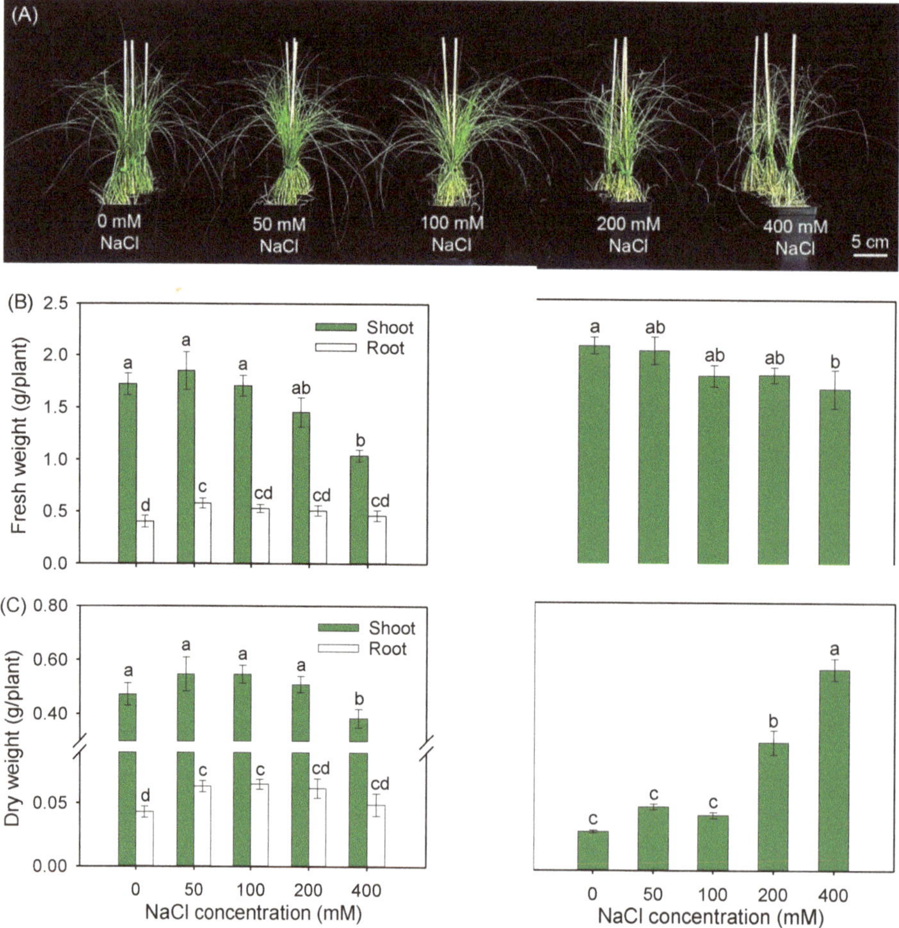

Figure 1. The growth responses (**A**), fresh weight (FW) (**B**), dry weight (DW) (**C**), tiller number (**D**) and leaf relative membrane permeability (RMP) (**E**) of *B. dactyloides* seedlings under different NaCl treatments for 8 days. Values in (**B–E**) are means $\pm$ SE ($n = 6$) and bars indicate SE. Columns with different letters indicate significant differences at $p < 0.05$ (Duncan test).

2.2. The Photosynthetic Responses of Buffalograss Exposed to Salinity

To investigate the photosynthetic capacity of buffalograss seedlings under saline conditions, the content of chlorophylls and photosynthetic physiological indexes were measured. The results showed that after 50 and 100 mM NaCl treatments, the content of chlorophyll a, chlorophyll b and total chlorophyll in leaves maintained stability compared to that of control plants (Figure 2), suggesting that these two salinities had no significant effect on the chlorophyll synthesis of buffalograss. However, 200 and 400 mM NaCl treatments resulted in an obviously decreased content of chlorophylls (Figure 2).

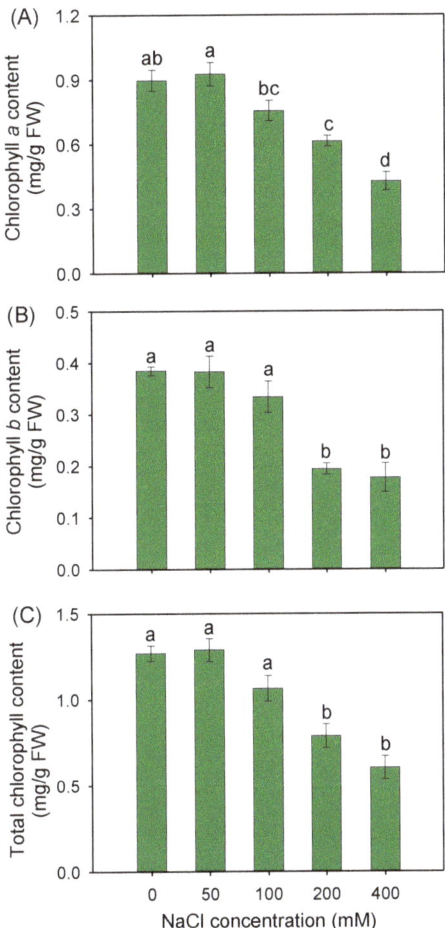

Figure 2. Chlorophyll *a* (**A**), chlorophyll *b* (**B**) and total chlorophyll (**C**) contents of *B. dactyloides* seedlings under different NaCl treatments for 8 days. Values are means $\pm$ SE ($n = 6$) and bars indicate SE. Columns with different letters indicate significant differences at $p < 0.05$ (Duncan test).

The net photosynthetic rate (Pn), transpiration rate (Tr), stomatal conductance (Gs) and intercellular CO_2 concentration (Ci) of buffalograss seedlings under 50 and 100 mM NaCl were comparable to those under the control condition (Figure 3A–D). However, 200 and 400 mM NaCl treatments significantly reduced Pn, Tr and Gs (Figure 3A–C), but increased Ci (Figure 3D). Notably, buffalograss could maintain a stable leaf water use efficiency (WUE) when plants were exposed to 50–200 mM NaCl treatments (Figure 3E).

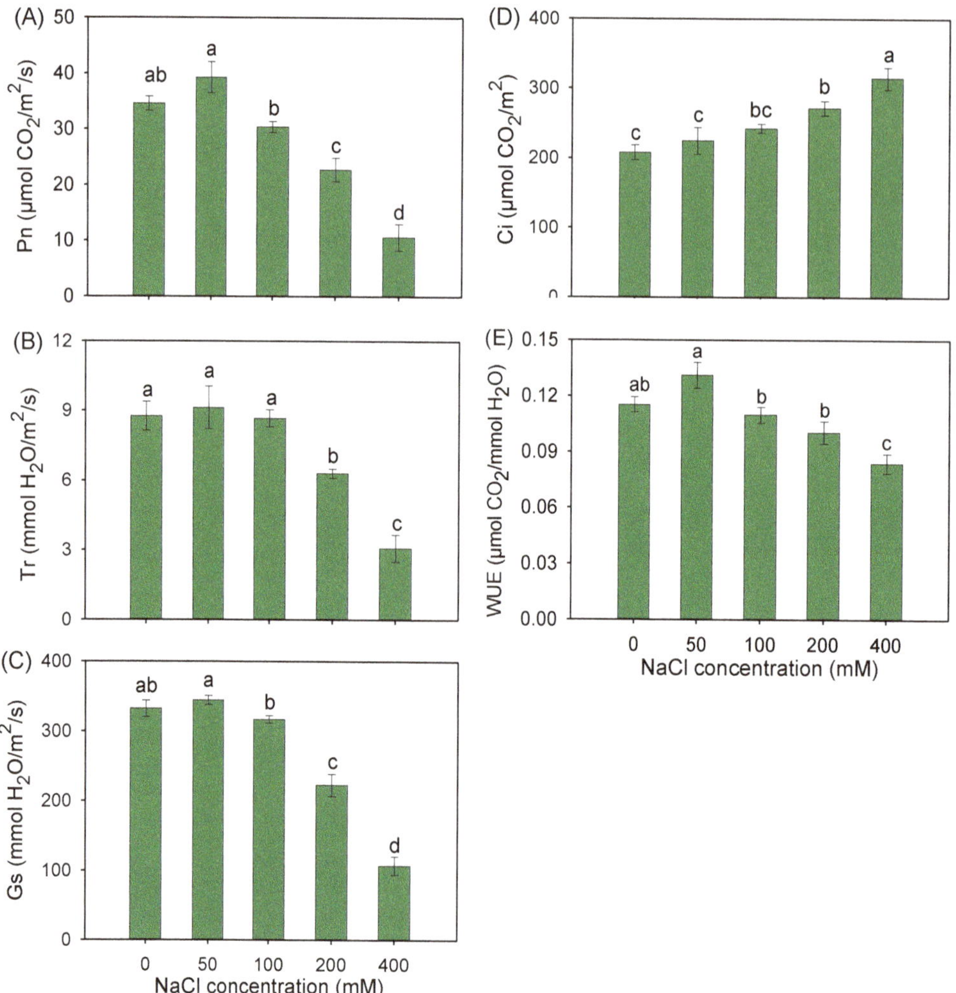

Figure 3. Net photosynthesis rate (Pn) (**A**), transpiration rate (Tr) (**B**), stomatal conductance (Gs) (**C**), intercellular CO_2 concentration (Ci) (**D**) and water use efficiency (WUE) (**E**) of *B. dactyloides* seedlings under different NaCl treatments for 8 days. Values are means ± SE (n = 6) and bars indicate SE. Columns with different letters indicate significant differences at $p < 0.05$ (Duncan test).

2.3. The Leaf Water Status and Root Activity of Buffalograss Seedlings Exposed to Salinity

To evaluate the water status of buffalograss under NaCl treatments, the leaf hydration and root activity were measured. As shown in Figure 4A,B, 50 and 100 mM NaCl had no significant effect on the leaf relative water content (RWC) and water saturation deficit (WSD). When treated with 200 and 400 mM NaCl, the RWCs were significantly decreased by 10.1% and 48.7%, respectively, and the WSDs were correspondingly significantly increased by 116.2% and 480.2%, respectively, compared with the those of the control condition (Figure 4A,B). Unexpectedly, in comparison with the control, 50, 100 and 200 mM NaCl treatments significantly increased the root activity by 5.4-, 4.2- and 2.0-fold, respectively (Figure 4C).

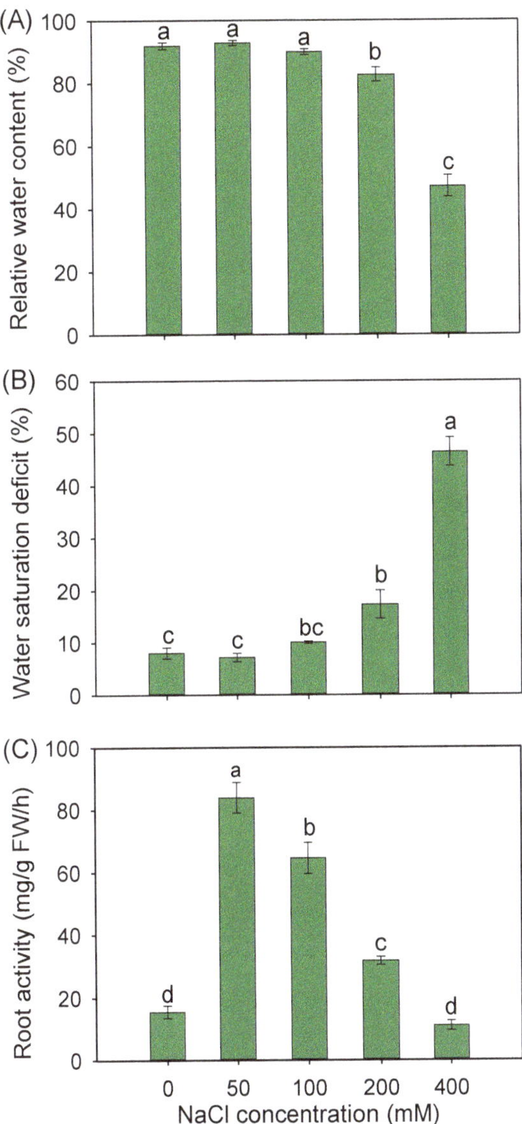

Figure 4. Leaf relative water content (**A**), leaf water saturation deficit (**B**) and root activity (**C**) in *B. dactyloides* seedlings under different NaCl treatments for 8 days. Values are means $\pm$ SE ($n = 6$) and bars indicate SE. Columns with different letters indicate significant differences at $p < 0.05$ (Duncan test).

2.4. The K^+/Na^+ Homeostasis in Buffalograss Seedlings Exposed to Salinity

With the increase in the external NaCl concentration, the amounts of Na^+ accumulated in different tissues of buffalograss seedlings were gradually elevated (Figure 5A–C). It was noticed that Na^+ contents in stems and leaves were lower than those in roots under all salt treatments (Figure 5A–C), indicating that this species could efficiently restrict the long-distance transport of Na^+ from roots to shoots. Although the K^+ accumulations in roots were significantly reduced with the increase in the external NaCl concentration, this parameter remained stable in stems, and unexpectedly, it gradually increased in leaves

(Figure 5D–F). The leaf K^+ concentrations in the presence of 100, 200 and 400 mM NaCl were increased by 35.5, 49.7 and 73.6% compared with those under the control condition, respectively (Figure 5F).

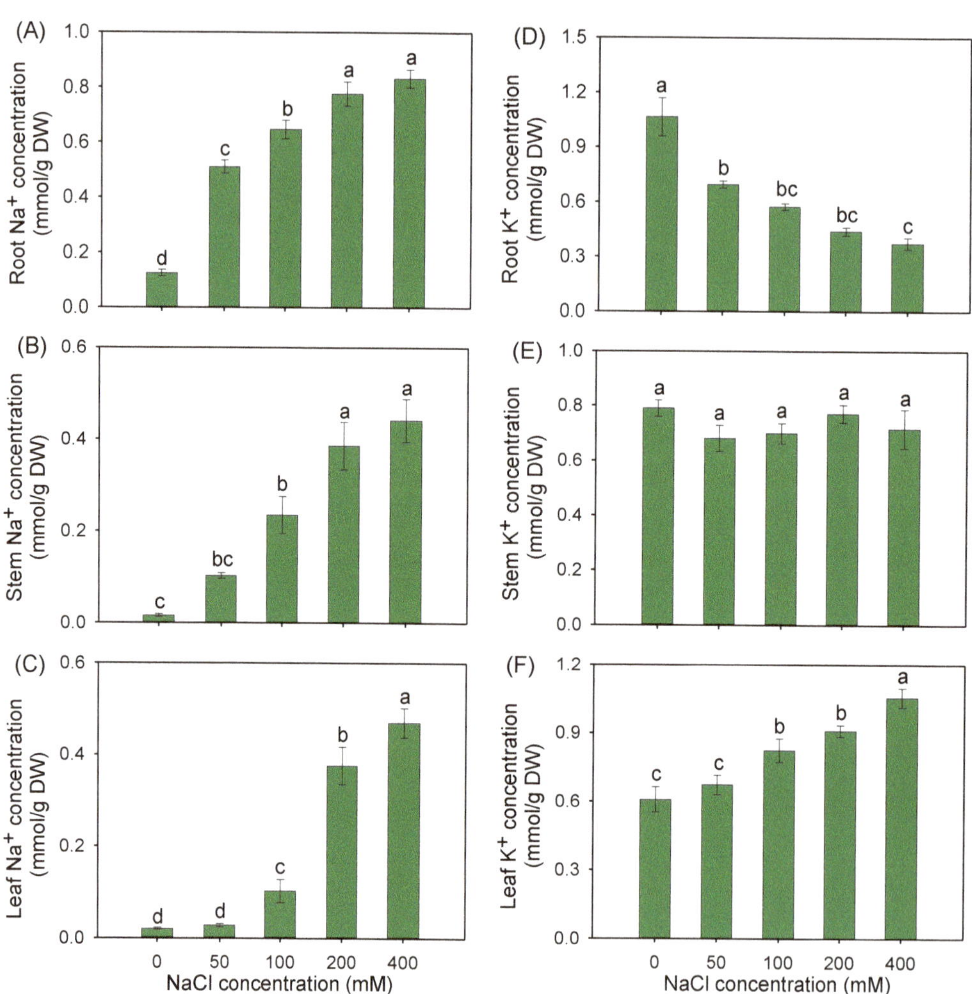

Figure 5. The Na^+ (**A–C**) and K^+ (**D–F**) concentrations in roots, stems and leaves of *B. dactyloides* seedlings under different NaCl treatments for 8 days. Values are means $\pm$ SE ($n = 6$) and bars indicate SE. Columns with different letters indicate significant differences at $p < 0.05$ (Duncan test).

Compared with the control, the leaf/root Na^+ ratio was significantly decreased by 70.4% under 50 mM NaCl, but sharply increased under 200 and 400 mM NaCl treatments (Figure 6A). In comparison with the control, all the NaCl treatments significantly decreased the K^+/Na^+ ratio in the roots (Figure 6B), while in the leaves, the K^+/Na^+ ratio under 100–400 mM NaCl treatments significantly declined, but it was maintained at the control level under a 50 mM NaCl treatment (Figure 6C). In addition, the K^+/Na^+ ratio in leaves under all salt treatments was greater than one, and was obviously higher than the root K^+/Na^+ ratio (Figure 6B,C).

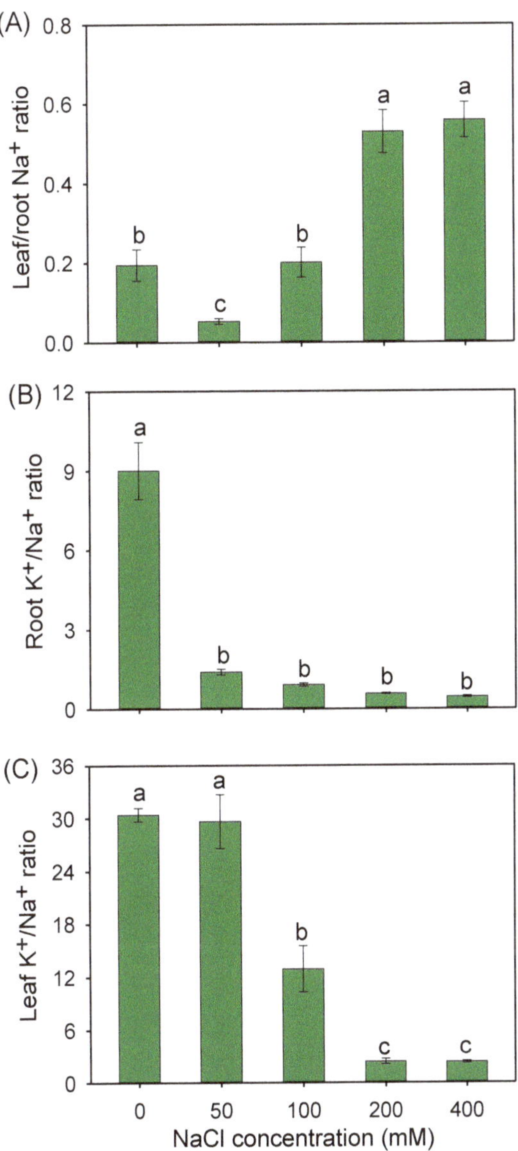

Figure 6. The leaf/root Na$^+$ ratio (**A**), root (**B**) and leaf (**C**) K$^+$/Na$^+$ ratio in *B. dactyloides* seedlings under different NaCl treatments for 8 days. Values are means ± SE (*n* = 6) and bars indicate SE. Columns with different letters indicate significant differences at *p* < 0.05 (Duncan test).

2.5. Effects of Salinity on Expression of Genes Related to Photosynthesis in Buffalograss

The expression of 12 genes related to photosynthesis, including 5 genes involved in chlorophyll synthesis (*BdUROD*, *BdCPO*, *BdProtox*, *BdChlH* and *BdCHLG*), 3 in photosynthetic electron transport (*BdLHCII*, *BdISP* and *BdFNR*) and 4 in CO$_2$ assimilation (*BdPEPC*, *BdNADP-ME*, *BdMDH* and *BdPPDK*), were analyzed in buffalograss seedlings in response to 50 and 200 mM NaCl treatments for 6 h. As shown in Figure 7, after a 50 mM NaCl treatment for 6 h, all genes tested were upregulated, and except for *BdCPO*, *BdPEPC* and *BdMDH*, the relative expression levels of the other 9 genes at 50 mM NaCl were more than two-fold higher than those under the control condition (salt treatment for 0 h). *BdProtox, BdLHCII* and *BdNADP-ME* were even increased by over ten-fold (Figure 7, Supplementary Table S2). When buffalograss seedlings were exposed to high salinity (200 mM NaCl), the expression levels of *BdLHCII, BdISP* and *BdFNR* were continuously upregulated; meanwhile, the expression of *BdNADP-ME, BdMDH, BdPEPC* and *BdPPDK* involved in C$_4$ carbon fixation remained at the control level. However, the expression levels of *BdProtox* and *BdChlH* significantly declined by 200 mM NaCl (Figure 7).

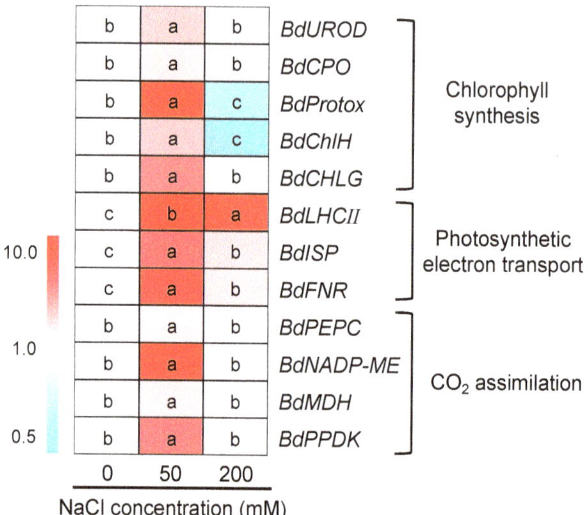

Figure 7. The expression of genes related to photosynthesis in *B. dactyloides* under 0, 50 and 200 mM NaCl for 6 h. Expression of the *B. dactyloides ACTIN* gene was used as an internal control for normalization. Different lowercase letters within each gene indicate significant differences ($p < 0.05$, Duncan test). UROD: uroporphyrinogen decarboxylase, CPO: coproporphyrinogen-III oxidase, Protox: protoporphyrinogen oxidase, ChlH: magnesium chelatase H subunit, CHLG: chlorophyll synthase, LHCII: LHCII type I chlorophyll a-b binding protein, ISP: cytochrome b6/f complex iron-sulfur subunit, FNR: ferrdoxin-NADP$^+$ reductase, PEPC: phosphoenolpyruvate carboxylase, NADP-ME: NADP$^+$-dependent malic enzyme, MDH: malate dehydrogenase, PPDK: pyruvate orthophosphate dikinase.

2.6. Effects of Salinity on Expression of Aquaporin Genes in Buffalograss

Buffalograss could significantly improve plant root activity to enhance water acquisition and simultaneously maintain the leaf water balance to adapt to saline environments (Figure 4). Therefore, the expression pattern of seven aquaporin genes (AQPs, including five plasma membrane-located *PIPs* and two tonoplast-located *TIPs*) was analyzed. In roots, the expression levels of three *PIP1s* (*BdPIP1;1 BdPIP1;2* and *BdPIP1;3*) were not induced by 50 mM NaCl and even showed slight downregulation, while two *PIP2s* (*BdPIP2;1* and *BdPIP2;2*) were significantly upregulated by 50 mM NaCl. After 200 mM NaCl treatment, the expression levels of *BdPIP1;3* and *BdPIP2;1* were upregulated two-fold more than those at control conditions (Figure 8A, Supplementary Table S3). Furthermore, the expression of *BdTIP1;2* in roots was upregulated under both 50 and 200 mM NaCl treatments (Figure 8A). In leaves, almost all tested *AQPs* (except for *BdPIP1;3*) were upregulated in response to 50 and/or 200 mM NaCl (Figure 8B). Among them, the expression of *BdPIP2;1* increased in both tissues after 50 and 200 mM NaCl treatments; its expression levels in the roots and leaves after 50 mM NaCl were 3.5- and 4.2-fold higher than those under the control condition, respectively (Figure 8, Supplementary Table S3).

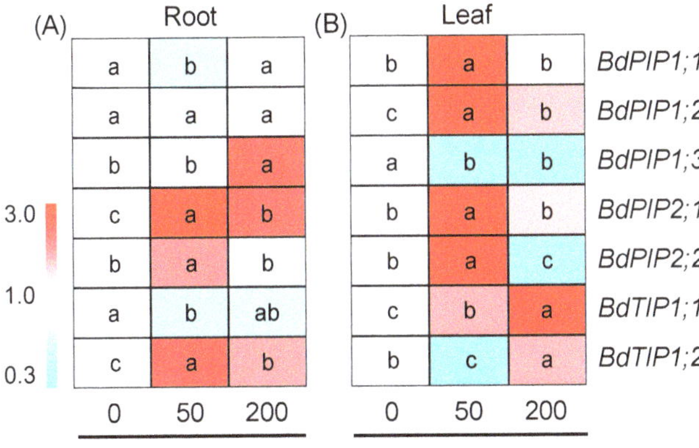

Figure 8. The expression of aquaporin genes in the roots (**A**) and leaves (**B**) of *B. dactyloides* under 0, 50 and 200 mM NaCl for 6 h. Expression of the *B. dactyloides ACTIN* gene was used as an internal control for normalization. Different lowercase letters within each gene indicate significant differences ($p < 0.05$, Duncan test). *PIP*: plasma membrane intrinsic protein, *TIP*: tonoplast intrinsic protein.

2.7. Effects of Salinity on Expression of Genes Related to Na^+ and K^+ Transport in Buffalograss

To understand how buffalograss seedlings maintained K^+/Na^+ homeostasis under saline conditions (Figures 5 and 6), we further analyzed the expression of key genes related to Na^+ and K^+ transport in roots in response to 50 and 200 mM NaCl treatments for 6 h. As shown in Figure 9 and Supplementary Table S4, when plants were exposed to moderate salinity (50 mM NaCl), except for the inwardly rectifying K^+ channel gene *BdAKT1*, other tested genes, encoding Na^+/H^+ antiporters (*BdSOS1* and *BdNHX1*), Na^+-selective transporters (*BdHKT1;4* and *BdHKT1;5*) and K^+ transporter/channel (*BdHAK5* and *BdSKOR*) were all upregulated (Figure 9). Under high salinity (200 mM NaCl), except for two *BdHKT1s* with no expression change, all the above genes were upregulated. In particular, the expression levels of *BdNHX1* and *BdHAK5* were 4.0- and 7.8-fold higher than those in control conditions, respectively (Figure 9, Supplementary Table S4).

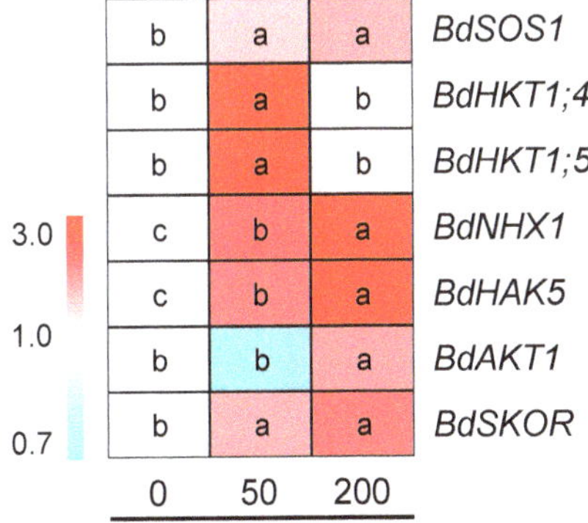

Figure 9. The expression of genes related to Na^+ and K^+ transport in the roots of *B. dactyloides* under 0, 50 and 200 mM NaCl for 6 h. Expression of the *B. dactyloides ACTIN* gene was used as an internal control for normalization. Different lowercase letters within each gene indicate significant differences ($p < 0.05$, Duncan test). SOS1: plasma membrane Na^+/H^+ antiporter, HKTs and HAK5: high-affinity K^+ transporter, NHX1: tonoplast Na^+/H^+ antiporter, AKT1: inwardly rectifying K^+ channel, SKOR: stelar K^+ outwardly rectifying channel.

3. Discussion

3.1. Buffalograss Maintains High Photosynthetic Capacity under Moderate Salinity by Stimulating the Expression of Photosynthesis-Related Genes

Salinity is generally harmful to the growth of most glycophytic species, but some special halophytes and xerophytes, such as *Suaeda* and *Atriplex* spp., can grow better at moderate salinity (within a 100 mM NaCl concentration) [41,42]. In the present work, under 50 and 100 mM NaCl treatments, the growth of buffalograss seedlings remained unaffected, and plant biomass, tillering ability and leaf relative membrane permeability were all comparable to plants under the control condition (Figure 1). Moreover, the root biomass was significantly increased under 50 and 100 mM NaCl treatments (Figure 1B,C). Differently, it has been reported that the growth of many glycophytes, especially traditional crops and turfgrasses in Poaceae such as wheat, rice, ryegrass and tall fescue, are inhibited by NaCl treatments lower than 100 mM [43–45]. Furthermore, although the leaf relative membrane permeability of buffalograss was significantly increased under the 200 mM NaCl treatment, the tissue biomass and tiller number were maintained at the same levels as those under the control condition (Figure 1), suggesting that buffalograss possesses a strong salt tolerance.

Biomass accumulation in higher plants directly depends on the photosynthetic capacity, while salinity reduced the photosynthetic capacity in most plants, resulting in a lower productivity [46,47]. The maintenance of a stable content of chlorophyll (Chl) is one of the most crucial physiological traits involved in the salt tolerance of plants, as it is directly linked to leaf photosynthetic capacity [48]. In this study, 50 and 100 mM NaCl had no obviously negative effect on the contents of Chl a and Chl b of buffalograss (Figure 2). Meanwhile, the Pn, Tr, Gs, Ci and WUE of buffalograss under 50 and 100 mM NaCl treatments were all maintained at the control levels (Figure 3). It was noticed that 200 and 400 mM NaCl treatments significantly decreased Chl contents and the Pn of buffalograss, which indicated that the external NaCl concentrations exceeding 200 mM would hamper the photosynthesis of this plant species.

It is known that salt stress influences Chl biosynthesis by restricting the expression levels of genes encoding relevant enzymes [44]. However, our results showed that moderate salinity (50 mM NaCl) stimulated the expression of genes associated with Chl biosynthesis in buffalograss, including four important rate-limiting enzymes, uroporphyrinogen decarboxylase (UROD), coproporphyrinogen-III oxidase (CPO), protoporphyrinogen oxidase (Protox) and magnesium chelatase H subunit (ChlH), involved in tetrapyrrole metabolism for the synthesis of Chl precursors [44,49,50], and a key enzyme (chlorophyll synthase, CHLG) catalyzing the last step of the Chl biosynthetic pathway [51] (Figure 7). Under high salinity (200 mM NaCl), the expressions of *BdProtox* (catalyzing the oxidation of protoporphyrinogen IX to protoporphyrin IX, the last enzyme of the common branch of the chlorophyll- and heme-synthesis pathway) [52] and *BdChlH* (involved in catalyzing the conversion of protoporphyrin IX to Mg-protoporphyrin IX in the chlorophyll pathway of tetrapyrrole metabolism) [50] in buffalograss were downregulated (Figure 7), which might be one of the crucial reasons why high salinity caused the reduction of Chl contents in buffalograss seedlings (Figure 2).

The expression of the *BdLHCII* gene coding an LHCII type I chlorophyll a-b binding protein (associated with PS II) was sharply upregulated by 11.9- and 24.9-fold under 50 and 200 mM NaCl, respectively (Figure 7, Supplementary Table S2), indicating that salinity might stimulate light energy harvest and transfer in the PS II complex of buffalograss. Additionally, 50 and 200 mM NaCl could upregulate the expression of *BdISP* (an iron-sulfur protein in the cytochrome b6/f complex) and *BdFNR* (chloroplast-targeted ferredoxin-NADP$^+$ reductase associated with PS I) to promote electron transfer between PS II and PS I (Figure 7) [53,54].

The CO_2 assimilation is the final stage in the oxygenic photosynthesis of higher plants, and C_4 plants share stronger CO_2 fixation capacity compared with C_3 plants based on a CO_2-concentrating mechanism [55]. Our study found that the expression levels of key enzymes in the C_4 pathway of buffalograss were upregulated by 50 mM NaCl, including phosphoenolpyruvate carboxylase (PEPC), NADP-dependent malic enzyme (NADP-ME), malate dehydrogenase (MDH) and pyruvate orthophosphate dikinase (PPDK), which play critical roles in providing sufficient Ci and, thus, improving the Pn of buffalograss seedlings to adapt to salinity (Figure 3A,D). Thus, these results indicate that buffalograss can maintain high Chl synthesis and photosynthetic capacity under moderate salinity by stimulating the expression of photosynthesis-related genes, which was positively correlated with the constant biomass of buffalograss plants.

3.2. Efficient Water Uptake and Transport Capacity Plays an Important Role in the Salt Tolerance of Buffalograss

Stable water status is essential for plants to adapt to saline environments [17]. In the present study, the leaf RWC and WSD of buffalograss seedlings was unaffected by 50 and 100 mM NaCl (Figure 4A,B), suggesting that this species was able to maintain the water balance of photosynthetic organs that might result in the healthy growth of plants under moderate salinity. Unexpectedly, the root activity of buffalograss seedlings sharply increased after salt treatments (except for 400 mM NaCl) (Figure 4C). Combined with the appearance of increased root biomass after salt treatments (Figure 1B,C), these results suggest that buffalograss possesses high water acquisition and retention capacities under saline conditions, which is conducive to promote mineral nutrient absorption, and provide adequate raw material for photosynthesis.

Aquaporins (AQPs) can effectively regulate water flow through the plant tissues and maintain cellular water homeostasis by mediating the rapid transmembrane transport of water [18]. Plasma membrane-located *PIPs* are involved in regulating most of the water absorption by roots and water transportation across cells in various tissues, whereas tonoplast-located *TIPs* are extremely vital for cellular osmoregulation because they mediate the water exchange between the vacuole and cytosol [56,57]. *PIPs* in plants are further categorized into *PIP1* and *PIP2* sub-groups [57]. In this study, an interesting phenomenon was found that in buffalograss roots, the expression levels of three *PIP1* genes (*BdPIP1;1 BdPIP1;2* and *BdPIP1;3*) all showed a slight downregulated tendency under 50 mM NaCl (Figure 8A). It has been established that the downregulation of *PIPs* in roots facilitates cellular water conservation by reducing the water permeability of the root cell membrane [58,59]. Given that the hydration status of buffalograss under 50 mM NaCl treatment was unaffected (Figure 4A), the downregulation of *BdPIP1;1*, *BdPIP1;2* and *BdPIP1;3* might also be involved in the root water conservation of buffalograss. In contrast, the expressions of the tested *PIP2* genes (*BdPIP2;1* and *BdPIP2;2*) in roots were upregulated after 50 mM NaCl treatment (Figure 8A), which were likely involved in water uptake to sustain plant water homeostasis at moderate salinity. Under 200 mM NaCl, the expression abundances of *BdPIP1;3* and *BdPIP2;1* in roots were upregulated (Figure 8A). Previous studies found that *PIP1;3* and *PIP2;1* are highly expressed in the epidermal cells of roots under salt and drought stresses [60,61]. Therefore, the upregulation of *BdPIP1;3* and *BdPIP2;1* under 50 and/or 200 mM NaCl treatments might be conducive to the water absorption of buffalograss roots and, hence, enhance root activity.

The *TIP* subfamily in plants was divided into five sub-groups (*TIP1-5*) [62], and *TIP1s* were found in the lytic vacuole membrane and abundantly expressed in salt-treated roots to regulate water transport [63,64]. In buffalograss roots, the expression of *BdTIP1;2* was upregulated by 50 and 200 mM NaCl treatments (Figure 8A), which may be involved in water accumulation in root cell vacuoles to improve water retention capacity. Furthermore, a study found that *AtTIP1;2* facilitates the emergence of new lateral root primordia [65]. We speculate that *BdTIP1;2* may have similar functions to stimulate lateral root development and thereby contribute to a higher root biomass and root activity in buffalograss plants after salt treatments (Figure 1B,C and Figure 4C).

In contrast, in buffalograss leaves, almost all tested aquaporin genes (except for *BdPIP1;3*) were upregulated in response to 50 and/or 200 mM NaCl (Figure 8B), suggesting that there was also a well-developed water transport system in buffalograss leaves to maintain leaf water homeostasis under saline environments. Meanwhile, *PIPs* can also transport CO_2 and are involve in CO_2 diffusion across the plasma membrane of mesophyll cells [66,67]; we speculate that the upregulation of *PIPs* in buffalograss leaves under NaCl treatments would increase the membrane permeability for CO_2 diffusion to contribute to maintaining a high Pn.

3.3. Maintaining K^+/Na^+ Homeostasis Is a Crucial Strategy for Buffalograss to Adapt to Saline Environments

Maintaining constant intracellular K^+ and Na^+ homeostasis is crucial for plants adapting to saline environments [12,68], which is mainly reflected in two aspects. One is the reduction in Na^+ accumulation in the cytoplasm of photosynthetic organs, another is the improvement in K^+ retention in the leaf mesophyll. In our study, the Na^+ concentration in the roots of buffalograss was higher than that in the stems and leaves (Figure 5A–C), indicative of the strong leaf Na^+ exclusion ability of buffalograss under salt stress. For most glycophytes, salinity stress seriously affects plant K^+ acquisition and the accumulation of K^+ in shoots [69–71]. Surprisingly, the K^+ concentrations were maintained in stems, and even significantly increased in the leaves of buffalograss under NaCl treatments (Figure 5E,F). In addition, 50 mM NaCl had no effect on leaf K^+/Na^+ ratio in buffalograss, and this parameter was far more than 1 (it is generally thought that leaf $K^+/Na^+ < 1$ will severely inhibit plant growth [27]) under 100–400 mM NaCl treatments (Figure 6C). These results suggested that buffalograss possesses an atypical ability to maintain K^+ and Na^+ homeostasis under salt stress.

To obtain the optimal K^+/Na^+ ratio under saline stress, K^+ absorption and K^+ transport from the roots to shoots through the xylem are required to be enhanced, and Na^+ influx and transport to the shoots should be limited [32,68]. Several proteins involved in root Na^+ transport, such as SOS1 and HKTs, play vital roles in restricting Na^+ overaccumulation in shoots under salt stress [29,72,73]. Our results showed that the expression of *BdSOS1* was upregulated in roots under two salinity levels (50 and 200 mM NaCl) (Figure 9), which should contribute to promote Na^+ exclusion from the root tip cells [29]. We also observed that two HKT1 genes (*BdHKT1;4* and *BdHKT1;5*) were upregulated by 50 mM NaCl (Figure 9). Previous studies demonstrated that *HKT1;4* and *HKT1;5* are highly expressed in roots under salt stress involved in Na^+ retrieval from the xylem into xylem parenchyma cells to limit excessive Na^+ accumulation in shoots [72–74], suggesting that BdHKT1;4 and BdHKT1;5 in buffalograss roots might play similar roles. The upregulation of *BdSOS1* and two *BdHKT1s* in roots by 50 mM NaCl is conducive to the leaf Na^+ exclusion of buffalograss, which closely correlated with holding the lower leaf/root Na^+ ratio at moderate salinity. Inversely, the lack of change in the expression of *BdHKT1s* at 200 mM NaCl may be one of the important reasons for the increase in leaf/root Na^+ ratio in leaves under high salinity. Additionally, the expression levels of another gene encoding for NHX1 protein function as the tonoplast Na^+/H^+ antiporter were increased 1.2- and 4.0-fold under 50 and 200 mM NaCl treatments for 6 h, respectively (Figure 6, Supplementary Table S4), which are capable of compartmentalizing more Na^+ into root vacuoles, thereby improving the osmotic adjustment ability to enhance water uptake and reducing Na^+ toxicity for root cells [75,76].

K^+ uptake capacity in the high-affinity range of concentrations is a prerequisite for plants to sustain K^+ homeostasis under saline environments, and HAK5 (high-affinity K^+ transporter) and AKT1 (inwardly rectifying K^+ channel) in plants have been shown to be the main proteins involved in this process [77,78]. The results of the present study showed that the expression of *BdHAK5* in buffalograss roots was significantly increased by two salinity; specifically it was 7.8-fold higher than that in control conditions at 200 mM NaCl (Figure 9, Supplementary Table S4). The expression of *BdAKT1* was slightly downregulated by 50 mM NaCl, whereas it was upregulated by 200 mM NaCl (Figure 9). These results suggest that *BdHAK5* in buffalograss may play a major role in root K^+ acquisition in the presence of moderate salinity, and that *BdAKT1* is primarily involved in K^+ uptake under high salinity. When K^+ absorbed by plant root epidermal cells is transported to the root xylem through the apoplast and symplast pathways, it needs to be loaded into xylem sap for translocation from roots to shoots, and SKOR (stelar K^+ outwardly rectifying channel) was a key protein reported to mediate the K^+ loading process [32]. In our study, *BdSKOR* expression levels in roots were increased by two salinity, and this upregulation enhanced K^+ loading into the xylem, which correlated with K^+ accumulation in the stems and leaves of buffalograss observed in the present results. In consequence, the increased expression of *BdHAK5* and *BdSKOR* as well as that of *BdSOS1* and *BdHKT1s* in buffalograss roots when plants were exposed to moderate salinity (50 mM NaCl) can account for the higher K^+/Na^+ ratio in leaves (Figure 6C).

4. Materials and Methods

4.1. Plant Growth Conditions and Salt Treatments

The seeds of buffalograss (*Buchloe dactyloides* (Nutt.) Engelm.) were obtained from Beijing Clover Ecotechnology Co., LTD., Beijing, China, in 2021. Considering the naturally low germination rate of buffalograss seeds, they were first sown in the experimental field of Northwest A&F University, and then the original stolon sections were excised from the middle of healthy buffalograss plants by using sharp scissors and their bases were placed into black plastic pots (10 cm high × 8 diameter, 4 cuttings/pot) filled with sterilized quartz sand (5–8 mm particle size). The pots were transferred into a greenhouse with a temperature of 28 °C/25 °C (day/night), a photoperiod of 16/8 h (light/dark, the flux density was about 800 μmol/m^2/s) and a relative humidity of approximately 65%. When the cuttings generated adventitious roots (about 10 days), uniform seedlings were chosen for further culture and irrigated with modified Hoagland nutrient solution [79].

After five weeks of preculture, buffalograss seedlings were treated with Hoagland nutrient solution containing additional 0 (control), 50, 100, 200 or 400 mM NaCl for 8 days, and 100, 200 and 400 mM NaCl treatment solutions were supplied incrementally by 50 mM each day to avoid salinity shock until the final concentrations were achieved. The treatment solutions were renewed every two days to keep a constant NaCl concentration. After 8 d, the plants were harvested for biomass measurement and physiological analysis, and six replicate seedlings ($n = 6$) were used for all measurements.

4.2. Measurement of Parameters Related to Growth

Tiller number was first counted. Then, the roots and shoots of individual seedlings were carefully separated, and tissue fresh weight (FW) was immediately measured after blotting off the surface water with filter paper. After that, all samples were oven-dried at 80 °C for 72 h to obtain dry weight (DW).

The leaf relative membrane permeability (RMP) was assessed according to the method described by Gibon et al. [80]. About 0.5 g of fresh mature leaves was cut into pieces and put in deionized water. After vacuuming for 30 min (10 min × 3 times) until the leaves were completely sunk, the leaf samples were shaken gently at 25 °C for 1.5 h, and then the initial electrolyte leakage (E1) was measured by using a conductivity meter (EC215, Hanna Instruments, Padovana, Italy). Finally, the leaf samples were incubated in a boiling water bath for 15 min, and the total electrolyte leakage (E2) was recorded after cooling. The RMP (%) was calculated as (E1/E2) × 100.

4.3. Measurement of Parameters for Photosynthesis Relations

The contents of chlorophylls were determined using fresh leaves according to the method described by Lichtenthaler [81]. Briefly, 0.1 g fresh leaf samples were cut into pieces and then immediately soaked in 10 mL of extraction solution, which was composed of 80% acetone and 95% ethyl alcohol at a volume ratio of 1:1. After extracting for 24 h in the dark, the supernatant was collected via centrifugation. Then, the absorbance at 645 nm and 663 nm was recorded using a UV spectrophotometer (UV-2102C, Unico Instrument

Co., Ltd., Shanghai, China). The contents of chlorophyll *a* (Chl a), chlorophyll *b* (Chl b) and total chlorophyll (total Chl) were calculated by using the equations described by Arnon [82].

The leaf net photosynthesis rate (Pn), transpiration rate (Tr), stomatal conductance (Gs) and intercellular CO_2 concentration (Ci) were measured in the greenhouse between 3 h and 5.5 h after the start of the photoperiod using a portable photosynthesis system (LI-6400XT, LI-COR, Lincoln, NE, USA). The intrinsic water use efficiency (WUE) was calculated as Pn/Gs [83]. Leaf areas were estimated using an Epson Perfection 4870 photo-scanner (Epson America, Inc., Long Beach, CA, USA).

4.4. Measurement of Parameters for Root Activity and Leaf Water Status

The root activity was determined using fresh roots according to the tetrazolium (TTC) staining method as described by Zhang et al. [84]. About 0.5 g of fresh tender roots was immersed in 10 mL of mixed solution of 0.4% TTC and phosphate buffer. After incubating at 37 °C in the dark for 3 h, 2 mL 1 M H_2SO_4 was immediately added to terminate the reaction. Subsequently, the root samples were dried with filter paper and then extracted with 10 mL of methyl alcohol. After being sealed in the dark for 3 h until the roots turned white, the absorbance of red extractant was recorded at 485 nm using a UV spectrophotometer. Root activity was calculated using the following formula: root activity (mg/g FW/h) = amount of TTC reduction (mg)/fresh root weight (g) × time (h).

The leaves were excised from buffalograss seedlings and the FW was weighed immediately. Then, the turgid weight (TW) was determined after soaking the leaves in distilled water at 4 °C overnight. Finally, the leaves were oven-dried at 80 °C for 72 h to obtain the DW. The leaf relative water content (RWC) and water saturation deficit (WSD) were calculated according to the following formulas: RWC (%) = 100 × (FW − DW)/(TW − DW); WSD (%) = 100 × (TW − FW)/(TW − DW) as described by Chen et al. [28].

4.5. Measurement of Na^+ and K^+ Concentrations in Tissues

The K^+ and Na^+ concentrations in tissues were measured according to the methods of Pan et al. [17]. Firstly, the roots, stems and leaves of seedlings were separated, and then the roots were rinsed with distilled water and placed in ice-cold 20 mM $LiNO_3$ solution for 8 min to exchange cell-wall-bound salts. After oven drying at 80 °C for 72 h, the K^+ and Na^+ were extracted from the dried tissues using 100 mM acetic acid at 90 °C for 2 h, and the K^+ and Na^+ concentrations were then determined with a flame spectrophotometer (Model 410, Sherwood Scientific, Ltd., Cambridge, UK).

4.6. RNA Extraction and cDNA Synthesis

According to the results of physiological indicators, buffalograss could adapt well to 50 mM NaCl and this salt concentration could even stimulate its growth, while 200 mM NaCl should be the critical concentration for this species to suffer from salt stress. Therefore, 50 and 200 mM NaCl were chosen as salt treatment concentrations in the gene expression analysis. Four-week-old buffalograss seedlings were treated with Hoagland nutrient solution supplemented with 0 (control), 50 or 200 mM NaCl. After treatments for 6 h, the roots and leaves of seedlings were collected, immediately frozen in liquid nitrogen and stored at −80 °C. After the frozen roots and leaves were ground into powder, the total RNA was extracted with an E.Z.N.A™ plant RNA kit (Omega Bio-Tek, Norcross, GA, USA). The integrity and purity of RNA were checked using 1% agarose gel electrophoresis and spectrophotometric analysis using a NanoDrop 2000 (Thermo Scientific, Pittsburgh, PA, USA), and then the first-strand cDNA was synthesized from 2 μg of total RNA using a PrimeScript™ RT Master Mix with gDNA eraser (Perfect Real Time) kit (TaKaRa, Biotech Co., Ltd., Dalian, China).

4.7. Gene Expression Analysis

We chose 12, 8 and 7 genes related to photosynthesis, aquaporin and ion transport, shown in Supplementary Table S1, respectively, to analyze their expression levels in response to different NaCl treatments. Quantitative real-time PCR (qRT-PCR) was used to estimate the expression levels of the genes. The primers used in qRT-PCR were designed with Primer 5.0 (Premier Biosoft International, Palo Alto, CA, USA) and are listed in Supplementary Table S1. Buffalograss *BdACTIN* (beta actin) was used as the reference gene for RNA normalization. SYBR Green Real-Time PCR Master Mix (Takara, Biotech Co., Ltd., Dalian, China) was used for qRT-PCR on a QuantStudio®5 real-time PCR system (Life Technologies, Waltham, MA, USA). Three biologically independent replicates were used to obtain the gene expression quantification data. All PCR reactions were performed with three replicates, and 3 μL first-strand cDNA after ten times dilution was used for each of the qRT-PCR reactions with a 20 μL reaction system. Finally, the relative expression levels of genes were calculated using the $2^{-\Delta\Delta Ct}$ method according to Duan et al. [85].

4.8. Statistical Analysis

SPSS 19.0 statistical software (SPSS Inc., Chicago, IL, USA) was used for statistical analyses. Data were subjected to a one-way ANOVA (Duncan's test, $p < 0.05$). Figures 1–6 were drawn with SigmaPlot 14.0 (Systat Software Inc., San Jose, CA, USA). All data in the physiological experiments are presented as mean $\pm$ SE ($n = 6$). Figures 7–9 were drawn using the MEV (Multiple Experiment Viewer) software, and the qRT-PCR data are presented in Supplementary Tables S2–S4 as mean $\pm$ SE ($n = 3$).

5. Conclusions

In conclusion, buffalograss possesses a prominent salt tolerance, as its growth, photosynthesis and water status were unaffected when external NaCl concentration was up to 200 mM. Moreover, this species could efficiently restrict Na^+ overaccumulation and increase K^+ concentration in leaves to maintain a high K^+/Na^+ ratio under salt stress. After NaCl treatments, buffalograss could maintain high leaf photosynthetic capacity and water balance by upregulating the expression of many genes associated with chlorophyll biosynthesis, photosynthetic electron transport, CO_2 assimilation as well as *PIP* and *TIP* aquaporins encoding genes. In addition, several genes related to Na^+ transport such as *BdSOS1*, *BdHKT1;4*, *BdHKT1;5* and *BdNHX1*, and genes related to K^+ transport such as *BdHAK5* and *BdSKOR*, should play key roles in the maintenance of K^+/Na^+ homeostasis in buffalograss leaves under salt stress. Further studies on the function of those genes would help to elucidate the salt tolerance mechanisms employed by buffalograss, and, thus, provide a theoretical basis for the cultivation of this species in salinized areas.

Supplementary Materials: The following supporting information can be downloaded at: https://www.mdpi.com/article/10.3390/plants12132459/s1. Table S1: The primers for qRT-PCR analysis; Table S2: The relative expression level of genes related to photosynthesis in *B. dactyloides* under 0 (Control), 50 and 200 mM NaCl for 6 h using qRT-PCR; Table S3: The relative expression level of aquaporin genes in the roots and leaves of *B. dactyloides* under 0 (Control), 50 and 200 mM NaCl for 6 h using qRT-PCR; Table S4: The relative expression level of genes related to Na^+ and K^+ transport in the roots of *B. dactyloides* under 0 (Control), 50 and 200 mM NaCl for 6 h using qRT-PCR.

Author Contributions: Conceptualization, Y.X. and H.G.; Data curation, H.G., Y.C., Z.L. and C.N.; Formal analysis, Z.L. and C.N.; Funding acquisition, H.G. and Y.X.; Investigation, H.G. and Y.C.; Methodology, H.G.; Project administration, Y.X. and H.G.; Resources, H.G.; Software, H.G., Y.C., Z.L. and C.N.; Supervision, H.G., Y.X. and T.H.; Validation, H.G. and Y.C.; Writing—original draft, H.G.; Writing—review and editing, Y.C., Y.X. and T.H. All authors have read and agreed to the published version of the manuscript.

Funding: This work was supported by the National Natural Science Foundation of China (No.32101253, No.32071888), the China Postdoctoral Science Foundation (No.2022M712610) and Shaanxi Province Key Research and Development Program General Project (No.2023-YBNY-046).

Institutional Review Board Statement: Not applicable.

Informed Consent Statement: Not applicable.

Data Availability Statement: The data that support the findings of this study are available on request from the corresponding author.

Conflicts of Interest: The authors declare no conflict of interest.

References

1. Zhao, C.; Zhang, H.; Song, C.; Zhu, J.K.; Shabala, S. Mechanisms of plant responses and adaptation to soil salinity. *Innovation* **2020**, *1*, 100017. [CrossRef] [PubMed]
2. Aroca, R.; Ruiz-Lozano, J.M.; Zamarreño, Á.M.; Paz, J.A.; García-Mina, J.M.; Pozo, M.J.; López-Ráez, J.A. Arbuscular mycorrhizal symbiosis influences strigolactone production under salinity and alleviates salt stress in lettuce plants. *J. Plant Physiol.* **2013**, *170*, 47–55. [CrossRef] [PubMed]
3. Deinlein, U.; Stephan, A.B.; Horie, T.; Luo, W.; Xu, G.; Schroeder, J.I. Plant salt-tolerance mechanisms. *Trends Plant Sci.* **2014**, *19*, 371–379. [CrossRef] [PubMed]
4. Nouri, M.Z.; Moumeni, A.; Komatsu, S. Abiotic stresses: Insight into gene regulation and protein expression in photosynthetic pathways of plants. *Int. J. Mol. Sci.* **2015**, *16*, 20392–20416. [CrossRef]
5. Chaves, M.M.; Costa, J.M.; Zarrouka, O.; Pinheiro, C.; Lopes, C.M.; Pereira, J.S. Controlling stomatal aperture in semi-arid regions: The dilemma of saving water or being cool? *Plant Sci.* **2016**, *251*, 54–64. [CrossRef]
6. Chaves, M.M.; Flexas, J.; Pinheiro, C. Photosynthesis under drought and salt stress: Regulation mechanisms from whole plant to cell. *Ann. Bot.* **2009**, *103*, 551–560. [CrossRef]
7. Sun, Z.W.; Ren, L.K.; Fan, J.W.; Li, Q.; Wang, K.J.; Guo, M.M.; Wang, L.; Li, J.; Zhang, G.X.; Yang, Z.Y.; et al. Salt response of photosynthetic electron transport system in wheat cultivars with contrasting tolerance. *Plant Soil Environ.* **2016**, *62*, 515–521. [CrossRef]
8. Furbank, R.T. Evolution of the C_4 photosynthetic mechanism: Are there really three C_4 acid decarboxylation types? *J. Exp. Bot.* **2011**, *62*, 3103–3108. [CrossRef]
9. Nevo, R.; Charuvi, D.; Tsabari, O.; Reich, Z. Composition, architecture and dynamics of the photosynthetic apparatus in higher plants. *Plant J.* **2012**, *70*, 157–176. [CrossRef]
10. Om, K.; Arias, N.N.; Jambor, C.C.; MacGregor, A.; Rezachek, A.N.; Haugrud, C.; Kunz, H.-H.; Wang, Z.; Huang, P.; Zhang, Q.; et al. Pyruvate, phosphate dikinase regulatory protein impacts light response of C_4 photosynthesis in *Setaria viridis*. *Plant Physiol.* **2022**, *190*, 1117–1133. [CrossRef]
11. Flowers, T.J.; Colmer, T.D. Salinity tolerance in halophytes. *New Phytol.* **2008**, *179*, 945–963. [CrossRef] [PubMed]
12. Tang, X.; Mu, X.; Shao, H.; Wang, H.; Brestic, M. Global plant-responding mechanisms to salt stress: Physiological and molecular levels and implications in biotechnology. *Crit. Rev. Biotechnol.* **2014**, *35*, 425–437. [CrossRef] [PubMed]
13. Yeo, A.R.; Lee, K.S.; Izard, P.; Boursier, P.J.; Flowers, T.J. Short and long-term effects of salinity on leaf growth in rice (*Oryza sativa* L.). *J. Exp. Bot.* **1991**, *42*, 881–889. [CrossRef]
14. Fricke, W.; Peters, W.S. The biophysics of leaf growth in salt-stressed barley. A study at the cell level. *Plant Physiol.* **2002**, *129*, 374–388. [CrossRef]
15. Taleisnik, E.; Rodríguez, A.A.; Bustos, D.; Erdei, L.; Ortega, L.; Senn, M.E. Leaf expansion in grasses under salt stress. *J. Plant Physiol.* **2009**, *166*, 1123–1140. [CrossRef]
16. Ma, Q.; Yue, L.J.; Zhang, J.L.; Wu, G.Q.; Bao, A.K.; Wang, S.M. Sodium chloride improves photosynthesis and water status in the succulent xerophyte *Zygophyllum xanthoxylum*. *Tree Physiol.* **2012**, *32*, 4–13. [CrossRef]
17. Pan, Y.Q.; Guo, H.; Wang, S.M.; Zhao, B.; Zhang, J.L.; Ma, Q.; Yin, H.J.; Bao, A.K. The photosynthesis, Na^+/K^+ homeostasis and osmotic adjustment of *Atriplex canescens* in response to salinity. *Front. Plant Sci.* **2016**, *7*, 848. [CrossRef]
18. Chaumont, F.; Tyerman, S.D. Aquaporins: Highly regulated channels controlling plant water relations. *Plant Physiol.* **2014**, *164*, 1600–1618. [CrossRef]
19. Hussain, S.; Ashraf, U.; Ali, M.F.; Zulfiqar, U.; Khaliq, A. Aquaporins: A Potential Weapon in Plants for Abiotic Stress Tolerance. In *Transporters and Plant Osmotic Stress*; Elsevier: Amsterdam, The Netherlands, 2021; pp. 63–76. [CrossRef]
20. Banerjee, A.; Roychoudhury, A. The Role of Aquaporins during Plant Abiotic Stress Responses. In *Plant Life Under Changing Environment*; Elsevier: Amsterdam, The Netherlands, 2020; pp. 643–661. [CrossRef]
21. Shafqat, W.; Jaskani, M.J.; Maqbool, R.; Chattha, W.S.; Ali, Z.; Naqvi, S.A.; Haider, M.S.; Khan, I.A.; Vincent, C.I. Heat shock protein and aquaporin expression enhance water conserving behavior of citrus under water deficits and high temperature conditions. *Environ. Exp. Bot.* **2021**, *181*, 104270. [CrossRef]
22. Ahmad, P.; Azooz, M.M.; Prasad, M.N.V. Salt Tolerance in Rice: Present Scenario and Future Prospects. In *Ecophysiology and Responses of Plants under Salt Stress*; Springer: Berlin/Heidelberg, Germany, 2013; pp. 203–211. [CrossRef]
23. Porcel, R.; Aroca, R.; Azcon, R.; Ruiz-Lozano, J.M. Regulation of cation transporter genes by the arbuscular mycorrhizal symbiosis in rice plants subjected to salinity suggests improved salt tolerance due to reduced Na^+ root-to-shoot distribution. *Mycorrhiza* **2016**, *26*, 673–684. [CrossRef]
24. Flowers, T.J.; Munns, R.; Colmer, T.D. Sodium chloride toxicity and the cellular basis of salt tolerance in halophytes. *Ann. Bot.* **2015**, *115*, 419–431. [CrossRef] [PubMed]
25. Shabala, S.; Cuin, T.A. Potassium transport and plant salt tolerance. *Physiol. Plant.* **2008**, *133*, 651–669. [CrossRef] [PubMed]

26. Wang, Y.; Wu, W.H. Genetic approaches for improvement of the crop potassium acquisition and utilization efficiency. *Curr. Opin. Plant Biol.* **2015**, *25*, 46–52. [CrossRef]
27. Munns, R.; Tester, M. Mechanisms of Salinity Tolerance. *Annu. Rev. Plant Biol.* **2008**, *59*, 651–681. [CrossRef] [PubMed]
28. Chen, J.; Zhang, H.; Zhang, X.; Tang, M. Arbuscular mycorrhizal symbiosis alleviates salt stress in black locust through improved photosynthesis, water status, and K^+/Na^+ homeostasis. *Front. Plant Sci.* **2017**, *8*, 1739. [CrossRef] [PubMed]
29. Shi, H.Z.; Ishitani, M.; Kim, C.; Zhu, J.K. The *Arabidopsis thaliana* salt tolerance gene SOS1 encodes a putative Na^+/H^+ antiporter. *Proc. Natl. Acad. Sci. USA* **2000**, *97*, 6896–6901. [CrossRef]
30. Fukuda, A.; Nakamura, A.; Tagiri, A.; Tanaka, H.; Miyao, A.; Hirochika, H.; Yoshiyuki, T. Function, intracellular localization and the importance in salt tolerance of a vacuolar Na^+/H^+ antiporter from rice. *Plant Cell Physiol.* **2004**, *45*, 146–159. [CrossRef]
31. Nieves-Cordones, M.; Rodenas, R.; Chavanieu, A.; Rivero, R.M.; Martinez, V.; Gaillard, I.; Rubio, F. Uneven HAK/KUP/KT protein diversity among angiosperms: Species distribution and perspectives. *Front. Plant Sci.* **2016**, *7*, 1–7. [CrossRef]
32. Assaha, D.V.; Ueda, A.; Saneoka, H.; Al-Yahyai, R.; Yaish, M.W. The role of Na^+ and K^+ transporters in salt stress adaptation in glycophytes. *Front. Physiol.* **2017**, *8*, 509. [CrossRef]
33. Marcum, K.B. Use of saline and non-potable water in the turfgrass industry: Constraints and developments. *Agric. Water Manag.* **2006**, *80*, 132–146. [CrossRef]
34. Qian, Y.Q.; Luo, D.; Gong, G.; Han, L.; Ju, G.S.; Sun, Z.Y. Effects of spatial scale of soil heterogeneity on the growth of a clonal plant producing both spreading and clumping ramets. *J. Plant Growth Regul.* **2014**, *33*, 214–221. [CrossRef]
35. Li, W.; Qian, Y.Q.; Han, L.; Liu, J.X.; Li, Z.J.; Ju, G.S.; Sun, Z.Y. Validation of candidate reference genes for gene expression normalization in *Buchloe dactyloides* using quantitative real-time RT-PCR. *Sci. Hortic-Amsterdam* **2015**, *197*, 99–106. [CrossRef]
36. Zhang, P.; Yang, P.; Zhang, Z.; Han, B.; Wang, W.; Wang, Y.; Cao, Y.; Hu, T. Isolation and characterization of a buffalograss (*Buchloe dactyloides*) dehydration responsive element binding transcription factor, BdDREB2. *Gene* **2014**, *536*, 123–128. [CrossRef] [PubMed]
37. Wu, F.; Chen, J.; Wang, J.; Wang, X.; Lu, Y.; Ning, Y.; Li, Y. Intra-population genetic diversity of *Buchloe dactyloides* (Nutt.) Engelm (buffalograss) determined using morphological traits and sequence-related amplified polymorphism markers. *3 Biotech* **2019**, *9*, 97. [CrossRef]
38. Alderman, E.J.; Hoyle, J.A.; Braun, R.C.; Reeves, J.A. Winter golf cart traffic, colorants, and overseeding influence winter green cover and spring recovery of buffalograss. *Crop Forage Turfgrass Manag.* **2020**, *6*, e20077. [CrossRef]
39. Hadle, J.J.; Russell, F.L.; Beck, J.B. Are buffalograss (*Buchloë dactyloides*) cytotypes spatially and ecologically differentiated? *Am. J. Bot.* **2019**, *106*, 1116–1125. [CrossRef]
40. Reid, S.D.; Koski, A.J.; Hughes, H.G. Buffalograss seedling screening invitro for Na chloride tolerance. *Hortic. Sci.* **1993**, *28*, 536.
41. Munns, R.; James, R.A.; Läuchli, A. Approaches to increasing the salt tolerance of wheat and other cereals. *J. Exp. Bot.* **2006**, *57*, 1025–1043. [CrossRef]
42. Song, J.; Wang, B. Using euhalophytes to understand salt tolerance and to develop saline agriculture: *Suaeda salsa* as a promising model. *Ann. Bot.* **2015**, *115*, 541–553. [CrossRef]
43. Guo, R.; Shi, L.; Yang, Y. Germination, growth, osmotic adjustment and ionic balance of wheat in response to saline and alkaline stresses. Soil Sci. *Plant Nutr.* **2009**, *55*, 667–679. [CrossRef]
44. Turan, S.; Tripathy, B.C. Salt-stress induced modulation of chlorophyll biosynthesis during de-etiolation of rice seedlings. *Physiol. Plantarum* **2015**, *153*, 477–491. [CrossRef] [PubMed]
45. Mutlu, S.S.; Atesoglu, H.; Selim, C.; Tokgöz, S. Effects of 24-epibrassinolide application on cool-season turfgrass growth and quality under salt stress. *Grassl. Sci.* **2017**, *63*, 61–65. [CrossRef]
46. Bose, J.; Munns, R.; Shabala, S.; Gilliham, M.; Pogson, B.; Tyerman, S. Chloroplast function and ion regulation in plants growing on saline soils: Lessons from halophytes. *J. Exp. Bot.* **2017**, *68*, 3129–3143. [CrossRef]
47. Hedrich, R.; Shabala, S. Stomata in a saline world. *Curr. Opin. Plant Biol.* **2018**, *46*, 87–95. [CrossRef] [PubMed]
48. Wang, Q.; Xie, W.; Xing, H.; Yan, J.; Meng, X.; Li, X.; Fu, X.; Xu, J.; Lian, X.; Yu, S.; et al. Genetic architecture of natural variation in rice chlorophyll content revealed by a genome-wide association study. *Mol. Plant* **2015**, *8*, 946–957. [CrossRef] [PubMed]
49. Keetman, U.; Mock, H.-P.; Grimm, B. Kinetics of antioxidative defence responses to photosensitisation in porphyrin-accumulating tobacco plants. *Plant Physiol. Bioch.* **2002**, *40*, 697–707. [CrossRef]
50. Li, Y.; Wang, X.; Zhang, Q.; Shen, Y.; Wang, J.; Qi, S.; Zhao, P.; Muhammad, T.; Islam, M.M.; Zhan, X.; et al. A mutation in *SlCHLH* encoding a magnesium chelatase H subunit is involved in the formation of yellow stigma in tomato (*Solanum lycopersicum* L.). *Plant Sci.* **2022**, *325*, 111466. [CrossRef]
51. Shalygo, N.; Czarnecki, O.; Peter, E.; Grimm, B. Expression of chlorophyll synthase is also involved in feedback-control of chlorophyll biosynthesis. *Plant Mol. Biol.* **2009**, *71*, 425–436. [CrossRef]
52. Kato, K.; Tanaka, R.; Sano, S.; Tanaka, A.; Hosaka, H. Identification of a gene essential for protoporphyrinogen IX oxidase activity in the cyanobacterium *Synechocystis* sp. PCC6803. *Proc. Natl. Acad. Sci. USA* **2010**, *107*, 16649–16654. [CrossRef]
53. Ceccarelli, E.A.; Arakaki, A.K.; Cortez, N.; Carrillo, N. Functional plasticity and catalytic efficiency in plant and bacterial ferredoxin-NADP (H) reductases. *Biochim. Biophys. Acta* **2004**, *1698*, 155–165. [CrossRef]
54. Wang, X.; Zhai, T.; Zhang, X.; Tang, C.; Zhuang, R.; Zhao, H.; Xu, Q.; Cheng, Y.; Wang, J.; Duplessis, S.; et al. Two stripe rust effectors impair wheat resistance by suppressing import of host Fe-S protein into chloroplasts. *Plant Physiol.* **2021**, *187*, 2530–2543. [CrossRef] [PubMed]

55. Wang, J.; Gao, H.; Guo, Z.; Meng, Y.; Yang, Q. Adaptation responses in C_4 photosynthesis of sweet maize (*Zea mays* L.) exposed to nicosulfuron. *Ecotoxicol. Environ. Safety* **2021**, *214*, 112096. [CrossRef] [PubMed]

56. Luu, D.T.; Maurel, C. Aquaporins in a challenging environment: Molecular gears for adjusting plant water status. *Plant Cell Environ.* **2005**, *28*, 85–96. [CrossRef]

57. Ahmed, S.; Kouser, S.; Asgher, M.; Gandhi, S.G. Plant aquaporins: A frontward to make crop plants drought resistant. *Physiol. Plantarum* **2021**, *172*, 1089–1105. [CrossRef] [PubMed]

58. Jang, J.Y.; Kim, D.G.; Kim, Y.O.; Kim, J.S.; Kang, H. An expression analysis of a gene family encoding plasma membrane aquaporins in response to abiotic stresses in *Arabidopsis thaliana*. *Plant Mol. Biol.* **2004**, *54*, 713–725. [CrossRef]

59. He, F.; Zhang, H.; Tang, M. Aquaporin gene expression and physiological responses of *Robinia pseudoacacia* L. to the mycorrhizal fungus *Rhizophagus irregularis* and drought stress. *Mycorrhiza* **2016**, *26*, 311–323. [CrossRef]

60. Wang, L.; Li, Q.; Lei, Q.; Feng, C.; Gao, Y.; Zheng, X.; Zhao, Y.; Wang, Z.; Kong, J. MzPIP2;1: An aquaporin involved in radial water movement in both water uptake and transportation, altered the drought and salt tolerance of transgenic Arabidopsis. *PLoS ONE* **2015**, *10*, e0142446. [CrossRef]

61. Liu, S.; Fukumoto, T.; Gena, P.; Feng, P.; Sun, Q.; Li, Q.; Matsumoto, T.; Kaneko, T.; Zhang, H.; Zhang, Y.; et al. Ectopic expression of a rice plasma membrane intrinsic protein (OsPIP1;3) promotes plant growth and water uptake. *Plant J.* **2020**, *102*, 779–796. [CrossRef]

62. Gattolin, S.; Sorieul, M.; Hunter, P.R.; Khonsari, R.H.; Frigerio, L. In vivo imaging of the tonoplast intrinsic protein family in Arabidopsis roots. *BMC Plant Biol.* **2009**, *9*, 133. [CrossRef]

63. Boursiac, Y.; Chen, S.; Luu, D.T.; Sorieul, M.; van den Dries, N.; Maurel, C. Early effects of salinity on water transport in Arabidopsis roots. Molecular and cellular features of aquaporin expression. *Plant Physiol.* **2005**, *139*, 790–805. [CrossRef]

64. Maurel, C.; Verdoucq, L.; Luu, D.T.; Santoni, V. Plant aquaporins: Membrane channels with multiple integrated functions. *Annu. Rev. Plant Biol.* **2008**, *59*, 595–624. [CrossRef] [PubMed]

65. Reinhardt, H.; Hachez, C.; Bienert, M.D.; Beebo, A.; Swarup, K.; Voß, U.; Bouhidel, K.; Frigerio, L.; Schjoerring, J.K.; Bennett, M.J.; et al. Tonoplast aquaporins facilitate lateral root emergence. *Plant Physiol.* **2016**, *170*, 1640–1645. [CrossRef] [PubMed]

66. Hanba, Y.T.; Shibasaka, M.; Hayashi, Y.; Hayakawa, T.; Kasamo, K.; Terashima, I.; Katsuhara, M. Overexpression of the barley aquaporin HvPIP2;1 increases internal CO_2 conductance and CO_2 assimilation in the leaves of transgenic rice plants. *Plant Cell Physiol.* **2004**, *45*, 521–529. [CrossRef]

67. Yang, S.; Cui, L. The action of aquaporins in cell elongation, salt stress and photosynthesis. *Chin. J. Biotechnol.* **2009**, *25*, 321–327. [CrossRef]

68. Shen, Y.; Shen, L.; Shen, Z.; Jing, W.; Ge, H.; Zhao, J.; Zhang, W. The potassium transporter OsHAK21 functions in the maintenance of ion homeostasis and tolerance to salt stress in rice. *Plant Cell Environ.* **2015**, *38*, 2766–2779. [CrossRef]

69. Rus, A.; Lee, B.H.; Munoz-Mayor, A.; Sharkhuu, A.; Miura, K.; Zhu, J.K.; Bressan, R.A.; Hasegawa, P.M. AtHKT1 facilitates Na^+ homeostasis and K^+ nutrition in planta. *Plant Physiol.* **2004**, *136*, 2500–2511. [CrossRef]

70. Himabindu, Y.; Chakradhar, T.; Reddy, M.C.; Kanygin, A.; Redding, K.E.; Chandrasekhar, T. Salt-tolerant genes from halophytes are potential key players of salt tolerance in glycophytes. *Environ. Exp. Bot.* **2016**, *124*, 39–63. [CrossRef]

71. Jaime-Pérez, N.; Pineda, B.; García-Sogo, B.; Atares, A.; Athman, A.; Byrt, C.S.; Olías, R.; Asins, M.J.; Gilliham, M.; Moreno, V.; et al. The sodium transporter encoded by the *HKT1;2* gene modulates sodium/potassium homeostasis in tomato shoots under salinity. *Plant Cell Environ.* **2017**, *40*, 658–671. [CrossRef] [PubMed]

72. Han, Q.Q.; Wang, Y.P.; Li, J.; Li, J.; Yin, X.C.; Jiang, X.Y.; Yu, M.; Wang, S.M.; Shabala, S.; Zhang, J.L. The mechanistic basis of sodium exclusion in *Puccinellia tenuiflora* under conditions of salinity and potassium deprivation. *Plant J.* **2022**, *112*, 322–338. [CrossRef]

73. James, R.A.; Blake, C.; Byrt, C.S.; Munns, R. Major genes for Na^+ exclusion, Nax1 and Nax2 (wheat HKT1;4 and HKT1;5), decrease Na^+ accumulation in bread wheat leaves under saline and waterlogged conditions. *J. Exp. Bot.* **2011**, *62*, 2939–2947. [CrossRef]

74. Chen, H.; Chen, X.; Gu, H.; Wu, B.; Zhang, H.; Yuan, X.; Cui, X. GmHKT1;4, a novel soybean gene regulating Na^+/K^+ ratio in roots enhances salt tolerance in transgenic plants. *Plant Growth Regul.* **2014**, *73*, 299–308. [CrossRef]

75. Wu, G.Q.; Xi, J.J.; Wang, Q.; Bao, A.K.; Ma, Q.; Zhang, J.L.; Wang, S.M. The *ZxNHX* gene encoding tonoplast Na^+/H^+ antiporter from the xerophyte *Zygophyllum xanthoxylum* plays important roles in response to salt and drought. *J. Plant Physiol.* **2011**, *168*, 758–767. [CrossRef] [PubMed]

76. Sahoo, D.P.; Kumar, S.; Mishra, S.; Kobayashi, Y.; Panda, S.K.; Sahoo, L. Enhanced salinity tolerance in transgenic mungbean overexpressing Arabidopsis antiporter (NHX1) gene. *Mol. Breeding* **2016**, *36*, 144. [CrossRef]

77. Rubio, F.; Nieves-Cordones, M.; Alemán, F.; Martínez, V. Relative contribution of AtHAK5 and AtAKT1 to K^+ uptake in the high-affinity range of concentrations. *Physiol. Plantarum* **2008**, *134*, 598–608. [CrossRef]

78. Nieves-Cordones, M.; Alemán, F.; Martínez, V.; Rubio, F. The *Arabidopsis thaliana* HAK5 K^+ transporter is required for plant growth and K^+ acquisition from low K^+ solutions under saline conditions. *Mol. Plant* **2010**, *3*, 326–333. [CrossRef] [PubMed]

79. Liu, B.; Zhang, X.; You, X.; Li, Y.; Long, S.; Wen, S.; Liu, Q.; Liu, T.; Guo, H.; Xu, Y. Hydrogen sulfide improves tall fescue photosynthesis response to low-light stress by regulating chlorophyll and carotenoid metabolisms. *Plant Physiol. Bioch.* **2022**, *170*, 133–145. [CrossRef] [PubMed]

80. Gibon, Y.; Bessieres, M.A.; Larher, F. Is glycine betaine a noncompatible solute in higher plants that do not accumulate it? *Plant Cell Environ.* **1997**, *20*, 329–340. [CrossRef]

81. Lichtenthaler, H.K. Chlorophylls and carotenoids: Pigments of photosynthetic biomembranes. *Methods Enzymol.* **1987**, *148*, 350–382. [CrossRef]
82. Arnon, D.I. Copper enzymes in isolated chloroplasts. Polyphenoloxidase in *Beta vulgaris*. *Plant Physiol.* **1949**, *24*, 1–15. [CrossRef]
83. Liu, F.; Andersen, M.N.; Jacobsen, S.E.; Jensen, C.R. Stomatal control and water use efficiency of soybean (*Glycine max* L. Merr.) during progressive soil drying. *Environ. Exp. Bot.* **2005**, *54*, 33–40. [CrossRef]
84. Zhang, X.; Huang, G.; Bian, X.; Zhao, Q. Effects of root interaction and nitrogen fertilization on the chlorophyll content, root activity, photosynthetic characteristics of intercropped soybean and microbial quantity in the rhizosphere. *Plant Soil Environ.* **2013**, *59*, 80–88. [CrossRef]
85. Duan, H.R.; Ma, Q.; Zhang, J.L.; Hu, J.; Bao, A.K.; Wei, L.; Wang, Q.; Luan, S.; Wang, S.M. The inward-rectifying K^+ channel SsAKT1 is a candidate involved in K^+ uptake in the halophyte *Suaeda salsa* under saline condition. *Plant Soil* **2015**, *395*, 173–187. [CrossRef]

Article

Transcriptome Analysis Reveals the Stress Tolerance Mechanisms of Cadmium in *Zoysia japonica*

Yi Xu [1,2], Yonglong Li [1], Yan Li [1], Chenyuan Zhai [1] and Kun Zhang [1,*]

[1] College of Grassland Science, Qingdao Agricultural University, Qingdao 266109, China; 20200202793@stu.qau.edu.cn (Y.X.); 20222115004@stu.qau.edu.cn (Y.L.); 20212103050@stu.qau.edu.cn (Y.L.); 20222215010@stu.qau.edu.cn (C.Z.)

[2] College of Life Sciences, Qingdao Agricultural University, Qingdao 266109, China

* Correspondence: zk61603@163.com

Abstract: Cadmium (Cd) is a severe heavy metal pollutant globally. *Zoysia japonica* is an important perennial warm-season turf grass that potentially plays a role in phytoremediation in Cd-polluted soil areas; however, the molecular mechanisms underlying its Cd stress response are unknown. To further investigate the early gene response pattern in *Z. japonica* under Cd stress, plant leaves were harvested 0, 6, 12, and 24 h after Cd stress (400 μM CdCl$_2$) treatment and used for a time-course RNA-sequencing analysis. Twelve cDNA libraries were constructed and sequenced, and high-quality data were obtained, whose mapped rates were all higher than 94%, and more than 601 million bp of sequence were generated. A total of 5321, 6526, and 4016 differentially expressed genes were identified 6, 12, and 24 h after Cd stress treatment, respectively. A total of 1660 genes were differentially expressed at the three time points, and their gene expression profiles over time were elucidated. Based on the analysis of these genes, the important mechanisms for the Cd stress response in *Z. japonica* were identified. Specific genes participating in glutathione metabolism, plant hormone signal and transduction, members of protein processing in the endoplasmic reticulum, transporter proteins, transcription factors, and carbohydrate metabolism pathways were further analyzed in detail. These genes may contribute to the improvement of Cd tolerance in *Z. japonica*. In addition, some candidate genes were highlighted for future studies on Cd stress resistance in *Z. japonica* and other plants. Our results illustrate the early gene expression response of *Z. japonica* leaves to Cd and provide some new understanding of the molecular mechanisms of Cd stress in *Zosia* and *Gramineae* species.

Keywords: transcriptome; cadmium; *Zoysia japonica*; differentially expressed genes; molecular mechanism

Citation: Xu, Y.; Li, Y.; Li, Y.; Zhai, C.; Zhang, K. Transcriptome Analysis Reveals the Stress Tolerance Mechanisms of Cadmium in *Zoysia japonica*. *Plants* **2023**, *12*, 3833. https://doi.org/10.3390/plants12223833

Academic Editors: Zhou Li and David Jespersen

Received: 18 October 2023
Revised: 9 November 2023
Accepted: 10 November 2023
Published: 12 November 2023

1. Introduction

With industrial development, Cadmium (Cd) has become a major heavy metal pollutant and extremely toxic to plants and humans [1]. Cd ions are often taken up into the airflow layer and released into the soil through atmospheric sedimentation, rainfall, and snowfall [2]. In addition, agricultural sewage irrigation, which contains a Cd solution, also increases the soil Cd ion content [3]. Previous strategies to remediate heavy metal pollution have mainly focused on physical and chemical methods [4]. However, these methods are neither durable nor reliable. Nowadays, phytoremediation is an eco-friendly approach that could remediate soil Cd contamination. It has multiple benefits, including a low cost, safety, reliability, and minimal environmental interference [5,6]. Actually, because of the food security problem for humans, not all plants are suitable for phytoremediation under Cd contamination aera. However, turfgrass could be a good candidate plant for phytoremediation in heavy metal-contaminated regions with great degradation ability and ornamental value [7,8].

Cd is an ion element that is nonessential for the growth and development of plants. Many studies have shown that Cd absorption can induce serious cellular damage, and that multiple physiological and metabolic processes interfere with plant growth [9]. Plants have evolved many detoxification mechanisms to reduce or resist Cd toxicity, including Cd ion absorption, chelation with metabolites, vacuolar storage, the activation of antioxidant systems, and related gene expression regulation networks [10]. When Cd enters the cell, cell plasma membrane proteins, such as ZRT, the IRT-like protein family protein, the natural resistance-associated macrophage protein, and yellow stripe-like (YSL) proteins, limit intracellular and extracellular Cd transportation [11]. The ATP-binding cassette transporter and heavy metal ATPase families of genes can be expressed in vacuoles, and Cd ions can be compartmentalized in roots and leaves to reduce Cd ion hyperaccumulation and protect plants [12]. Metallothionein, phytochelatin, glutathione, and pectin are the major heavy metal chelators in plants [13]. When the Cd ion concentration in the cytoplasm increases, these metabolites fix the Cd ions to chelate them in vacuoles or cell walls. In addition, antioxidant systems, including superoxide dismutase, catalase, ascorbate peroxidase, glutathione peroxidase, dehydroascorbate reductase, and glutathione reductase, play important roles in scavenging reactive oxygen species during Cd stress [14]. Meanwhile, the above-mentioned physiological and metabolic processes, which are regulated by a complex network system involving a large number of genes, are the core components of the plant Cd stress detoxification [15].

Transcription factors play an important regulatory role in plant development and their response to various environmental stresses. Under Cd stress conditions, transcription factors participate in the plant Cd stress response and tolerance by controlling the expression of downstream genes [16]. Numerous transcription factors, such as bHLH, MYB, and WRKY, have been reported to be involved in plant Cd stress tolerance. For example, *AtMYB4* overexpression can improve plant Cd resistance by specifically activating phytochelatin synthases and metallothionein 1C gene expression and promoting phytochelatin and metallothionein synthesis for plant detoxification [17]. Cd stress can upregulate the expression of the OBP3-responsive gene (ORG), which is a bHLH transcription factor in soybean (*Glycine max*), and *GmORG3* overexpression enhances cadmium tolerance via increasing iron and reducing cadmium uptake from roots to shoots [18]. Cd stress could also induce the expression of *WRKY45*. Overexpressing *WRKY45* confers Cd tolerance via promoting the expression of phytochelatin-synthesis-related genes, *PCS1* and *PCS2*, respectively, in Arabidopsis [19]. In addition, other transcription factors have been reported to play important roles in the signal transduction of a Cd stress response by regulating a series of genes involved in Cd uptake, transport, and tolerance in different plant such as tea plants (*Camellia sinensis*) [20], muskmelon [21], and the golden rain tree (*Koelreuteria paniculata*) [22].

Zoysia japonica is an important perennial warm-season turf grass with a large biomass and excellent abiotic stress resistance [23]. A previous study has shown the potential ability of *Z. japonica* as a bioindicator in heavy metal pollution areas [24], but its molecular mechanisms regarding Cd tolerance are not yet well understood, which provides a limitation for *Z. japonica* utilization in soil remediation in Cd-contaminated areas. In the present study, we characterized the differentially expressed genes in *Z. japonica* leaves under Cd stress at different time points and explored the tolerance mechanisms of *Z. japonica* in response to Cd toxicity. Our results are an important addition to plant Cd stress molecular mechanism studies and may contribute to improvements in the phytoremediation and environmental protection of soils polluted with heavy metals. Moreover, the novel candidate genes identified in our study may serve as molecular markers for further improvement of Cd stress tolerance in *Zoysia* and other grass species.

2. Results

2.1. Transcriptome Analysis in Z. japonica Leaves after Cd Treatment

In the present study, *Z. japonica* leaves were harvested 0, 6, 12, and 24 h after Cd stress treatment and subjected to RNA sequencing (RNA-seq) analysis. Twelve cDNA libraries were constructed from *Z. japonica* leaves and sequenced following Cd stress treatment. The RNA-seq data were uploaded to the NCBI SRA database (no. PRJNA1008854). For each library, 39,850,416–46,067,222 and 39,253,016–45,438,370 bp of raw and clean reads were generated, respectively (Table 1). The total mapped rate was higher than 94%, with more than 6,017,412,816 bp of sequence. The Q20 and Q30 values exceeded 97% and 93%, respectively. In addition, a strong correlation was observed between the different samples (Figure S1). These results indicated that the RNA-seq data were of high quality.

Table 1. Summary of RNA-seq results in *Z. japonica* under Cd stress treatment.

Sample	Reads No.	Clean Reads No.	Total Mapped	Bases (bp)	Q20 (%)	Q30 (%)
L_0h_1	46,067,222	45,438,370	43,480,195 (95.69%)	6,956,150,522	98.00	94.14
L_0h_2	41,800,042	41,147,778	39,405,918 (95.77%)	6,311,806,342	97.74	93.55
L_0h_3	43,795,346	43,206,630	41,483,551 (96.01%)	6,613,097,246	97.92	93.89
L_6h_1	45,520,700	44,950,190	42,576,734 (94.72%)	6,873,625,700	98.01	94.14
L_6h_2	44,767,754	44,113,820	42,032,965 (95.28%)	6,759,930,854	97.90	93.95
L_6h_3	40,830,954	40,260,512	38,151,253 (94.76%)	6,165,474,054	97.97	94.08
L_12h_1	39,850,416	39,253,016	37,335,934 (95.12%)	6,017,412,816	97.80	93.64
L_12h_2	44,664,070	43,988,038	41,964,721 (95.40%)	6,744,274,570	97.78	93.59
L_12h_3	45,168,508	44,507,012	42,304,429 (95.05%)	6,820,444,708	97.93	94.02
L_24h_1	40,804,266	40,169,930	38,163,038 (95.00%)	6,161,444,166	97.81	93.71
L_24h_2	44,778,440	44,073,644	41,890,835 (95.05%)	6,761,544,440	97.90	93.99
L_24h_3	44,738,228	44,142,912	42,173,022 (95.54%)	6,755,472,428	98.02	94.19

2.2. Differential Gene Expression Analysis in Z. japonica Leaves under Cd Treatment

Differentially expressed genes were screened using the standard cut-off metrics of $|\log2$ fold change$| > 1$ and a p-value < 0.05. Volcano plots identified 5321, 6526, and 4016 genes expressed 6, 12, and 24 h after Cd stress treatment, respectively (Figure 1). This indicates that a strong induction of gene expression occurred 12 h after Cd stress treatment. Between L_6 h (leaves sampled 6 h after Cd treatment) and L_0 h (control), 2242 and 3097 upregulated and downregulated genes, respectively, were identified (Figure 1A); between L_12 h (leaves sampled 12 h after Cd treatment) and L_0 h, 2605 and 3921 upregulated and downregulated genes, respectively, were identified (Figure 1B); and between L_24 h (leaves sampled 24 h after Cd treatment) and L_0 h, 1450 and 2566 upregulated and downregulated genes, respectively, were identified (Figure 1C).

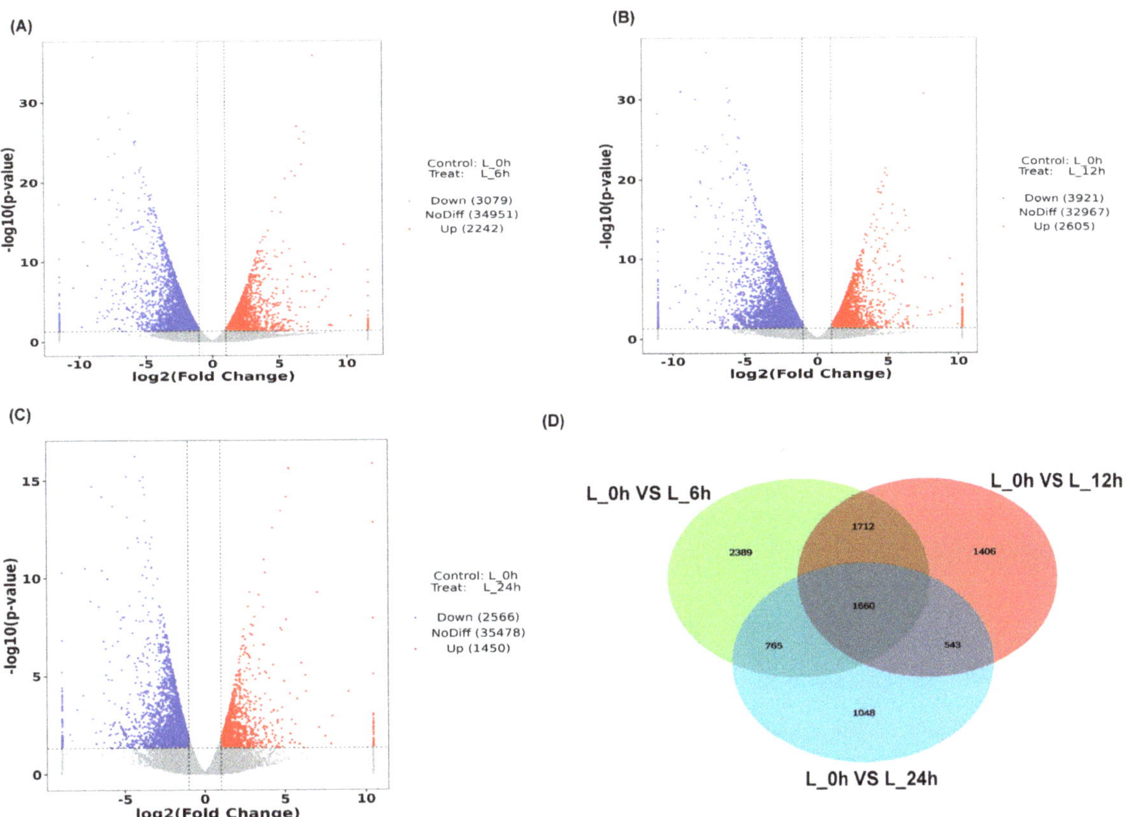

Figure 1. The profile of differentially expressed genes at different time points. (**A**) Volcano plot of the L_6 h vs. L_0 h group; (**B**) Volcano plot of the L_12 h vs. L_0 h group; (**C**) Volcano plot of the L_24 h vs. L_0 h group; (**D**) Venn diagram for all differentially expressed genes.

For L_6 h vs. L_0 h group, a gene ontology (GO) analysis showed that the chloroplast, plastid, thylakoid, plastid envelope, plastid stroma, plastid thylakoid, chloroplast thylakoid, chloroplast stroma, plastid membrane, and chloroplast thylakoid membrane were the top ten GO terms in the cellular component category ($p < 0.05$). Photosynthesis, photosynthesis light reaction, photosynthetic electron transport chain, plastid organization, generation of precursor metabolites and energy, chloroplast organization, response to abiotic stimulus, response to light intensity, electron transport chain, and photosynthetic electron transport in photosystem I were the top ten GO terms in biological processes. For the molecular function category, phosphoenolpyruvate carboxykinase activity, protein-domain-specific binding, snoRNA binding, sugar-phosphatase activity, inorganic molecular entity transmembrane transporter activity, starch synthase activity, carbon–carbon lyase activity, inorganic diphosphate phosphatase activity, fructose 1,6-bisphosphate 1-phosphatase activity, and transmembrane transporter activity were highly enriched in the *Z. japonica* leaves under Cd stress treatment (Figure S2A).

For the L_12 h vs. L_0 h group GO analysis, the plastid stroma, chloroplast stroma, chloroplast, plastid, plastid thylakoid membrane, chloroplast thylakoid membrane, thylakoid, photosynthetic membrane, thylakoid membrane, and plastid envelope were the top ten GO terms in the cellular component category ($p < 0.05$). Photosynthesis, the response to abiotic stimulus, photosynthesis light reaction, glucose metabolic process, gluconeogenesis, hexose biosynthetic process, response to karrikin, response to light stimulus, hexose metabolic process, and rRNA metabolic process were the top ten GO terms in the biological process category. For the molecular function category, snoRNA binding, phosphoenolpyruvate carboxykinase activity, fructose 1,6-bisphosphate 1-phosphatase activity, oxidoreductase activity, carbon–carbon lyase activity, pigment binding, sugar-phosphatase activity, transmembrane transporter activity, catalytic activity, inorganic molecular entity, and transmembrane transporter activity were highly enriched in *Z. japonica* leaves under Cd stress treatment (Figure S2B).

For the L_24 h vs. L_0 h group GO analysis, the chloroplast, plastid, plastid envelope, plastid stroma, chloroplast stroma, thylakoid, organelle envelope, envelope, chloroplast envelope, and plastid membrane were the top ten GO terms in the cellular component category ($p < 0.05$). Photosynthesis, the photosynthesis light reaction, plastid organization, chloroplast organization, generation of precursor metabolites and energy, glucose metabolic process, starch metabolic process, photosynthetic electron transport chain, cellular glucan metabolic process, and hexose biosynthetic process were the top ten GO terms in the biological process category. For the molecular function category, phosphoenolpyruvate carboxykinase activity, fructose 1,6-bisphosphate 1-phosphatase activity, sugar-phosphatase activity, transmembrane transporter activity, inorganic molecular entity transmembrane, transporter activity, carbon–carbon lyase activity, oxidoreductase activity, oxidoreductase activity, acting on other nitrogenous compounds as donors, and ion transmembrane transporter activity were highly enriched in the *Z. japonica* leaves under Cd stress treatment (Figure S2C).

For the KEGG pathway analysis, the pathways associated with carbon fixation in photosynthetic organisms; photosynthesis; photosynthesis-antenna proteins; nitrogen metabolism; pyruvate metabolism; glyoxylate and dicarboxylate metabolism; pentose phosphate pathway; glycolysis/gluconeogenesis; glycine, serine, and threonine metabolism; and alanine, aspartate, and glutamate metabolism were the ten most enriched in the L_6 h vs. L_0h group (Figure S3A). For the L_12 h vs. L_0h group, carbon fixation in photosynthetic organisms, photosynthesis-antenna proteins, glycolysis/gluconeogenesis, pyruvate metabolism, ribosome biogenesis in eukaryotes, photosynthesis, pentose phosphate pathway, nitrogen metabolism, flavonoid biosynthesis, and glyoxylate and dicarboxylate metabolism were the ten most enriched KEGG pathways (Figure S3B). For the L_24 h vs. L_0h group, carbon fixation in photosynthetic organisms, starch and sucrose metabolism, pyruvate metabolism, nitrogen metabolism, glycolysis/gluconeogenesis, pentose phosphate pathway, phenylalanine metabolism, photosynthesis, flavonoid biosynthesis, and ubiquinone and other terpenoid-quinone biosyntheses were the ten most enriched KEGG pathways (Figure S3C).

2.3. Common Differentially Expressed Genes Identified in the Time Course Analysis

The Venn diagram in Figure 1D shows that there were 1660 genes differentially expressed in the L_6 h vs. L_0 h, L_12 h vs. L_0 h, and L_24 h vs. L_0 h groups. These common differentially expressed genes play an important role in the response of *Z. japonica* to Cd stress and were chosen for further analysis (Table S2). The variation in the common differentially expressed genes is displayed in Figure 2A. The dynamic expression patterns of the differentially expressed genes were analyzed. The 1660 differentially expressed genes were divided into nine clusters (Figure 2B), including three upregulated clusters (clusters 1, 2, and 3) and four downregulated clusters (clusters 6, 7, 8, 9).

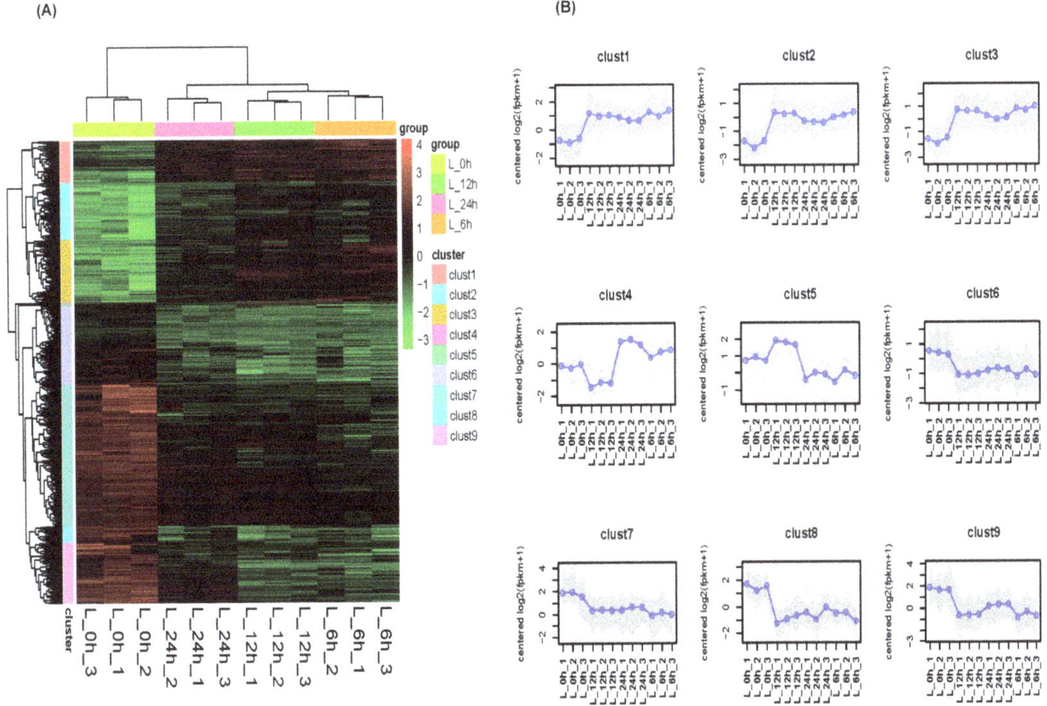

Figure 2. Expression profiles of the common genes at different time points. (**A**) Heatmap of all differentially expressed genes; (**B**) cluster analysis of differentially expressed gene profiles.

The common differentially expressed genes were subjected to a GO analysis (Figure 3A, Table S3). These genes were mainly involved in the chloroplasts, plastids, plastid stroma, chloroplast stroma, plastid envelope, thylakoid, chloroplast thylakoid membrane, organelle envelope, envelope, and plastid thylakoid membrane in the cellular component category ($p < 0.05$). The glucose metabolic process, gluconeogenesis, the hexose biosynthetic process, the hexose metabolic process, the starch metabolic process, photosynthesis, the starch biosynthetic process, the photosynthesis light reaction, the response to light intensity, and the generation of precursor metabolites and energy were the top ten GO terms in the biological process category. In the molecular function category, phosphoenolpyruvate carboxykinase activity, oxidoreductase activity, acting on other nitrogenous compounds as donors, fructose 1,6-bisphosphate 1-phosphatase activity, transmembrane transporter activity, transporter activity, phosphoenolpyruvate carboxykinase (ATP) activity, sugar-phosphatase activity, oxidoreductase activity, cytochrome as an acceptor, and nitrite reductase (NO-forming) activity were highly enriched in the *Z. japonica* leaves under Cd stress treatment. The common differentially expressed genes were also subjected to a KEGG pathway analysis. As shown in Figure 3B and Table S4, the pathways associated with carbon fixation in photosynthetic organisms; nitrogen metabolism; glycolysis/gluconeogenesis; pyruvate metabolism; the pentose phosphate pathway; starch and sucrose metabolism; alanine, aspartate, and glutamate metabolism; monobactam biosynthesis; and galactose metabolism were the top ten significantly enriched pathways in the *Z. japonica* leaves under Cd stress treatment.

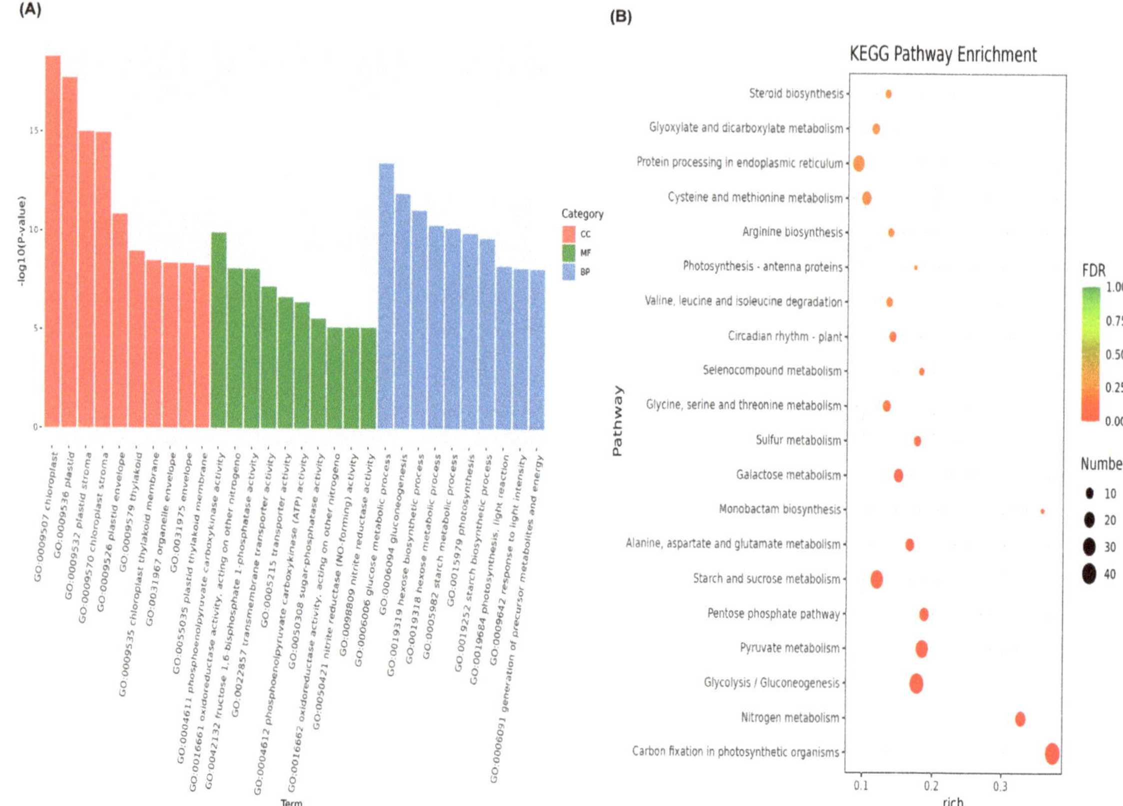

Figure 3. The GO and KEGG enrichment analysis of commonly expressed genes. (**A**) GO analysis; (**B**) KEGG enrichment analysis.

2.4. The Molecular Mechanism of the Response to Cd Treatment in Z. japonica Leaves

Briefly, the molecular mechanism was composited by the following processes. Fourteen glutathione-metabolism-related genes were commonly differentially expressed in the L_6 h vs. L_0 h, L_12 h vs. L_0 h, and L_24 h vs. L_0 h groups (Figure 4A, Table S5). Most of the differentially expressed genes were downregulated. However, one glutathione transferase, *GST23* (Zjn_sc00068.1.g02430.1.am.mkhc) (3.95-, 2.94, and 2.95-fold), and one probable glutathione S-transferase (Zjn_sc00068.1.g02420.1.sm.mk) (4.23-, 3.57-, and 2.03-fold) were upregulated in the L_6 h vs. L_0 h, L_12 h vs. L_0 h, and L_24 h vs. L_0 h groups, respectively, indicating that these two genes have potential roles in Cd chelation mediated by glutathione.

A total of 23 plant hormone-signal-transduction-related genes (10 upregulated and 13 downregulated) were commonly differentially expressed between the L_6 h vs. L_0h, L_12 h vs. L_0 h, and L_24 h vs. L_0 h groups (Figure 4B, Table S6). Among these genes, auxin-responsive proteins (Zjn_sc00007.1.g01680.1.am.mk, Zjn_sc00066.1.g01480.1.sm.mkhc, Zjn_sc00024.1.g02230.1.am.mkhc, Zjn_sc00107.1.g00840.1.sm.mkhc) (*IAA9*, *IAA15*, and *IAA21*), ethylene receptor 3 (Zjn_sc00007.1.g00600.1.am.mkhc), the abscisic acid receptor (Zjn_sc00016.1.g02200.1.sm.mkhc) (*PYL5*), and the probable protein phosphatase 2C 9 (Zjn_sc00048.1.g00330.1.am.mk) were upregulated, suggesting that these hormone signal transduction pathways play roles in Cd detoxification.

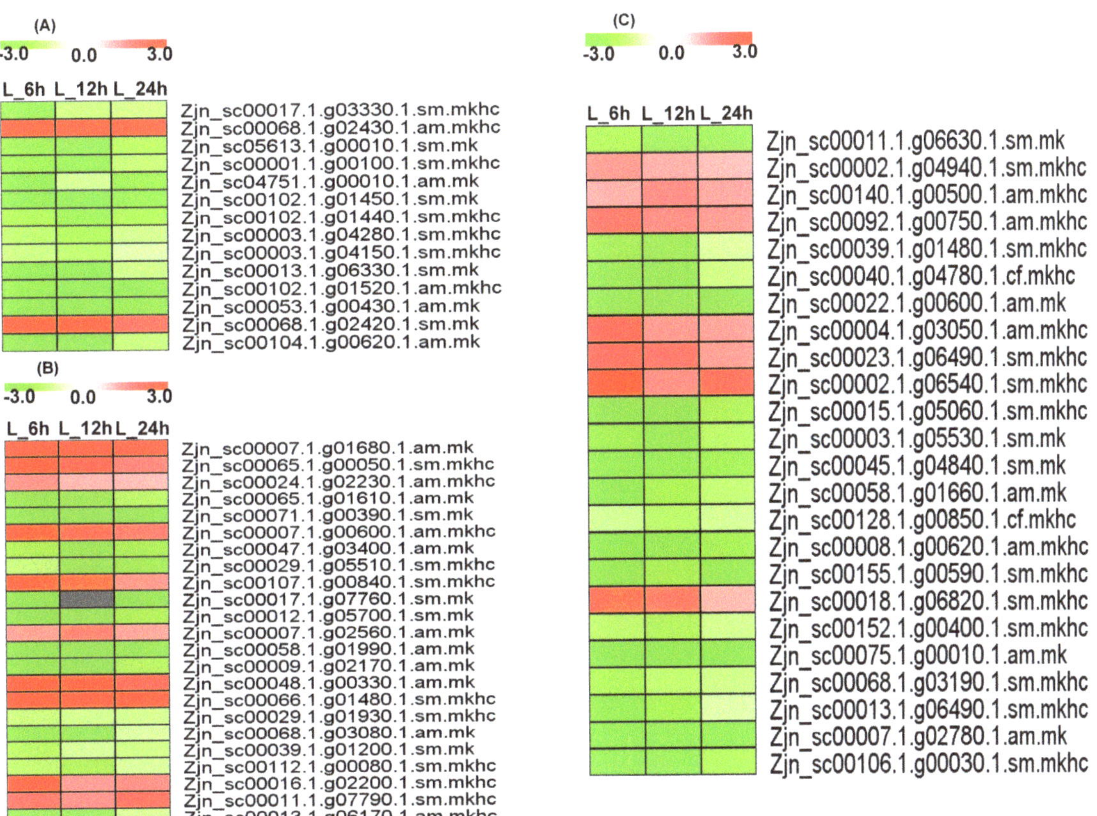

Figure 4. Expression profiles of commonly expressed genes related to glutathione metabolism, plant hormone signal and transduction, and transcription regulation pathways. (**A**) Glutathione metabolism pathway; (**B**) plant hormone signal and transduction pathway; (**C**) Protein processing in endoplasmic reticulum pathway.

Twenty-four differentially expressed genes (7 upregulated and 17 downregulated) were involved in protein processing in the endoplasmic reticulum pathway (Figure 4C, Table S7). These included the probable ubiquitin receptor (Zjn_sc00018.1.g06820.1.sm.mkhc) (*RAD23*); the dolichyl-diphosphooligosaccharide-protein glycosyltransferase subunit (Zjn_sc00023.1.g06490.1.sm.mkhc) (*STT3B*); the eIF-2-alpha kinase (Zjn_sc00004.1.g03050.1.am.mkhc) (*GCN2*); protein disulfide isomerase (Zjn_sc00140.1.g00500.1.am.mkhc), and the eukaryotic translation initiation factor 2 subunit alpha homolog (Zjn_sc00002.1.g04940.1.sm.mkhc), which were upregulated, suggesting that different protein or enzyme metabolism pathways are involved in Cd stress tolerance.

A total of 75 differentially expressed genes (14 upregulated and 61 downregulated) were identified with transport functions (Figure 5A, Table S8), including a sugar transporter ERD6-like 4/5 (Zjn_sc00027.1.g01490.1.sm.mkhc, Zjn_sc00043.1.g01470.1.am.mkhc); probable metal-nicotianamine transporters (Zjn_sc00056.1.g02740.1.sm.mkhc, Zjn_sc00056.1.g02750.1.sm.mkhc, Zjn_sc00056.1.g02760.1.sm.mkhc) (*YSL12*, *YSL13*); a probable mitochondrial adenine nucleotide transporter (Zjn_sc00015.1.g05910.1.sm.mkhc) (*BTL1*); sugar transport protein 14 (Zjn_sc00107.1.g01620.1.sm.mkhc); oligopeptide transporter 1 (Zjn_sc00106.1.g00260.1.am.mk); equilibrative nucleotide transporter 3 (Zjn_sc00008.1.g11390.1.sm.mkhc); the adenine nucleotide transporter (Zjn_sc00086.1.g01550.1.am.mk) (*BT1*); and folate transporter 1 (Zjn_sc00014.1.g03250.1.sm.mkhc), indicating that the transportation of Cd ions and small-molecule metabolic substances was aggregated in *Z. japonica*.

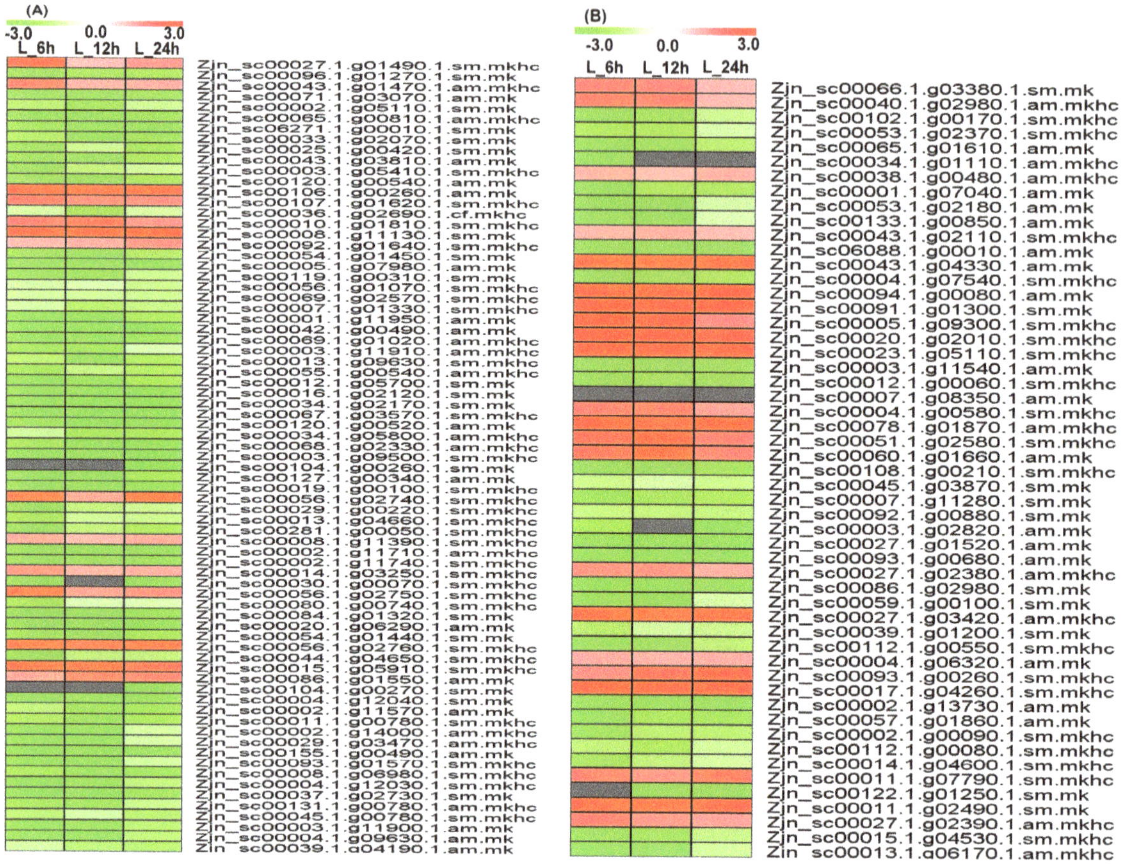

(A)

-3.0 0.0 3.0
L_6h L_12h L_24h

(B)

-3.0 0.0 3.0
L_6h L_12h L_24h

Figure 5. Expression profiles of common expressed genes related to transportation and protein processing in endoplasmic reticulum pathway. (**A**) Transporter protein genes; (**B**) transcription factor.

Fifty-three transcription-factor-related genes (23 upregulated and 30 downregulated) were commonly differentially expressed between the L_6 h vs. L_0 h, L_12 h vs. L_0 h, and L_24 h vs. L_0 h groups (Figure 5B, Table S9). These genes included basic helix–loop–helix transcription factors (Zjn_sc00038.1.g00480.1.am.mkhc, Zjn_sc00007.1.g08350.1.am.mk) (*bHLH74*, *bHLH96*), ethylene-responsive transcription factors (Zjn_sc00004.1.g06320.1.am.mk, Zjn_sc00043.1.g04330.1.am.mk, Zjn_sc00027.1.g03420.1.am.mkhc, Zjn_sc00078.1.g01870.1.am.mkhc) (*ERF1A*, *CRF1*, WRI1, *ABR1*), heat stress transcription factors (Zjn_sc00020.1.g02010.1.sm.mkhc, Zjn_sc00017.1.g04260.1.sm.mkhc, Zjn_sc00011.1.g02490.1.sm.mk) (*HSFA3*, *HSFC1B*), transcription factor *ICE1* (Zjn_sc00094.1.g00080.1.am.mk) (*SCRM*), a MADS-box transcription factor (Zjn_sc00091.1.g01300.1.sm.mk) (*MADS23*), and transcription factor *GAMYB* (Zjn_sc00040.1.g02980.1.am.mkhc) (*GAM1*), which were upregulated and suggested to play important roles in transcriptional regulation in response to Cd stress.

In addition, 91 carbohydrate-metabolism-related genes (six upregulated and eighty-five downregulated) were identified (Table S10), but only six of these genes, including benzaldehyde dehydrogenase (Zjn_sc00007.1.g05950.1.sm.mkhc), plastidial pyruvate kinase 2 (Zjn_sc00036.1.g00110.1.am.mkhc), dihydrolysine residue acetyltransferase component 4 of the pyruvate dehydrogenase complex (Zjn_sc00091.1.g01210.1.sm.mkhc), and phosphoglycerate mutase-like protein 4 (Zjn_sc00028.1.g03020.1.sm.mkhc), were upregulated after the Cd stress treatment, indicating that the carbohydrate-related pathway was severely inhibited by Cd toxicity.

Figure 6 shows a simple graph of the Cd tolerance mechanism in *Z. japonica* leaves. When Cd ions are transported into the cell cytoplasm, transcriptional regulation begins, and different metabolic processes, such as protein metabolism, glutathione metabolism, and carbohydrate metabolism, are triggered. Hormones, such as auxin, ethylene, and abscisic acid, participate in Cd detoxification.

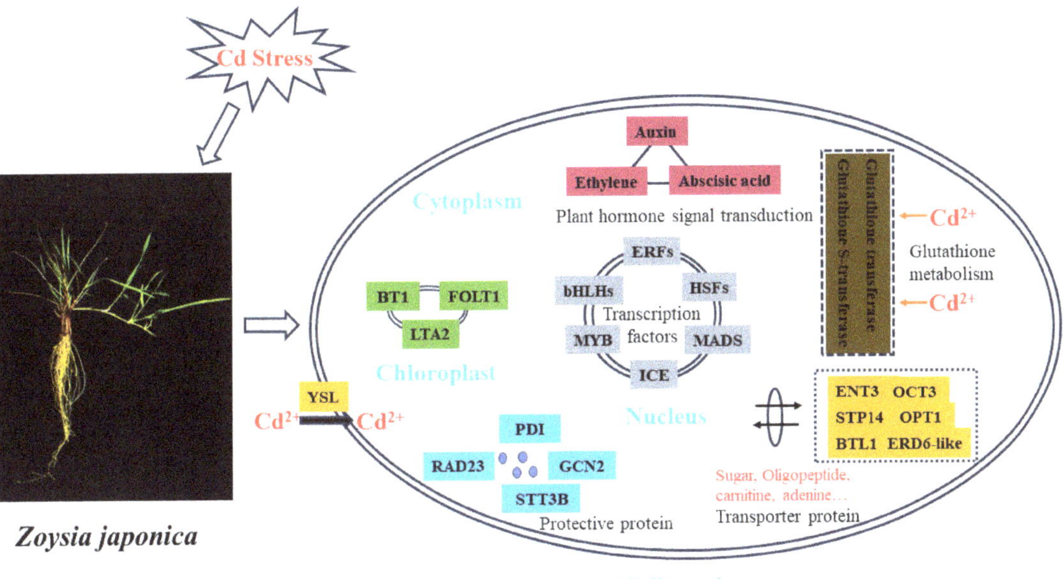

Figure 6. A simple diagram associated with the Cd tolerance mechanism in *Z. japonica* leaves. Basic helix–loop–helix transcription factors, bHLH; adenine nucleotide transporter, BT1; probable mitochondrial adenine nucleotide transporter, BTL1; Equilibrative nucleotide transporter 3, ENT3; ethylene-responsive transcription factors, ERFs; Sugar transporter ERD6-like, ERD6-like; Folate transporter 1, FOLT1; eIF-2-alpha kinase, GCN2; heat stress transcription factors, HSFs; ICE-like transcription factor, ICE; dihydrolipoyllysine-residue acetyltransferase component 4 of the pyruvate dehydrogenase complex, LTA2; MADS-box transcription factor, MADS; organic cation/carnitine transporter 3, OCT3; oligopeptide transporter 1, OPT1; protein disulfide isomerase, PDI; probable ubiquitin receptor, RAD23; sugar transport protein 14, STP14; dolichyl-diphosphooligosaccharide-protein glycosyltransferase subunit, STT3B; probable metal-nicotianamine transporters, YSLs.

2.5. Validation of RNA-seq Results with qRT-PCR

To verify the RNA-seq results, we randomly selected nine differentially expressed genes and performed a quantitative reverse transcription polymerase chain reaction (qRT-PCR) analysis of the L_6 h vs. L_0 h, L_12 h vs. L_0 h, and L_24 h vs. L_0 h groups. The relative expression levels of the differentially expressed genes are shown in Table S11. We then performed a correlation analysis of the RNA-seq and qRT-PCR results (Figure 7). A strong correlation (R^2 = 0.8319) was observed, confirming the reliability of the RNA-seq data.

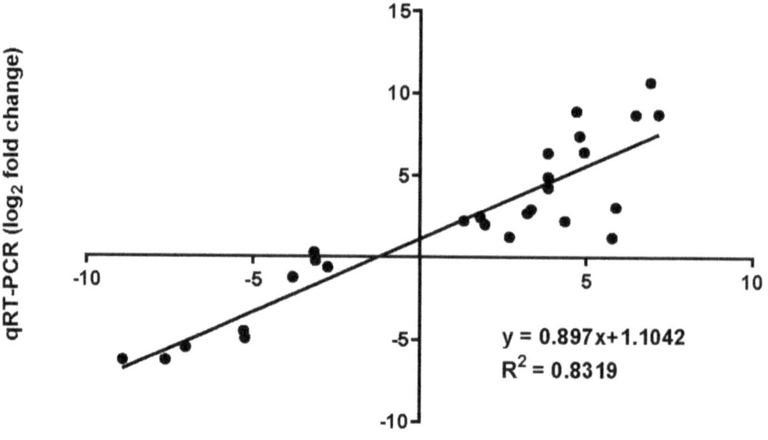

Figure 7. Validation of RNA-seq results with qRT-PCR.

3. Discussion

RNA-seq has been widely used to explore the molecular mechanisms underlying Cd stress in plants, especially in Cd-accumulating species. For example, 110.07 Gb of clean data and 63,957 unigenes were acquired from the hyperaccumulator *Phytolacca americana* after the Cd stress treatment [25]. In the Cd accumulator *Erigeron canadensis*, 12 cDNA libraries were constructed, and a total of 89.51 Gb of clean reads were obtained after 0.5 mmol/L of Cd treatment [26]. In the hyperaccumulator plant *Noccaea caerulescens*, transcriptomes were assembled using the Trinity de novo assembler and annotated along with the predicted protein sequences [27]. Turf grass is an important phytoremediation plant as it successfully mitigates soil Cd levels during the cutting and regrowth processes [28]. However, few studies have performed an RNA-seq analysis of Cd stress tolerance in turf grass species, and only in seashore paspalum (*Paspalum vaginatum*) and tall fescue (*Festuca arundinacea*) [29,30]. In this study, we obtained high-quality transcriptome data that can provide new insights into the plant Cd stress tolerance mechanisms in turfgrass species.

Glutathione plays an important role in the phytochelatin pathways in plants that cope with cadmium stress [31]. Under Cd stress, the expression of glutathione-metabolism-related genes is induced, and this promotes the biosynthesis of phytochelatin to chelate free Cd to generate a Cd–phytochelatin complex, which is then transported to the vacuole to isolate and complete the detoxification of Cd [32]. Increased glutathione metabolism has been reported in *Solanum nigrum* and clove basil *(Ocimum gratissimum)* under Cd stress conditions [33,34]. One glutathione transferase and one probable glutathione S-transferase were found to be upregulated in the present study, suggesting that these genes are closely associated with Cd detoxification in *Z. japonica* leaves.

Under Cd stress, plant hormones (such as ABA and IAA) undergo significant changes, and the expression levels of antioxidant synthesis and transport protein genes are affected [35]. ABA treatment counteracts Cd-induced fluctuations in non-enzymatic proteins and antioxidant enzymes [36]. The abscisic acid receptor PYL5 and probable protein phosphatase 2C 9 were upregulated after Cd stress, suggesting that the ABA signaling pathway was activated. Previous studies have shown that there are significant differences in the IAA concentration and distribution in the primary root tips and cotyledons of Cd-treated plants [37]. IAA content is significantly reduced after Cd treatment. In rice (*Oryza sativa*) shoots, GO and KEGG enrichment analyses showed that genes encoding the auxin-responsive protein IAA were upregulated after Cd stress [38]. Genes encoding the auxin-responsive proteins IAA9, 15, and 21 were upregulated in the present study, suggesting that auxins are involved in the response of Z. *japonica* leaves to Cd stress.

The endoplasmic reticulum is a complex metabolic site and an important organelle for the regulation of protein synthesis, signal transduction, and calcium homeostasis [39]. The synthesis and folding of all secreted proteins and most membrane proteins, as well as the modification and processing of proteins (mainly glycosylation, hydroxylation, acylation, and disulfide bond formation), are carried out in the endoplasmic reticulum. When plants are stressed by heavy metals, large amounts of unfolded or misfolded proteins accumulate in cells, resulting in endoplasmic reticulum stress [40]. The probable ubiquitin receptor RAD23 has been reported to be a principal shuttle of ubiquitin conjugates and to contribute to Arabidopsis UV tolerance [41]; the dolichyl-diphosphooligosaccharide-protein glycosyltransferase subunit STT3B could encode an essential subunit of the oligosaccharyltransferase complex and control adaptive responses to stress in plants [42]; the eIF-2-alpha kinase GCN2 functions as a translation of mRNAs and plays a role in plants' stress responses [43]; protein disulfide isomerase had been reported to participate in plant abiotic and biotic stress resistance [44]. These genes, which relate to protein processing in the endoplasmic reticulum pathway, were upregulated after the Cd treatment in the present study. However, the potential role of these genes in regulating Cd stress tolerance requires further analyses.

The oligopeptide transporter (OPT) family plays diverse roles in metal homeostasis, nitrogen mobilization, and sulfur distribution [45]. In *Arabidopsis*, 17 OPT members have been identified and phylogenetically divided into two subfamilies: OPT and YSL [46]. Many OPTs have been shown to be involved in the uptake and translocation of Fe and Cd [47]. OPT1 was also identified in the present study, suggesting that it plays an important role in metal homeostasis. The YSL family plays an important role in the transport of metal ions and in chelating metals (Fe, Cu, Mn, and Cd) [48]. Feng et al. discovered that SnYSL3 encodes a local plasma transporter that transports multiple metal–nicotinamide complexes and that SnYSL3 plays an important role in Cd stress [49]. In blueberries, VcYSL6 is upregulated by Cd stress, and its overexpression increases the Cd accumulation in the leaves [50]. Two probable metal–nicotianamine transporters were found to be upregulated in our study, suggesting an important role for nicotianamine chelation via *YSL* genes in Z. *japonica*. Sugar transporters mediate sugar transport in plants and play crucial roles in various physiological processes [51]. Here, we found that the sugar transporters ERD6-like 4/5 and sugar transport protein 14 showed increased expression levels after Cd stress, which is consistent with the KEGG analysis that showed that sugar-related pathways were the most enriched pathways under Cd stress conditions.

Transcription factors bind to DNA and are involved in the activity of hormonal and antioxidant enzymes, thereby regulating the absorption, transport, and accumulation of Cd. *ERF* gene expression plays a crucial role in the Cd stress response in plants. For example, in *Hibiscus syriacus*, numerous ERF-family transcription factors were induced after Cd stress [52]. In creeping bentgrass (*Agrostis stolonifera*), ERF1B, ERF110, ERF7, ERF113, and ERF15 are upregulated in response to Cd stress [53]. In *Iris lacteal*, the overexpression of *IlAP2*, an AP2/ERF superfamily gene, reduced Cd toxicity by hindering Cd transport [54]. In the present study, ethylene-responsive transcription factors (*ERF1A*, *CRF1*, *WRI1*, and *ABR1*) were upregulated after Cd treatment, suggesting that they may play critical roles in the Cd stress tolerance of Z. *japonica*. Heat shock transcription factors (HSFs) regulate HSP expression, participate in various aspects of protein homeostasis (such as the refolding, assembly, and transport of damaged proteins), and maintain protein stability [55]. Previous studies have proven that Cd stress could lead to the upregulation of *HSF* family genes and metallothionein genes in wheat and rice roots [56]. HSF3A and HSFC1B have been widely identified as playing roles in plant oxidative stress tolerance, and HSFC1B is induced by salt and heat stress [57]. HSF3A and HSFC1B from the Z. *japonica* leaves were found to be upregulated after the Cd stress treatment in this study, implying their potentially novel functions in Cd tolerance in Z. *japonica*. bHLH transcription factors are involved in the regulation of plant growth and development [58]. bHLH74 is required for leaf and root development in *Arabidopsis* [59]. In *Neolamarckia cadamba*, bHLH74 has been reported to function in cell elongation and is upregulated after gibberellin treatment [60]. In tomatoes, SlbHLH96 interacts with SlERF4 and represses the expression of ABA catabolic genes, thereby promoting drought tolerance [61]. In rice, OsbHLH96 influences brown planthopper resistance by regulating the expression of pathogen-related genes [62]. In the present study, bHLH74 and bHLH96 were found to be upregulated after Cd stress treatment, implying that these two transcription factors may positively regulate the Cd stress response in *Z. japonica* leaves by affecting plant growth, development, and stress-related gene expression.

Previous studies have shown that Cd stress affects the chlorophyll content and carbon metabolism enzyme activity in a plant's photosynthetic organs and destroys the photosynthetic structure of plants, thereby reducing their photosynthetic rate. Cd reduces the photosynthetic efficiency of plants by affecting photoreactions, dark reactions, and photochemical efficiencies [63,64]. Similar results were observed in this study, which found that carbon fixation in photosynthetic organisms was the most enriched pathway in the KEGG analysis, and most carbohydrate-metabolism-related genes were downregulated after the Cd stress treatment, suggesting that the photosynthetic process was inhibited.

In summary, this study revealed the molecular mechanisms of the response of *Z. japonica* leaves to Cd stress at different time points. A series of specific genes in glutathione metabolism, plant hormone signaling and transduction, and those involved in protein processing in the endoplasmic reticulum, transporter proteins, transcription factors, and carbohydrate metabolism pathways were highlighted in *Z. japonica*. Candidate genes, including the *bHLH*, *ERF*, and *HSF* family genes; *OPT* genes; *GST* genes; and auxin-responsive protein genes, which are shown in Figure 6, are novel for a plant Cd stress tolerance study. A similar abiotic stress regulation function of these novel genes has been reported in Arabidopsis or other model plants, which suggests their potential and special roles during Cd detoxification in *Z. japonica*. Therefore, further experiments on how the function of these genes varies in plant Cd stress tolerance should be carried out. Overall, our results enrich the understanding of the molecular mechanisms of Cd stress in *Z. japonica* and also provide important clues for Cd tolerance genetic breeding in other turfgrass and Gramineae species.

4. Materials and Methods

4.1. Plant Materials and Methods

Z. japonica seeds (Shi Ji Tian Yuan Ecological Technology Co., Zhengzhou, China) were sown in fritted clay and watered with a half-strength Hoagland nutrient solution. After approximately 3 months of cultivation in a growth chamber (GXZ-500, Jiangnan, China) (25/20 °C (day/night temperature), 16 h light (1200 μmol m^{-2}s^{-1})/8 h dark), uniform *Z. japonica* seedlings were selected and transferred to hydroponic conditions with half-strength Hoagland. Plants were cultivated for 2 months and then subjected to Cd treatment. The Cd concentration was 400 μM, supplied as $CdCl_2$ based on the previous study [65]. Leaf tissues were sampled 0, 6, 12, and 24 h after Cd treatment for physiological measurements, RNA isolation, and RNA-seq analysis. All time points had three biological replicates.

4.2. Total RNA Isolation and Sequencing Library Generation

Total RNA was isolated using a MiniBEST Plant RNA Extraction Kit (Takara, Kusatsu, Japan). RNA concentration, quality, and integrity were determined using a NanoDrop spectrophotometer (Thermo Fisher Scientific, Waltham, MA, USA). Three micrograms of RNA were used for sequencing library generation using a previously published protocol. The library was sequenced using a NovaSeq 6000 platform (Illumina, San Diego, CA, USA).

4.3. Quality Control and Read Mapping

The original (raw) sequencing data were exported in an FASTQ format. The raw data contained a number of adaptor sequences and low-quality reads; therefore, Cutadapt (v1.15) software was used to filter the sequencing data to obtain high-quality (clean) reads for further analysis. The reference genome and gene annotation files were downloaded from the *Zoysia* Genome Database (http://zoysia.kazusa.or.jp/ (accessed on 1 June 2023)). The filtered reads were mapped to the reference genome using HISAT2 v2.0.5.

4.4. Differential Expression Analysis

HTSeq (0.9.1) statistics were used to compare the read count values for each gene with the original gene expression data and then to standardize the expression levels. Differences in gene expression levels were analyzed using DESeq (version 1.30.0) with screening conditions as follows: expression difference multiple |log2 fold change| > 1 and *p*-value < 0.05. An R language software package, Pheatmap (1.0.8), was used to perform a bidirectional clustering analysis of all the differentially expressed genes in the samples. Heatmaps were generated according to the expression levels of the same gene in different samples and the expression patterns of different genes in the same sample, using the Euclidean method to calculate the distance and the complete linkage method for clustering. The dynamic expression patterns of the differentially expressed genes were analyzed using Short Time-series Expression Miner software (v1.3.8) [66].

4.5. GO and KEGG Enrichment Analyses

The mapped genes were further analyzed using the GO database, and the number of differentially expressed genes in each term was calculated. topGO was used to perform a GO enrichment analysis of the differentially expressed genes, and the *p*-value was calculated using the hypergeometric distribution method (the standard of significant enrichment is a *p*-value < 0.05). The GO terms significantly enriched for the differentially expressed genes were used to determine the main biological functions performed by the genes. ClusterProfiler (v3.4.4) software was used to perform a KEGG pathway enrichment analysis of the differentially expressed genes, focusing on significantly enriched pathways with a *p*-value < 0.05.

4.6. qRT-PCR Analysis

Total RNA was used for cDNA synthesis using a PrimeScript RT Reagent Kit with a gDNA Eraser (Takara, Kusatsu, Japan). qRT-PCR was performed using a CFX96 Touch Real-Time PCR Detection System (Bio-Rad, Hercules, CA, USA) with a TB Green Premix Ex Taq (Tli RNase H Plus; Takara). Nine differentially expressed genes (Zjn_sc00066.1. g01480.1.sm.mkhc, Zjn_sc00094.1.g00080.1.am.mk, Zjn_sc00002.1.g13730.1.am.mk, Zjn_ sc00011.1.g02490.1.sm.mk, Zjn_sc00106.1.g00260.1.am.mk, Zjn_sc00003.1.g09500.1.sm.mkhc, Zjn_sc00008.1.g11130.1.sm.mk, Zjn_sc00058.1.g01990.1.am.mk, and Zjn_sc00023.1.g06490.1. sm.mkhc) were selected for the qRT-PCR analysis.

The qRT-PCR primers used are listed in Supplementary Table S1. Relative gene expression was calculated using the $2^{-\Delta\Delta Ct}$ method [67]. Each qRT-PCR analysis was performed in triplicate.

Supplementary Materials: The following supporting information can be downloaded at https://www.mdpi.com/article/10.3390/plants12223833/s1. Figure S1: The correlation between different samples in RNA-seq results; Figure S2: The GO analysis of differentially expressed genes at different time points; Figure S3: The KEGG analysis of differentially expressed genes at different time points; Table S1 Primer sequences for qRT-PCR analysis; Table S2 All the identified common differentially expressed genes; Table S3 GO analysis of common differentially expressed genes; Table S4 KEGG enrichment analysis of common differentially expressed genes; Table S5 Glutathione metabolism related common differentially expressed genes; Table S6 Plant hormone signal transduction related common differentially expressed genes; Table S7 Protein processing in the endoplasmic reticulum pathway related common differentially expressed genes; Table S8 Transport function related common differentially expressed genes; Table S9 Transcription factor related common differentially expressed genes; Table S10 Carbohydrate metabolism related common differentially expressed genes; Table S11 Relative gene expression of selected differentially expressed genes

Author Contributions: K.Z. was responsible for the experiment's design. Y.X., Y.L. (Yonglong Li), Y.L. (Yan Li). and C.Z. were responsible for performing the experiment. K.Z. and Y.X. were responsible for writing and reviewing the manuscript. K.Z. was responsible for the funding support. All authors have read and agreed to the published version of the manuscript.

Funding: This research was funded by the Natural Science Foundation of Shandong Province, China (no. ZR2021QC037), and the Start-up Foundation for High Talents of Qingdao Agricultural University (no. 665/1121011).

Data Availability Statement: Data are contained within the article and Supplementary Materials.

Conflicts of Interest: The authors declare no conflict of interest.

References

1. Khanna, K.; Kohli, S.K.; Ohri, P.; Bhardwaj, R.; Ahmad, P. Agroecotoxicological Aspect of Cd in Soil-Plant System: Uptake, Translocation and Amelioration Strategies. *Environ. Sci. Pollut. Res. Int.* **2022**, *29*, 30908–30934. [CrossRef] [PubMed]
2. Kubier, A.; Wilkin, R.T.; Pichler, T. Cadmium in soils and groundwater: A review. *Appl. Geochem. J. Int. Assoc. Geochem. Cosmochem.* **2019**, *108*, 104388. [CrossRef] [PubMed]
3. Asgher, M.; Khan, M.I.; Anjum, N.A.; Khan, N.A. Minimising toxicity of cadmium in plants—Role of plant growth regulators. *Protoplasma* **2015**, *252*, 399–413. [CrossRef] [PubMed]
4. Qin, G.; Niu, Z.; Yu, J.; Li, Z.; Ma, J.; Xiang, P. Soil heavy metal pollution and food safety in China: Effects, sources and removing technology. *Chemosphere* **2021**, *267*, 129205. [CrossRef] [PubMed]
5. Yan, A.; Wang, Y.; Tan, S.N.; Mohd Yusof, M.L.; Ghosh, S.; Chen, Z. Phytoremediation: A Promising Approach for Revegetation of Heavy Metal-Polluted Land. *Front. Plant Sci.* **2020**, *11*, 359. [CrossRef]
6. Bhat, S.A.; Bashir, O.; Ul Haq, S.A.; Amin, T.; Rafiq, A.; Ali, M.; Américo-Pinheiro, J.H.P.; Sher, F. Phytoremediation of heavy metals in soil and water: An eco-friendly, sustainable and multidisciplinary approach. *Chemosphere* **2022**, *303*, 134788. [CrossRef]
7. Li, F.; Jin, H.; Wu, X.; Liu, Y.; Chen, X.; Wang, J. Remediation for trace metals in polluted soils by turfgrass assisted with chemical reagents. *Chemosphere* **2022**, *295*, 133790. [CrossRef]
8. Liu, H.T.; Guo, X.X. Hydroxyapatite reduces potential Cadmium risk by amendment of sludge compost to turf-grass grown soil in a consecutive two-year study. *Sci. Total Environ.* **2019**, *661*, 48–54. [CrossRef]
9. Zhu, T.; Li, L.; Duan, Q.; Liu, X.; Chen, M. Progress in our understanding of plant responses to the stress of heavy metal cadmium. *Plant Signal. Behav.* **2021**, *16*, 1836884. [CrossRef]
10. Haider, F.U.; Liqun, C.; Coulter, J.A.; Cheema, S.A.; Wu, J.; Zhang, R.; Wenjun, M.; Farooq, M. Cadmium toxicity in plants: Impacts and remediation strategies. *Ecotoxicol. Environ. Saf.* **2021**, *211*, 111887. [CrossRef]
11. Song, Y.; Jin, L.; Wang, X. Cadmium absorption and transportation pathways in plants. *Int. J. Phytoremed.* **2017**, *19*, 133–141. [CrossRef] [PubMed]
12. Khan, N.; You, F.M.; Datla, R.; Ravichandran, S.; Jia, B.; Cloutier, S. Genome-wide identification of ATP binding cassette (ABC) transporter and heavy metal associated (HMA) gene families in flax (*Linum usitatissimum* L.). *BMC Genom.* **2020**, *21*, 722. [CrossRef] [PubMed]
13. Carrillo, J.T.; Borthakur, D. Methods for metal chelation in plant homeostasis: Review. *Plant Physiol. Biochem.* **2021**, *163*, 95–107. [CrossRef] [PubMed]
14. Kang, Y.; Yang, L.; Dai, H.; Xie, M.; Wang, Y.; Peng, J.; Sun, H.; Ao, T.; Chen, W. Antioxidant system response, mineral element uptake and safe utilization of *Polygonatum sibiricum* in cadmium-contaminated soil. *Sci. Rep.* **2021**, *11*, 18737. [CrossRef]
15. Li, Y.; Ding, L.; Zhou, M.; Chen, Z.; Ding, Y.; Zhu, C. Transcriptional Regulatory Network of Plant Cadmium Stress Response. *Int. J. Mol. Sci.* **2023**, *24*, 4378. [CrossRef]
16. Moravčíková, D.; Žiarovská, J. The Effect of Cadmium on Plants in Terms of the Response of Gene Expression Level and Activity. *Plants* **2023**, *12*, 1848. [CrossRef]
17. Agarwal, P.; Mitra, M.; Banerjee, S.; Roy, S. MYB4 transcription factor, a member of R2R3-subfamily of MYB domain protein, regulates cadmium tolerance via enhanced protection against oxidative damage and increases expression of PCS1 and MT1C in Arabidopsis. *Plant Sci. Int. J. Exp. Plant Biol.* **2020**, *297*, 110501. [CrossRef]
18. Xu, Z.; Liu, X.; He, X.; Xu, L.; Huang, Y.; Shao, H.; Zhang, D.; Tang, B.; Ma, H. The Soybean Basic Helix-Loop-Helix Transcription Factor ORG3-Like Enhances Cadmium Tolerance via Increased Iron and Reduced Cadmium Uptake and Transport from Roots to Shoots. *Front. Plant Sci.* **2017**, *8*, 1098. [CrossRef]
19. Li, F.; Deng, Y.; Liu, Y.; Mai, C.; Xu, Y.; Wu, J.; Zheng, X.; Liang, C.; Wang, J. Arabidopsis transcription factor WRKY45 confers cadmium tolerance via activating PCS1 and PCS2 expression. *J. Hazard. Mater.* **2023**, *460*, 132496. [CrossRef]
20. Liu, S.; Peng, X.; Wang, X.; Zhuang, W. Transcriptome Analysis Reveals Differentially Expressed Genes Involved in Cadmium and Arsenic Accumulation in Tea Plant (*Camellia sinensis*). *Plants* **2023**, *12*, 1182. [CrossRef]

21. Cheng, Y.; Qiu, L.; Shen, P.; Wang, Y.; Li, J.; Dai, Z.; Qi, M.; Zhou, Y.; Zou, Z. Transcriptome studies on cadmium tolerance and biochar mitigating cadmium stress in muskmelon. *Plant Physiol. Biochem.* **2023**, *197*, 107661. [CrossRef]
22. He, Q.; Zhou, T.; Sun, J.; Wang, P.; Yang, C.; Bai, L.; Liu, Z. Transcriptome Profiles of Leaves and Roots of Goldenrain Tree (*Koelreuteria paniculata* Laxm.) in Response to Cadmium Stress. *Int. J. Environ. Res. Public Health* **2021**, *18*, 12046. [CrossRef]
23. Wang, J.; An, C.; Guo, H.; Yang, X.; Chen, J.; Zong, J.; Li, J.; Liu, J. Physiological and transcriptomic analyses reveal the mechanisms underlying the salt tolerance of Zoysia japonica Steud. *BMC Plant Biol.* **2020**, *20*, 114. [CrossRef]
24. Hu, C.; Yang, X.; Gao, L.; Zhang, P.; Li, W.; Dong, J.; Li, C.; Zhang, X. Comparative analysis of heavy metal accumulation and bioindication in three seagrasses: Which species is more suitable as a bioindicator? *Sci. Total Environ.* **2019**, *669*, 41–48. [CrossRef]
25. Zhao, L.; Zhu, Y.H.; Wang, M.; Ma, L.G.; Han, Y.G.; Zhang, M.J.; Li, X.C.; Feng, W.S.; Zheng, X.K. Comparative transcriptome analysis of the hyperaccumulator plant Phytolacca americana in response to cadmium stress. *3Biotech* **2021**, *11*, 327. [CrossRef]
26. Gan, C.; Liu, Z.; Pang, B.; Zuo, D.; Hou, Y.; Zhou, L.; Yu, J.; Chen, L.; Wang, H.; Gu, L.; et al. Integrative physiological and transcriptome analyses provide insights into the Cadmium (Cd) tolerance of a Cd accumulator: Erigeron canadensis. *BMC Genom.* **2022**, *23*, 778. [CrossRef]
27. Blande, D.; Halimaa, P.; Tervahauta, A.I.; Aarts, M.G.; Kärenlampi, S.O. De novo transcriptome assemblies of four accessions of the metal hyperaccumulator plant Noccaea caerulescens. *Sci. Data* **2017**, *4*, 160131. [CrossRef]
28. Rabêlo, F.H.S.; Vangronsveld, J.; Baker, A.J.M.; van der Ent, A.; Alleoni, L.R.F. Are Grasses Really Useful for the Phytoremediation of Potentially Toxic Trace Elements? A Review. *Front. Plant Sci.* **2021**, *12*, 778275. [CrossRef]
29. Xu, L.; Zheng, Y.; Yu, Q.; Liu, J.; Yang, Z.; Chen, Y. Transcriptome Analysis Reveals the Stress Tolerance to and Accumulation Mechanisms of Cadmium in Paspalum vaginatum Swartz. *Plants* **2022**, *11*, 2078. [CrossRef]
30. Zhu, H.; Ai, H.; Cao, L.; Sui, R.; Ye, H.; Du, D.; Sun, J.; Yao, J.; Chen, K.; Chen, L. Transcriptome analysis providing novel insights for Cd-resistant tall fescue responses to Cd stress. *Ecotoxicol. Environ. Saf.* **2018**, *160*, 349–356. [CrossRef]
31. Hasanuzzaman, M.; Nahar, K.; Anee, T.I.; Fujita, M. Glutathione in plants: Biosynthesis and physiological role in environmental stress tolerance. *Physiol. Mol. Biol. Plants Int. J. Funct. Plant Biol.* **2017**, *23*, 249–268. [CrossRef]
32. Chen, F.; Wang, F.; Wu, F.; Mao, W.; Zhang, G.; Zhou, M. Modulation of exogenous glutathione in antioxidant defense system against Cd stress in the two barley genotypes differing in Cd tolerance. *Plant Physiol. Biochem.* **2010**, *48*, 663–672. [CrossRef] [PubMed]
33. Li, Z.; Guan, W.; Yang, L.; Yang, Y.; Yu, H.; Zou, L.; Teng, Y. Insight into the Vacuolar Compartmentalization Process and the Effect Glutathione Regulation to This Process in the Hyperaccumulator Plant *Solanum nigrum* L. BioMed Res. Int. **2022**, *2022*, 4359645. [CrossRef]
34. Wang, B.; Wang, Y.; Yuan, X.; Jiang, Y.; Zhu, Y.; Kang, X.; He, J.; Xiao, Y. Comparative transcriptomic analysis provides key genetic resources in clove basil (Ocimum gratissimum) under cadmium stress. *Front. Genet.* **2023**, *14*, 1224140. [CrossRef] [PubMed]
35. Pacenza, M.; Muto, A.; Chiappetta, A.; Mariotti, L.; Talarico, E.; Picciarelli, P.; Picardi, E.; Bruno, L.; Bitonti, M.B. In Arabidopsis thaliana Cd differentially impacts on hormone genetic pathways in the methylation defective ddc mutant compared to wild type. *Sci. Rep.* **2021**, *11*, 10965. [CrossRef] [PubMed]
36. Chen, K.; Chen, P.; Qiu, X.; Chen, J.; Gao, G.; Wang, X.; Zhu, A.; Yu, C. Regulating role of abscisic acid on cadmium enrichment in ramie (*Boehmeria nivea* L.). *Sci. Rep.* **2021**, *11*, 22045. [CrossRef]
37. Guo, J.; Qin, S.; Rengel, Z.; Gao, W.; Nie, Z.; Liu, H.; Li, C.; Zhao, P. Cadmium stress increases antioxidant enzyme activities and decreases endogenous hormone concentrations more in Cd-tolerant than Cd-sensitive wheat varieties. *Ecotoxicol. Environ. Saf.* **2019**, *172*, 380–387. [CrossRef]
38. Sun, L.; Wang, J.; Song, K.; Sun, Y.; Qin, Q.; Xue, Y. Transcriptome analysis of rice (*Oryza sativa* L.) shoots responsive to cadmium stress. *Sci. Rep.* **2019**, *9*, 10177. [CrossRef]
39. Xu, H.; Xu, W.; Xi, H.; Ma, W.; He, Z.; Ma, M. The ER luminal binding protein (BiP) alleviates Cd(2+)-induced programmed cell death through endoplasmic reticulum stress-cell death signaling pathway in tobacco cells. *J. Plant Physiol.* **2013**, *170*, 1434–1441. [CrossRef]
40. Beaupere, C.; Labunskyy, V.M. (Un)folding mechanisms of adaptation to ER stress: Lessons from aneuploidy. *Curr. Genet.* **2019**, *65*, 467–471. [CrossRef]
41. Lahari, T.; Lazaro, J.; Schroeder, D.F. RAD4 and RAD23/HMR Contribute to Arabidopsis UV Tolerance. *Genes* **2017**, *9*, 8. [CrossRef]
42. Koiwa, H.; Li, F.; McCully, M.G.; Mendoza, I.; Koizumi, N.; Manabe, Y.; Nakagawa, Y.; Zhu, J.; Rus, A.; Pardo, J.M.; et al. The STT3a subunit isoform of the Arabidopsis oligosaccharyltransferase controls adaptive responses to salt/osmotic stress. *Plant Cell* **2003**, *15*, 2273–2284. [CrossRef] [PubMed]
43. Lokdarshi, A.; von Arnim, A.G. Review: Emerging roles of the signaling network of the protein kinase GCN2 in the plant stress response. *Plant Sci. Int. J. Exp. Plant Biol.* **2022**, *320*, 111280. [CrossRef] [PubMed]
44. Li, T.; Wang, Y.H.; Huang, Y.; Liu, J.X.; Xing, G.M.; Sun, S.; Li, S.; Xu, Z.S.; Xiong, A.S. A novel plant protein-disulfide isomerase participates in resistance response against the TYLCV in tomato. *Planta* **2020**, *252*, 25. [CrossRef] [PubMed]
45. Lubkowitz, M. The oligopeptide transporters: A small gene family with a diverse group of substrates and functions? *Mol. Plant* **2011**, *4*, 407–415. [CrossRef]
46. Koh, S.; Wiles, A.M.; Sharp, J.S.; Naider, F.R.; Becker, J.M.; Stacey, G. An oligopeptide transporter gene family in Arabidopsis. *Plant Physiol.* **2002**, *128*, 21–29. [CrossRef]

47. Wang, C.; Wang, X.; Li, J.; Guan, J.; Tan, Z.; Zhang, Z.; Shi, G. Genome-Wide Identification and Transcript Analysis Reveal Potential Roles of Oligopeptide Transporter Genes in Iron Deficiency Induced Cadmium Accumulation in Peanut. *Front. Plant Sci.* **2022**, *13*, 894848. [CrossRef]

48. Tao, J.; Lu, L. Advances in Genes-Encoding Transporters for Cadmium Uptake, Translocation, and Accumulation in Plants. *Toxics* **2022**, *10*, 411. [CrossRef]

49. Feng, S.; Tan, J.; Zhang, Y.; Liang, S.; Xiang, S.; Wang, H.; Chai, T. Isolation and characterization of a novel cadmium-regulated Yellow Stripe-Like transporter (SnYSL3) in Solanum nigrum. *Plant Cell Rep.* **2017**, *36*, 281–296. [CrossRef]

50. Chen, S.; Liu, Y.; Deng, Y.; Liu, Y.; Dong, M.; Tian, Y.; Sun, H.; Li, Y. Cloning and functional analysis of the VcCXIP4 and VcYSL6 genes as Cd-regulating genes in blueberry. *Gene* **2019**, *686*, 104–117. [CrossRef]

51. Julius, B.T.; Leach, K.A.; Tran, T.M.; Mertz, R.A.; Braun, D.M. Sugar Transporters in Plants: New Insights and Discoveries. *Plant Cell Physiol.* **2017**, *58*, 1442–1460. [CrossRef] [PubMed]

52. Li, X.; Liu, L.; Sun, S.; Li, Y.; Jia, L.; Ye, S.; Yu, Y.; Dossa, K.; Luan, Y. Physiological and transcriptional mechanisms associated with cadmium stress tolerance in Hibiscus syriacus L. *BMC Plant Biol.* **2023**, *23*, 286. [CrossRef] [PubMed]

53. Yuan, J.; Bai, Y.; Chao, Y.; Sun, X.; He, C.; Liang, X.; Xie, L.; Han, L. Genome-wide analysis reveals four key transcription factors associated with cadmium stress in creeping bentgrass (*Agrostis stolonifera* L.). *PeerJ* **2018**, *6*, e5191. [CrossRef] [PubMed]

54. Wang, Z.; Ni, L.; Liu, L.; Yuan, H.; Gu, C. IlAP2, an AP2/ERF Superfamily Gene, Mediates Cadmium Tolerance by Interacting with IlMT2a in Iris lactea var. chinensis. *Plants* **2023**, *12*, 823. [CrossRef]

55. Guo, J.; Wu, J.; Ji, Q.; Wang, C.; Luo, L.; Yuan, Y.; Wang, Y.; Wang, J. Genome-wide analysis of heat shock transcription factor families in rice and Arabidopsis. *J. Genet. Genom.* **2008**, *35*, 105–118. [CrossRef] [PubMed]

56. Shim, D.; Hwang, J.U.; Lee, J.; Lee, S.; Choi, Y.; An, G.; Martinoia, E.; Lee, Y. Orthologs of the class A4 heat shock transcription factor HsfA4a confer cadmium tolerance in wheat and rice. *Plant Cell* **2009**, *21*, 4031–4043. [CrossRef]

57. Seni, S.; Kaur, S.; Malik, P.; Yadav, I.S.; Sirohi, P.; Chauhan, H.; Kaur, A.; Chhuneja, P. Transcriptome based identification and validation of heat stress transcription factors in wheat progenitor species Aegilops speltoides. *Sci. Rep.* **2021**, *11*, 22049. [CrossRef]

58. Zhou, X.; Liao, Y.; Kim, S.U.; Chen, Z.; Nie, G.; Cheng, S.; Ye, J.; Xu, F. Genome-wide identification and characterization of bHLH family genes from Ginkgo biloba. *Sci. Rep.* **2020**, *10*, 13723. [CrossRef]

59. Bao, M.; Bian, H.; Zha, Y.; Li, F.; Sun, Y.; Bai, B.; Chen, Z.; Wang, J.; Zhu, M.; Han, N. miR396a-Mediated basic helix-loop-helix transcription factor bHLH74 repression acts as a regulator for root growth in Arabidopsis seedlings. *Plant Cell Physiol.* **2014**, *55*, 1343–1353. [CrossRef]

60. Li, L.; Wang, J.; Chen, J.; Wang, Z.; Qaseem, M.F.; Li, H.; Wu, A. Physiological and Transcriptomic Responses of Growth in Neolamarckia cadamba Stimulated by Exogenous Gibberellins. *Int. J. Mol. Sci.* **2022**, *23*, 11842. [CrossRef]

61. Liang, Y.; Ma, F.; Li, B.; Guo, C.; Hu, T.; Zhang, M.; Liang, Y.; Zhu, J.; Zhan, X. A bHLH transcription factor, SlbHLH96, promotes drought tolerance in tomato. *Hortic. Res.* **2022**, *9*, uhac198. [CrossRef]

62. Wang, M.; Yang, D.; Ma, F.; Zhu, M.; Shi, Z.; Miao, X. OsHLH61-OsbHLH96 influences rice defense to brown planthopper through regulating the pathogen-related genes. *Rice* **2019**, *12*, 9. [CrossRef]

63. Zhang, H.; Xu, Z.; Guo, K.; Huo, Y.; He, G.; Sun, H.; Guan, Y.; Xu, N.; Yang, W.; Sun, G. Toxic effects of heavy metal Cd and Zn on chlorophyll, carotenoid metabolism and photosynthetic function in tobacco leaves revealed by physiological and proteomics analysis. *Ecotoxicol. Environ. Saf.* **2020**, *202*, 110856. [CrossRef]

64. Parmar, P.; Kumari, N.; Sharma, V. Structural and functional alterations in photosynthetic apparatus of plants under cadmium stress. *Bot. Stud.* **2013**, *54*, 45. [CrossRef]

65. Liu, J.X.; Sun, Z.Y.; Ju, G.S.; Han, L.; Qian, Y.Q. Physiological response of Zoysia japonica to Cd2+. *Acta Ecol. Sin.* **2011**, *31*, 6149–6156. (In Chinese)

66. Ernst, J.; Bar-Joseph, Z. STEM: A tool for the analysis of short time series gene expression data. *BMC Bioinform.* **2006**, *7*, 191. [CrossRef]

67. Rao, X.; Huang, X.; Zhou, Z.; Lin, X. An improvement of the 2^(-delta delta CT) method for quantitative real-time polymerase chain reaction data analysis. *Biostat. Bioinform. Biomath.* **2013**, *3*, 71–85.

MDPI
St. Alban-Anlage 66
4052 Basel
Switzerland
www.mdpi.com

Plants Editorial Office
E-mail: plants@mdpi.com
www.mdpi.com/journal/plants